Operator Theory
Advances and Applications
Vol. 103

Editor:
I. Gohberg

Recent Progress in Operator Theory

International Workshop on Operator Theory and Applications, IWOTA 95, in Regensburg, July 31–August 4, 1995

I. Gohberg
R. Mennicken
C. Tretter
Editors

Springer Basel AG

Authors:

I. Gohberg
School of Mathematical Sciences
Raymond and Beverly Sackler
Faculty of Exact Sciences
Tel Aviv University
Ramat Aviv 69978
Israel
e-mail: gohberg@math.tau.ac.il

R. Mennicken
NWFI-Mathematik
Universität Regensburg
D-93040 Regensburg
Germany
e-mail: reinhard.mennicken@mathematik.uni-regensburg.de

C. Tretter
NWFI-Mathematik
Universität Regensburg
D-93040 Regensburg
Germany
e-mail: christiane.tretter@mathematik.uni-regensburg.de

A CIP catalogue record for this book is available from the
1991 Mathematics Subject Classification 47-06, 93-06

A CIP catalogue record for this book is available from the
Library of Congress, Washington D.C., USA

Deutsche Bibliothek Cataloging-in-Publication Data
Recent progress in operator theory / International Workshop on Operator Theory
and Applications, IWOTA 95, in Regensburg, July 31 – August 4, 1995.
I. Gohberg ... ed.. – Basel ; Boston ; Berlin : Birkhäuser, 1998
(Operator theory ; Vol. 103)
ISBN 978-3-0348-9776-1 ISBN 978-3-0348-8793-9 (eBook)
DOI 10.1007/978-3-0348-8793-9

Originally published by Birkhäuser Verlag in 1998
Softcover reprint of the hardcover 1st edition 1998
Printed on acid-free paper produced from chlorine-free pulp. TCF ∞
Cover design: Heinz Hiltbrunner, Basel

ISBN 978-3-0348-9776-1

9 8 7 6 5 4 3 2 1

Table of contents

On the normal solvability of cohomological equations on compact topological spaces

G. Belitskii and Yu. Lyubich

On nonnegative realizations of rational matrix functions and nonnegative input–output systems

K.-H. Förster and B. Nagy

On the geometric structure of regular dilations

D. Gaşpar and N. Suciu

On generalized interpolation and shift invariant maximal semidefinite subspaces

A. Gheondea

The sum of matrix Nevanlinna functions and self-adjoint extensions in exit spaces

S. Hassi, M. Kaltenbäck and H.S.V. de Snoo

Editorial introduction

This and the previous volume of the OT series contain the proceedings of the Workshop on Operator Theory and its Applications, IWOTA 95, which was held at the University of Regensburg, Germany, July 31 to August 4, 1995. It was the eigth workshop of this kind. Following is a list of the seven previous workshops with reference to their proceedings:

1981 Operator Theory
(Santa Monica, California, USA)

1983 Applications of Linear Operator Theory to Systems and Networks
(Rehovot, Israel), OT 12

1985 Operator Theory and its Applications
(Amsterdam, The Netherlands), OT 19

1987 Operator Theory and Functional Analysis
(Mesa, Arizona, USA), OT 35

1989 Matrix and Operator Theory
(Rotterdam, The Netherlands), OT 50

1991 Operator Theory and Complex Analysis
(Sapporo, Japan), OT 59

1993 Operator Theory and Boundary Eigenvalue Problems
(Vienna, Austria), OT 80

IWOTA 95 offered a rich programme on a wide range of latest developments in operator theory and its applications. The programme consisted of 6 invited plenary lectures, 54 invited special topic lectures and more than 100 invited session talks. About 180 participants from 25 countries attended the workshop, more than a third came from Eastern Europe.

The conference covered different aspects of linear and nonlinear spectral problems, starting with problems for abstract operators up to spectral theory of ordinary and partial differential operators, pseudodifferential operators, and integral operators. The workshop was also focussed on operator theory in spaces with indefinite metric, operator functions, interpolation and extension problems. The applications concerned problems in mathematical physics, hydrodynamics, magnetohydrodynamics, quantum mechanics, astrophysics as well as the theory of networks and systems. The papers in the proceedings bring the readers up to date on recent achievements in these areas.

This volume contains the contributions to different aspects of operator theory and its applications. A second volume is focussed especially on differential and integral operators.

The financial support of the following institutions for IWOTA 95 is highly appreciated:

Bayerisches Staatsministerium für Unterricht, Kultus, Wissenschaft und Kunst,
Deutsche Forschungsgemeinschaft (DFG),
Regensburger Universitätsstiftung Hans Vielberth,
Gesellschaft für Angewandte Mathematik und Mechanik (GAMM),
Akademie Verlag, Berlin,
Birkhäuser Verlag, Basel.

We also thank the University of Regensburg for supporting the workshop in many ways, in particular by providing lecture rooms and other facilities.

I.C. Gohberg, R. Mennicken and C. Tretter

List of participants

1	M.S. Agranovich	30	E. Grinshpun	59	W.N. Everitt	88	I.S. Kac
2	K. Seddighi	31	I. Koltracht	60	R. Vonhoff	89	S.G. Pyatkov
3	Ch. Davis	32	N. Suciu	61	C. Sadosky	90	J. Kos
4	B. Fritzsche	33	A. Kozhevnikov	62	M.A. Kaashoek	91	A.C.M. Ran
5	B. Kirstein	34	B. Silbermann	63	M. Moeller	92	S.N. Naboko
6	S. Roch	35	B. Gramsch	64	A.K. Motovilov	93	V. Hardt
7	S. Treil	36	M.P.H. Wolff	65	A.B. Mingarelli	94	H. Gail
8	S.M. Verduyn Lunel	37	S. Behm	66	D.Z. Arov	95	M. Weber
9	K. Lorentz	38	P. Kurasov	67	I.M. Spitkovsky	96	A. Hartmann
10	H.S.V. de Snoo	39	L. Rodman	68	J.A. Dubinskii	97	D. Gaspar
11	A. Ben-Artzi	40	H.-G. Leopold	69	V. Matsaev	98	E.W. Lytvynov
12	G. Schlüchtermann	41	E. Litsyn	70	V. Liskevich	99	W.G. Litvinov
13	R. Lauter	42	H. Langer	71	A.A. Nudelman	100	E.R. Tsekanovskii
14	S. Albeverio	43	I. Marek	72	G. Litvinchuk	101	Yu.V. Egorov
15	J. Zemanek	44	A. Gheondea	73	D. Bakic	102	L.A. Sakhnovich
16	A.S. Markus	45	H. Triebel	74	B. Najman	103	V. Peller
17	M. Solomyak	46	Ch.R. Johnson	75	V.M. Adamjan	104	M. Markin
18	Z.G. Sheftel	47	A.F. dos Santos	76	D. Pik	105	P. Gaspar
19	L.R. Volevich	48	R. Duduchava	77	G. Heinig	106	K.-H. Förster
20	M. Malamud	49	F.-O. Speck	78	V. Strauss	107	R. Gaspar
21	E. Meister	50	Ya. Roitberg	79	M. Taghavi	108	P. Jonas
22	D.V. Yakubovich	51	K. Makarov	80	V. Pivovarchik	109	Yu.M. Berezansky
23	V. Kapustin	52	B.-W. Schulze	81	F. Ali Mehmeti	110	R. Mennicken
24	E. Schrohe	53	M. Demuth	82	N. Gopal Reddy	111	I.C. Gohberg
25	R. Denk	54	H.R. Beyer	83	K. Naimark	112	C. Tretter
26	J. Brasche	55	I. Roitberg	84	B.M. Levitan	113	V. Kondratiev
27	M.A. Nudelman	56	V. Mikhailets	85	A.A. Shkalikov	114	B.A. Plamenevskii
28	A. Kulesko	57	I. Suciu	86	R.O. Griniv	115	J. Saurer
29	M. Rakowski	58	V.S. Rabinovich	87	A.I. Kozhanov		

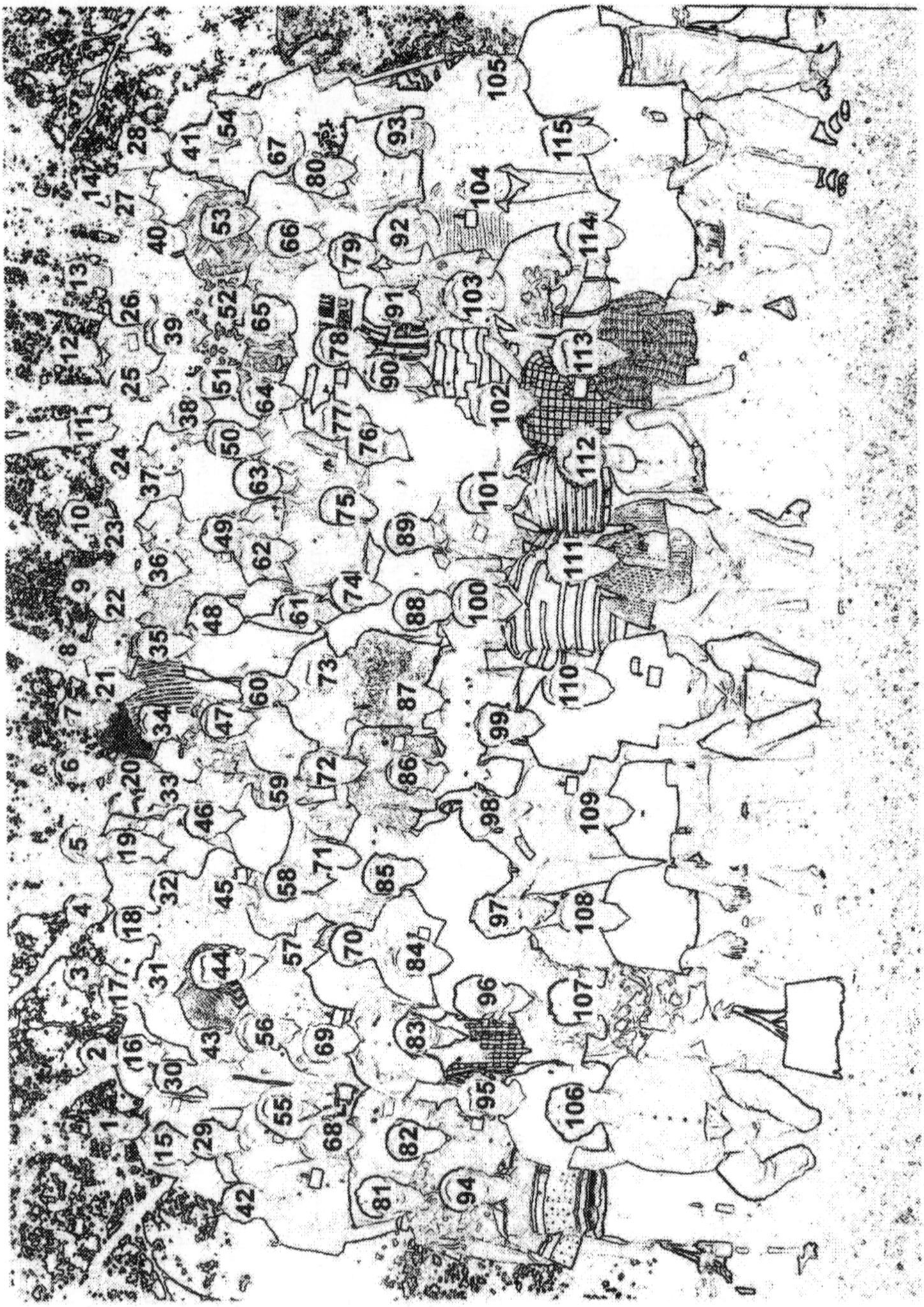

Operator Theory:
Advances and Applications, Vol. 103

Inversion formulas for compressions of block–Toeplitz operators

D.Z. AROV

Inversion formulas are obtained for invertible compressions $T(f) = P_{K_b}M_f|K_b$ of block-Toeplitz operators to a left shift invariant subspace $K_b = H_n^2 \ominus bH_m^2$ of the Hardy space H_n^2 of vector functions. These results are compared with known inversion formulas for block-Toeplitz matrices, block-Pick matrices and for Toeplitz integral operators in $L_n^2(0,a)$.

1. Introduction

1.1. This work is a development of the investigations of inversion formulas (Gohberg-Semencul, and Gohberg-Krupnik, A. Sakhnovich, L. Sakhnovich, G. Heinig, L. Lerer, T. Kailath, and others; see references in [7], [2], [5], [6], [9], [8]) for matrices with special structure (block-Toeplitz, block-Pick, etc.), their continuous analogues for integral operators in $L_n^2(0,a)$ with difference kernels and other structured operators [4].

1.2. Let $H_{n\times m}^p$ denote the Hardy space of $n \times m$ matrix functions (m.f.s) $h(z)$ analytic in the disk $\mathcal{D} = \{z : |z| < 1\}$ with $\|h\|_p < +\infty$, where

$$\|h\|_p = \sup\{\|h(z)\| : z \in \mathcal{D}\} \quad \text{for} \ \ p = +\infty,$$

$$\|h\|_p^p = \sup_{0<r<1} \frac{1}{2\pi} \int_{\partial\mathcal{D}} (Sp\ h^*(r\zeta)h(r\zeta))^{\frac{p}{2}} |d\zeta| \quad \text{for} \ \ 1 \le p < +\infty.$$

By identifying m.f.s $h(z) = \Sigma_0^\infty h_k z^k$ ($\in H_{n\times m}^p$) with their boundary values $h(\zeta) = \Sigma_0^\infty h_k \zeta^k$ ($\zeta \in \partial\mathcal{D}$ a.e.) one can consider $H_{n\times m}^p$ as a subspace of the space $L_{n\times m}^p$ of $n \times m$ m.f.s $h(\zeta)$ measurable on the circle $\partial\mathcal{D} = \{\zeta : |\zeta| = 1\}$ with $\|h\|_p < \infty$, where

$$\|h\|_p = \text{ess sup}\{\|h(\zeta)\| : \ \zeta \in \partial\mathcal{D}\} \quad \text{for} \ \ p = +\infty,$$

$$\|h\|_p^p = \frac{1}{2\pi} \int_{\partial\mathcal{D}} (Sp\ h^*(\zeta)h(\zeta))^{\frac{p}{2}} |d\zeta| \quad \text{for} \ \ 1 \le p < +\infty.$$

We consider $L_{n\times m}^2$ as a Hilbert space; moreover, for $F \in L_{n\times m}^2$ and $G \in L_{n\times r}^2$ we define

$$\langle F, G\rangle := \frac{1}{2\pi} \int_{\partial\mathcal{D}} G^*(\zeta)F(\zeta)\,|d\zeta|.$$

Then $H^2_{n\times m} \subset L^2_{n\times m}$, and we define $K^2_{n\times m} := L^2_{n\times m} \ominus H^2_{n\times m}$, $H^2_n := H^2_{n\times 1}$, $K^2_n := K^2_{n\times 1}$.

Let $[X, Y]$ denote the space of bounded linear operators acting from the Hilbert space X into the Hilbert space Y, $[X] = [X, X]$, let $P_{\mathcal{L}}$ be the orthoprojection onto the subspace $\mathcal{L}$, and let $A|\mathcal{L}$ be the restriction of the operator A on the subspace $\mathcal{L}$.

1.3. Let $b(z)$ be a fixed inner m.f. of order $n \times m$, i.e., $b \in H^\infty_{n\times m}$, $\|b\|_\infty \leq 1$ and $b^*(\zeta)b(\zeta) = I_m$ a.e. on $\partial\mathcal{D}$. We consider the left shift invariant subspace K_b of the space H^2_n, which is defined by the m.f. b,

$$K_b = H^2_n \ominus bH^2_m,$$

and let S be the left shift operator in the space K_b,

$$(Sh)(z) = z^{-1}(h(z) - h(0)), \quad h \in K_b. \tag{1.1}$$

For a m.f. $f(\zeta)$ from $L_{n\times n}$, let M_f be the operator of multiplication by f in the space L^2_n,

$$(M_f h)(\zeta) = f(\zeta)h(\zeta), \quad h \in L^2_n\ , \qquad T(f) := P_{K_b} M(f)|K_b.$$

Let $\mathbb{K}(b)$ denote the set of operators $T \in [K_b]$ which are quasicommutative with respect to the shift S, i.e., for which

$$\operatorname{rank}(TS - ST) < \infty.$$

In Section 2 we shall prove that for any $f \in L^\infty_{n\times n}$, the compression $T(f)$ belongs to $\mathbb{K}(b)$ and we shall obtain inversion formulas for the invertible compressions. Earlier, in [10], we obtained inversion formulas for strictly positive compressions $T(f)$ (> 0, $f(\zeta) > 0$, $f^{\pm 1} \in L^\infty_{n\times n}$) on the space K_b when $b(z) = b_0(z)I_n$ and b_0 is a scalar inner function. This was a generalization of the well-known Gohberg-Semencul formula for positive block-Toeplitz matrices [8], [3]. Following the author's lecture on this result in M.A. Kaashoek's Summer Operator Theory Seminar (July, 1994), I.C. Gohberg conjectured that one could obtain similar inversion formulas for invertible operators which are skew-Toeplitz in the sense of the work [1], i.e., for operators T_ω of the form

$$T_\omega = \sum_{j,k=1}^{N} (S^*)^j T(c_{jk}) S^k,$$

where c_{jk}, $1 \leq j, k \leq N$, are constant matrices of order n. These operators belong to $\mathbb{K}(b)$.

By the method considered in this paper we have also obtained inversion formulas for the invertible operators T in the wider class $\mathbb{K}(b)$. This result will be published elsewhere.

2. Main results

2.1. Let $\chi(z) = zI_p$ ($p = m$ or $p = n$),

$$e_1(z) = I_n - b(z)b^*(0), \qquad e_2(z) = z^{-1}(b(z) - b(0)). \tag{2.1}$$

For $h \in K_b$ one has

$$\begin{aligned} Sh &= T(\chi^*)h, \qquad T(\chi)h = \zeta h(\zeta) - P_{bH^2_m} M_\chi h, \\ S^*h &= \zeta h(\zeta) - b(\zeta) P_{H^2_m} b^* M_\chi h \\ &= \zeta h(\zeta) - b(\zeta) P_{H^2_m} M_\chi b^* h. \end{aligned}$$

Because $b^*h \in K^2_m$, one has

$$P_{H^2_m} M_\chi b^* h = \frac{1}{2\pi} \int_{\partial\mathcal{D}} M_\chi b^* h \, |d\zeta| = \langle h, M^*_\chi b\rangle = \langle h, e_2\rangle.$$

Thus,

$$(S^*h)(\zeta) = \zeta h(\zeta) - b(\zeta)\langle h, e_2\rangle. \tag{2.2}$$

Using the formulas (1.1) and (2.2), one has that

$$(I - S^*S)h = e_1(\zeta)\langle h, e_1\rangle \quad \text{and} \quad (I - SS^*)h = e_2(\zeta)\langle h, e_2\rangle.$$

These formulas can be rewritten in the form

$$I - S^*S = e_1\langle\,\cdot\,, e_1\rangle \quad \text{and} \quad I - SS^* = e_2\langle\,\cdot\,, e_2\rangle.$$

Let $f_1(z)$ and $f_2(z)$ be m.f.s from $H^2_{n\times n}$ and $H^2_{n\times m}$ with columns from K_b which are defined for a fixed m.f. $f \in L^\infty_{n\times n}$ by the formulas

$$f_1 = P_{K^n_b} M^*_\chi f, \qquad f_2 = P_{K^m_b} f^* b. \tag{2.3}$$

Lemma 2.1. *If $f \in L^\infty_{n\times n}$, then the compression $T = T(f)$ belongs to $\mathbb{K}(b)$ and*

$$R := TS - ST = e_2\langle\,\cdot\,, f_2\rangle - f_1\langle\,\cdot\,, e_1\rangle, \tag{2.4}$$

where e_1, e_2 and f_1, f_2 are defined by the formulas (2.1) and (2.3), respectively.

Proof. We have

$$\begin{aligned} Rh &= (TS - ST)h = (P_{K_b} M_f P_{K_b} M^*_\chi - P_{K_b} M^*_\chi P_{K_b} M_f)h \\ &= P_{K_b} M^*_\chi P_{bH^2_m} M_f h - P_{K_b} M_f M^*_\chi h(0) \\ &= P_{K_b} M^*_\chi P_{b\mathbb{C}^m} M_f h - f_1(\zeta)\langle h, e_1\rangle \\ &= e_2(\zeta)\langle h, f_2\rangle - f_1(\zeta)\langle h, e_1\rangle, \end{aligned}$$

and hence the assertion. □

2.2. The basic formula

$$h(z) = \langle (I - zS)^{-1}h, e_1\rangle, \qquad h \in K_b, \tag{2.5}$$

is easily deduced from the Cauchy formula

$$h(z) = \frac{1}{2\pi i}\oint_{\partial\mathcal{D}} \frac{h(\zeta)}{\zeta - z}\,d\zeta = \frac{1}{2\pi}\int_{\partial\mathcal{D}} \frac{h(\zeta)}{1 - z\bar{\zeta}}\,|d\zeta|,$$

which is valid for $h \in H^2_n$.

For an operator $A \in [X, Y]$ we define the operator $\mathbb{A} \in [X^p, Y^p]$ by the rule

$$\mathbb{A}[x_1, \ldots, x_p] = [Ax_1, \ldots, Ax_p]$$

for every choice of vectors $x_k \in X$, $1 \leq k \leq p$. If $[x_1, \ldots, x_p]$ is a m.f. from K^p_b ($\subset H^2_{n\times p}$) and $A \in [K_b]$, then $\mathbb{A} \in [K^p_b]$.

Now, upon applying this notation to the operators T and S, and invoking formula (2.5), we obtain

$$H(z) = \langle (I - z\mathbb{S})^{-1}H, e_1\rangle, \qquad H \in K^p_b\ , \tag{2.6}$$

and

$$\begin{aligned} (T^{-1}h)(z) &= \langle (I - zS)^{-1}T^{-1}h, e_1\rangle, &\qquad h \in K_b, \\ (\mathbb{T}^{-1}h)(z) &= \langle (I - z\mathbb{S})^{-1}\mathbb{T}^{-1}H, e_1\rangle, &\qquad H \in K^p_b, \end{aligned} \tag{2.7}$$

if $T^{-1} \in [K_b]$.

2.3. The main result of the paper is the following theorem.

Theorem 2.2. *Let b be an inner matrix function of order $n \times m$, let $f \in L^\infty_{n\times n}$, and let $T = T(f)$ be an invertible compression on K_b. Let e_1, e_2 and f_1, f_2 be defined by the formulas (2.1) and (2.3), and let $x\,(\in K^n_b)$, $y\,(\in K^m_b)$, $u\,(\in K^n_b)$, $v\,(\in K^m_b)$ be the solutions of the equations*

$$\mathbb{T}^*x = e_1, \qquad \mathbb{T}y = e_2, \qquad \mathbb{T}^*u = f_2, \qquad \mathbb{T}v = f_1. \tag{2.8}$$

Then the following inversion formula holds:

$$(T^{-1}h)(z) = (I_n - zv(z))\frac{1}{2\pi i}\oint_{\partial\mathcal{D}} \frac{x^*(\zeta)h(\zeta)}{\zeta - z}d\zeta + zy(z)\frac{1}{2\pi i}\oint_{\partial\mathcal{D}} \frac{u^*(\zeta)h(\zeta)}{\zeta - z}d\zeta \tag{2.9}$$

for $h \in K_b$.

Proof. By Lemma 2.1, we have formula (2.4) for the commutant R of the operators $T = T(f)$ and S. Using the identity

$$(I - zS)^{-1}T^{-1} - T^{-1}(I - zS)^{-1} = z(I - zS)^{-1}T^{-1}RT^{-1}(I - zS)^{-1}$$

and the formula (2.7), we have that

$$(T^{-1}h)(z) = \langle T^{-1}(I - zS)^{-1}h, e_1\rangle + z\langle (I - zS)^{-1}T^{-1}RT^{-1}(I - zS)^{-1}h, e_1\rangle.$$

But

$$\begin{aligned}\langle T^{-1}(I - zS)^{-1}h, e_1\rangle &= \langle (I - zS)^{-1}h, \mathbb{T}^{*-1}e_1\rangle \\ &= \langle (I - zS)^{-1}h, x\rangle \\ &= \frac{1}{2\pi i}\oint_{\partial\mathcal{D}} \frac{x^*(\zeta)h(\zeta)}{\zeta - z}d\zeta\end{aligned}$$

and

$$\begin{aligned}&\langle (I - zS)^{-1}T^{-1}RT^{-1}(I - zS)^{-1}h, e_1\rangle \\ &= \langle (I - zS)^{-1}T^{-1}(e_2\langle T^{-1}(I - zS)^{-1}h, f_2\rangle - f_1\langle T^{-1}(I - zS)^{-1}h, e_1\rangle), e_1\rangle \\ &= \langle (I - zS)^{-1}\mathbb{T}^{-1}e_2, e_1\rangle\langle (I - zS)^{-1}h, \mathbb{T}^{*-1}f_2\rangle \\ &\quad - \langle (I - zS)^{-1}\mathbb{T}^{-1}f_1, e_1\rangle\langle (I - zS)^{-1}h, \mathbb{T}^{*-1}e_1\rangle \\ &= y(z)\frac{1}{2\pi i}\oint_{\partial\mathcal{D}} \frac{u^*(\zeta)h(\zeta)}{\zeta - z}d\zeta - v(z)\frac{1}{2\pi i}\oint_{\partial\mathcal{D}} \frac{x^*(\zeta)h(\zeta)}{\zeta - z}d\zeta.\end{aligned}$$

In the last equality we have used formula (2.6) for $y(z)$ and $v(z)$. □

Remark 2.3. It is easy to check that in the scalar case, when $n = 1$, we have $e_1(\zeta) = \zeta^{-1}b(\zeta)e_2^*(\zeta)$, $f_2(\zeta) = \zeta^{-1}b(\zeta)f_1^*(\zeta)$, $x(\zeta) = \zeta^{-1}b(\zeta)y^*(\zeta)$ and $u(\zeta) = \zeta^{-1}b(\zeta)v^*(\zeta)$. Because of this, it is enough to consider only two of the equations in the system (2.8), e.g., the pair $\mathbb{T}y = e_2$ and $\mathbb{T}v = f_2$.

2.4. In the next theorem we present a simpler inversion formula which holds under additional assumptions.

Theorem 2.4. *Suppose that in addition to the conditions imposed in Theorem 2.2 $b(z) = b_0(z)I_n$ where $b_0(z)$ is a scalar inner function and the solutions x, y, u, v of the equations (2.8) are m.f.s from $H^\infty_{n\times n}$. Then for $T = T(f)$ we have the inversion formula*

$$T^{-1} = T^*(x) + S^*(T(y)T^*(u) - T(v)T^*(x)). \tag{2.10}$$

Proof. If $b(z) = b_0(z)I_n$ where b_0 is a scalar inner function, then $H^*(\zeta)h(\zeta) \perp bH^2_n$ for $h \in K_b$ and $H \in H^\infty_{n\times n}$. Consequently,

$$T(H^*)h = P_{K_b}M_{H^*}h = P_{H^2_n}M_{H^*}h$$

and

$$(T(H^*)h)(z) = \frac{1}{2\pi i}\oint_{\partial\mathcal{D}} \frac{H^*(\zeta)h(\zeta)}{\zeta - z}d\zeta = \langle (I - zS)^{-1}h, H\rangle.$$

If x, y, u, v are m.f.s from $H^\infty_{n\times n}$, then

$$\langle (I - zS)^{-1}h, x\rangle = (T(x^*)h)(z) = (T^*(x)h)(z)$$

and

$$\langle (I - zS)^{-1}h, u\rangle = (T^*(u)h)(z).$$

Thus, because of the additional conditions assumed in Theorem 2.4, we can rewrite formula (2.9) in the form

$$(T^{-1}h)(\zeta) = (T^*(x)h)(\zeta) + \zeta y(\zeta)(T^*(u)h)(\zeta) - \zeta v(\zeta)(T^*(x)h)(\zeta).$$

By the orthoprojection of both parts of this identity on K_b we have

$$(T^{-1}h)(\zeta) = (T^*(x)h)(\zeta) + (S^*T(y)T^*(u)h)(\zeta) - (S^*T(v)T^*(x)h)(\zeta).$$

This completes the proof. □

2.5. Let $f = a + c^*$ where $a, c \in H^\infty_{n\times n}$, and let $b(z) = b_0 I_n$ where b_0 is a scalar inner function. Then

$$\begin{aligned} T &:= T(f) = T(a) + T(c^*), \qquad ST(c^*) = T(c^*)S, \\ R &= TS - ST = T(a)S - ST(a) \\ &= (I - SS^*)T(a)S - ST(a)(I - S^*S) \\ &= e_2\langle\cdot, \mathbb{S}^*\mathbb{T}^*(a)e_2\rangle - \mathbb{S}\mathbb{T}(a)e_1\langle\cdot, e_1\rangle \\ &= e_2\langle\cdot, f_2\rangle - f_1\langle\cdot, e_1\rangle, \end{aligned}$$

i.e., we have formula (2.4) with

(2.11) $$f_1 = \mathbb{S}\mathbb{T}(a)e_1 \quad \text{and} \quad f_2 = \mathbb{S}^*\mathbb{T}^*(a)e_2.$$

Let $a_0 = P_{K_b^n} a$. Then

$$f_1(\zeta) = \zeta^{-1}(a_0(\zeta) - a_0(0)), \qquad f_2(\zeta) = b_0(\zeta)(a_0^*(\zeta) - a_0^*(0)).$$

In the next sections we consider three cases:

(2.12) $$\begin{aligned} &b_0(z) = z^N, \\ &b_0(z) = \Pi_1^N (z - z_k)/(1 - \bar{z}_k z) \quad \text{and} \\ &b_0(z) = \exp(-a(1-z)/(1+z)), \quad a > 0. \end{aligned}$$

For these choices of $b_0(\zeta)$ we can obtain from formula (2.10) inversion formulas for block-Toeplitz matrices, block-Pick matrices and block-Toeplitz integral operators in $L^2_n(0, a)$ $(a < +\infty)$, respectively.

3. Inversion formulas for block-Toeplitz and block-Pick matrices

3.1. Inversion formulas for block-Toeplitz matrices

We now consider the formula (2.10) when $b(z) = z^N I_n$. In this case

$$K_b = \left\{ h(z) : h(z) = \sum_{k=0}^{N-1} h_k z^k,\ h_k \in \mathbb{C}^n \right\},$$

i.e., K_b is unitarily equivalent to the space

$$\mathbb{C}^{nN} = \{h : h = \mathrm{col}[h_k]_1^N,\ h_k \in \mathbb{C}^n\}.$$

For the compression $T = T(f)$ with symbol $f(\zeta) = \Sigma_{-\infty}^{\infty} c_k \zeta^k \in L^{\infty}_{n\times n}$ we have

$$(Th)(\zeta) = \sum_{j=0}^{N-1} \left(\sum_{k=0}^{N-1} c_{j-k} h_k \right) \zeta^j, \qquad h(\zeta) = \sum_{k=0}^{N-1} h_k \zeta^k.$$

The corresponding operator in the space $\mathbb{C}^{nN}$ is defined by the block-Toeplitz matrix

$$T_N = [c_{j-k}]_{j,k=0}^{N-1}.$$

For m.f.s e_1, e_2 and f_1, f_2 we now have:

$$e_1(z) = I_n, \quad e_2(z) = z^{N-1} I_n, \quad f_1(z) = \sum_{k=0}^{N-1} c_{k+1} z^k, \quad f_2(z) = \sum_{k=0}^{N-1} c^*_{N-k} z^k,$$

where $c_N = 0$. The system of equations (2.8) for

$$x = \sum_{k=0}^{N-1} x_k z^k, \quad y = \sum_{k=0}^{N-1} y_k z^k, \quad u = \sum_{k=0}^{N-1} u_k z^k, \quad v = \sum_{k=0}^{N-1} v_k z^k$$

is equivalent to the algebraic system of equations

$$\mathbb{T}_N^* \boldsymbol{x} = \begin{bmatrix} I_n \\ 0_{n\times n} \\ \vdots \\ 0_{n\times n} \end{bmatrix}, \qquad \mathbb{T}_N \boldsymbol{y} = \begin{bmatrix} 0_{n\times n} \\ 0_{n\times n} \\ \vdots \\ I_n \end{bmatrix},$$

$$\mathbb{T}_N^* \boldsymbol{u} = \begin{bmatrix} c_1 \\ c_2 \\ \vdots \\ c_{N-1} \\ 0_{n\times n} \end{bmatrix}, \qquad \mathbb{T}_N \boldsymbol{v} = \begin{bmatrix} 0_{n\times n} \\ c^*_{N-1} \\ \vdots \\ c^*_1 \end{bmatrix}, \tag{3.1}$$

where $\boldsymbol{x} = \mathrm{col}[x_k]_0^{N-1}$, $\boldsymbol{y} = \mathrm{col}[y_k]_0^{N-1}$, $\boldsymbol{u} = \mathrm{col}[u_k]_0^{N-1}$, $\boldsymbol{v} = \mathrm{col}[v_k]_0^{N-1}$.

In this case formula (2.10) gives the following inversion formula for an invertible block-Toeplitz matrix T_N:

$$T_N^{-1} = \begin{bmatrix} 0 & 0 & \cdots & 0 & 0 \\ y_0 & 0 & \cdots & 0 & 0 \\ y_1 & y_0 & \cdots & 0 & 0 \\ \vdots & \vdots & & \vdots & \vdots \\ y_{N-2} & y_{N-3} & \cdots & y_0 & 0 \end{bmatrix} \begin{bmatrix} u_0^* & u_1^* & \cdots & u_{N-1}^* \\ 0 & u_0^* & \cdots & u_{N_2}^* \\ \vdots & \vdots & & \vdots \\ 0 & 0 & \cdots & u_0^* \end{bmatrix}$$

$$- \begin{bmatrix} -I_n & 0 & \cdots & 0 & 0 \\ v_0 & -I_n & \cdots & 0 & 0 \\ v_1 & v_0 & \cdots & 0 & 0 \\ \vdots & \vdots & & \vdots & \vdots \\ v_{N-2} & v_{N-3} & \cdots & v_0 & -I_n \end{bmatrix} \begin{bmatrix} x_0^* & x_1^* & \cdots & x_{N-1}^* \\ 0 & x_0^* & \cdots & x_{N-2}^* \\ \vdots & \vdots & & \vdots \\ 0 & 0 & \cdots & x_0^* \end{bmatrix}.$$

In the scalar case, when $n = 1$, we have $x_k^* = y_{N-k-1}$ and $u_k^* = v_{N-k-1}$. This inversion formula is known, see, e.g., [5], [9] and [4].

3.2. Inversion formulas for block-Pick matrices

Let $b(z) = b_0(z)I_n$ where $b_0(z)$ is a finite Blaschke product with simple zeros z_k $(\in \mathcal{D})$, $1 \le k \le N$. Then

$$K_b = \left\{ h(\zeta) : h(\zeta) := b_0(\zeta) \sum_{k=1}^{N} h_k(\zeta - z_k)^{-1},\ h_k \in \mathbb{C}^n \right\}.$$

Let U $(\in [\mathbb{C}^{nN}, K_b])$, V $(\in [K_b, \mathbb{C}^{nN}])$, and B $(\in [\mathbb{C}^{nN}])$ be the operators which are defined by the formulas

$$U \operatorname{col}[h_k]_{k=1}^N = b_0(\zeta) \sum_{k=1}^{n} h_k(\zeta - z_k)^{-1}, \qquad Vh(\zeta) = \operatorname{col}[h(z_k)]_{k=1}^N,$$

and

$$B \operatorname{col}[h_k]_{k=1}^N = \operatorname{col}[(b_0'(z_k))^{-1} h_k],$$

respectively. The operator B is defined by the block-diagonal matrix B with diagonal blocks $(b_0'(z_k))^{-1} I_n$ where

$$b_0'(z_k) = (1 - |z_k|^2)^{-1} b_k(z_k) \quad \text{and} \quad b_k(z) = b_0(z) \frac{1 - \bar{z}_k z}{z - z_k}, \quad 1 \le k \le N.$$

Then

$$U^{-1} = BV \quad \text{and} \quad V^{-1} = UB.$$

Let $\mathcal{P}(\boldsymbol{a};\boldsymbol{c})$, $\mathcal{P}(\boldsymbol{a})$ and $\mathcal{P}$ denote the block-Pick matrices

$$\mathcal{P}(\boldsymbol{a};\boldsymbol{c}) = \left[\frac{a_k + c_j^*}{1 - z_k\bar{z}_j}\right]_{j,k=1}^N, \quad \mathcal{P}(\boldsymbol{a}) = \left[\frac{a_k}{1 - z_k\bar{z}_j}\right]_{j,k=1}^N, \quad \mathcal{P} = \left[\frac{I_n}{1 - z_k\bar{z}_j}\right]_{j,k=1}^N,$$

respectively, where $\boldsymbol{a} = \mathrm{col}[a_k]_1^N$, $a_k \in \mathbb{C}^{n\times n}$, $\boldsymbol{c} = \mathrm{col}[c_k]_1^N$, $c_k \in \mathbb{C}^{n\times n}$.

Lemma 3.5. *If $a \in H^\infty_{n\times n}$, $c \in H^\infty_{n\times n}$, $a_k = a(z_k)$ and $c_k = c(z_k)$, $1 \le k \le N$, then*

(3.3) $$\mathcal{P}(\boldsymbol{a};\boldsymbol{c}) = U^*T(a + c^*)U.$$

Proof. If $a \in H^\infty_{n\times n}$ and $a(z_k) = a_k$, $1 \le k \le N$, then

$$\begin{aligned} T(a)U\ \mathrm{col}[h_k]_1^N &= b_0(\zeta)\sum_{K=1}^N a_k h_k(\zeta - z_k)^{-1}, \\ (T(a)\,U\,\mathrm{col}[h_k]_1^N, U\,\mathrm{col}[g_k]_1^N)_{K_b} &= \sum_{j=1}^n\sum_{k=1}^n\left(\frac{a_k h_k}{1 - z_k\bar{z}_j}, g_j\right)_{\mathbb{C}^n} \\ &= (\mathcal{P}(\boldsymbol{a})\,\mathrm{col}[h_k], \mathrm{col}[g_k])_{\mathbb{C}^{nN}}. \end{aligned}$$

So we have

$$\mathcal{P}(\boldsymbol{a}) = U^*T(a)U.$$

Because

$$\mathcal{P}(\boldsymbol{a};\boldsymbol{c}) = \mathcal{P}(\boldsymbol{a}) + \mathcal{P}^*(\boldsymbol{c}), \qquad \mathcal{P}^*(\boldsymbol{c}) = U^*T^*(c)U$$

and

$$T(a + c^*) = T(a) + T(c^*) = T(a) + T^*(c),$$

we have the relation (3.3). In particular,

(3.4) $$\mathcal{P} = U^*U.$$

For $\boldsymbol{a} = \mathrm{col}[a_k]_1^N$ $(\in \mathbb{C}^{nN\times n})$ there exists a unique m.f. a such that $a \in K_b^n$ and $a(z_k) = a_k$, $1 \le k \le N$. This m.f. is given by the formula

(3.5) $$a(z) = (\mathbb{V}^{-1}\boldsymbol{a} =)\, b_0(z)\sum_{k=1}^N (b_0'(z_k))^{-1}a_k(z - z_k)^{-1}.$$

Thus for the block-Pick matrix $\mathcal{P}(\boldsymbol{a};\boldsymbol{c})$ there exist m.f.s $a = \mathbb{V}^{-1}\boldsymbol{a}$ and $c = \mathbb{V}^{-1}\boldsymbol{c}$ such that $\mathcal{P}(\boldsymbol{a};\boldsymbol{c})$ and the compression $T(a+c^*)$ are linked by (3.3). This relation permits us to obtain the following inversion formulas for invertible block-Pick matrices $\mathcal{P}(\boldsymbol{a};\boldsymbol{c})$ via the inversion formula (2.10) for the corresponding compression $T = T(a+c^*)$:

$$\begin{aligned}\mathcal{P}^{-1}(\boldsymbol{a};\boldsymbol{c}) &= \mathcal{P}^{-1}\mathcal{P}^*(\boldsymbol{x})\mathcal{P}^{-1}+\mathcal{P}^{-1}\mathcal{P}(\widetilde{\boldsymbol{y}})\mathcal{P}^{-1}\mathcal{P}^*(\boldsymbol{u})\mathcal{P}^{-1}\\ &\quad -\mathcal{P}^{-1}\mathcal{P}(\widetilde{\boldsymbol{v}})\mathcal{P}^{-1}\mathcal{P}^*(\boldsymbol{x})\mathcal{P}^{-1},\end{aligned} \tag{3.6}$$

where

$$\boldsymbol{x} = \operatorname{col}[x_k]_1^N = \mathbb{V}x, \quad x_k = x(z_k), \quad \widetilde{\boldsymbol{y}} = \operatorname{col}[z_k y_k]_1^N, \quad y_k = y(z_k),$$
$$\boldsymbol{u} = \operatorname{col}[u_k]_1^N = \mathbb{V}u, \quad u_k = u(z_k), \quad \widetilde{\boldsymbol{v}} = \operatorname{col}[z_k v_k]_1^N, \quad v_k = v(z_k),$$

x, y, u, v are the solutions of the equations (2.8) for the corresponding $T = T(a+c^*)$ $(a = \mathbb{V}^{-1}\boldsymbol{a},\ c = \mathbb{V}^{-1}\boldsymbol{c})$.
We would like to obtain a formula for $\mathcal{P}^{-1}$ and the system of algebraic equations for $\boldsymbol{x},\ \boldsymbol{y}\ (= \operatorname{col}[y_k]_1^N),\ \boldsymbol{u},\ \boldsymbol{v}\ (= \operatorname{col}[v_k]_1^N)$ which is equivalent to the system (2.8). At first we will obtain a formula for $\mathcal{P}^{-1}$. Let $\mathcal{P}\xi = \eta$, $\xi = \operatorname{col}[\xi_k]_1^N\ (\in \mathbb{C}^{nN})$, $\eta = \operatorname{col}[\eta_k]_1^N (\in \mathbb{C}^{nN})$. Then

$$\sum_{k=1}^N \frac{\xi_k}{1 - z_k\bar{z}_j} = \eta_j, \qquad 1 \le j \le N,$$
$$\sum_{k=1}^N \frac{1}{2\pi i}\oint_{\partial\mathcal{D}} \frac{\xi_k d\zeta}{(\zeta - z_k)(1-\bar{z}_j\zeta)} = \eta_j, \qquad 1 \le j \le N,$$
$$\sum_{k=1}^N \frac{1}{2\pi i}\oint_{\partial\mathcal{D}} \frac{\bar{\xi}_k d\zeta}{(1-\bar{z}_k\zeta)(\zeta - z_j)} = \bar{\eta}_j, \qquad 1 \le j \le N.$$

This system of equations can be written in the form

$$\bar{\xi}(z_j) = \bar{\eta}_j, \quad 1 \le j \le N, \qquad \text{where } \bar{\xi}(z) = \sum_{k=1}^N \frac{\bar{\xi}_k}{1 - \bar{z}_k z} \ (\in K_b).$$

If $\bar{\eta} = \operatorname{col}[\bar{\eta}_j]_1^N$, then $\bar{\xi}(z) = \mathbb{V}^{-1}\bar{\eta}$. Consequently,

$$\bar{\xi}(z) = b_0(z)\sum_{k=1}^N (b_0'(z_k))^{-1}\bar{\eta}_j(z - z_k)^{-1},$$
$$\xi(z) := \overline{\bar{\xi}(\frac{1}{\bar{z}})} = \sum_{k=1}^N \frac{\xi_k z}{z - z_k} = b_0^{-1}(z)\sum_{k=1}^N \overline{(b_0'(z_k))^{-1}}\eta_k z(1 - \bar{z}_k z)^{-1}.$$

From this we have

$$\xi_j = (b_0'(z_j))^{-1}\sum_{k=1}^N \overline{(b_0'(z_k))^{-1}}(1 - \bar{z}_k z_j)^{-1}\eta_k.$$

Thus

$$\mathcal{P}^{-1} = B\mathcal{P}'B^*, \tag{3.7}$$

where

$$\mathcal{P}' = (\overline{\mathcal{P}} =) \left[\frac{I_n}{1-\bar{z}_k z_j}\right]_{j,k=1}^N \quad \text{and} \quad B = \operatorname{diag}\left[(b_0'(z_1))^{-1}I_n, \dots, (b_0'(z_N))^{-1}I_n\right].$$

The system (2.8) is equivalent to the algebraic system

$$\begin{aligned} \mathbb{V}\mathbb{T}^*\mathbb{V}^{-1}\boldsymbol{x} &= \mathbb{V}e_1, & \mathbb{V}\mathbb{T}\mathbb{V}^{-1}\boldsymbol{y} &= \mathbb{V}e_2, \\ \mathbb{V}\mathbb{T}^*\mathbb{V}^{-1}\boldsymbol{u} &= \mathbb{V}f_2, & \mathbb{V}\mathbb{T}\mathbb{V}^{-1}\boldsymbol{v} &= \mathbb{V}f_1. \end{aligned} \tag{3.8}$$

Let

$$\boldsymbol{e}_1 = \mathbb{V}e_1, \qquad \boldsymbol{e}_2 = \mathbb{V}e_2, \qquad \boldsymbol{f}_1 = \mathbb{V}f_1, \qquad \boldsymbol{f}_2 = \mathbb{V}f_2.$$

Because

$$\begin{aligned} e_1(z) &= (1-b_0(z)\overline{b_0(0)})I_n, & e_2(z) &= z^{-1}(b_0(z)-b_0(0))I_n, \\ f_1(z) &= z^{-1}(a(z)-a(0)), & f_2(z) &= b_0(z)\left(a^*\left(\frac{1}{\bar{z}}\right) - a^*(0)\right), \end{aligned}$$

where $a(z)$ is defined by formula (3.5), we have

$$\begin{aligned} \boldsymbol{e}_1 &= \operatorname{col}[e_1(z_k)]_1^N, & e_1(z_k) &= I_n, \\ \boldsymbol{e}_2 &= \operatorname{col}[e_2(z_k)]_1^N, & e_2(z_k) &= -z_k^{-1}b_0(0)I_n \end{aligned}$$

$$(e_2(z_k) = b_0'(z_k)I_n \ \text{ if } \ z_k = 0),$$

$$\begin{aligned} \boldsymbol{f}_1 &= \operatorname{col}[f_1(z_k)]_1^N, & \boldsymbol{f}_1(z_k) &= z_k^{-1}(a_k - a(0)), \\ \boldsymbol{f}_2 &= \operatorname{col}[f_2(z_k)]_1^N, & \boldsymbol{f}_2(z_k) &= z_k\sum_{j=1}^N \overline{(b_0'(z_j))}^{-1}\frac{1}{1-\bar{z}_j z_k}a_j^* \end{aligned}$$

$$(\boldsymbol{f}_2 = \mathcal{D}\mathcal{P}'B^*\operatorname{col}[a_j^*]_1^N, \qquad \mathcal{D} := \operatorname{diag}[z_1 I_n, \dots, z_N I_n]),$$

$$\boldsymbol{f}_2 = \mathcal{D}^{-1}(I + b_0(0)B\mathcal{D}^{-1})\operatorname{col}[a_k].$$

(If $b_0(0) = 0$ we can consider $z_1 = 0$. Then $e_2(z_1) = b_0'(z_1)I_n$, $e_2(z_k) = 0_{n\times n}$ for $k > 1$, $b_0'(z_1) = (-1)\Pi_{k=2}^{N-1}z_k$, $a(0) = a_1$, $f_2(z_1) = 0$ and $f_1(z_1) = a'(0)$.) Furthermore,

$$\mathbb{V}\mathbb{T}\mathbb{V}^{-1} = \mathbb{V}\mathbb{U}^{*-1}\mathbb{P}(\boldsymbol{a};\mathbf{c})\mathbb{U}^{-1}\mathbb{V}^{-1},$$

$$\mathbb{U}^{-1}\mathbb{V}^{-1} = \mathbb{B},$$

and

$$\mathbb{V}\mathbb{U}^{*-1} = \mathbb{V}\mathbb{U}\mathbb{P}^{-1} = \mathbb{B}^{-1}\mathbb{P}^{-1}.$$

Thus the system (3.8) is equivalent to the system

$$\begin{aligned} &\mathbb{P}^*(\boldsymbol{a};\boldsymbol{c})\mathbb{B}\boldsymbol{x} = \mathbb{P}\mathbb{B}\boldsymbol{e}_1, &&\mathbb{P}(\boldsymbol{a};\boldsymbol{c})\mathbb{B}\boldsymbol{y} = \mathbb{P}\mathbb{B}\boldsymbol{e}_2,\\ &\mathbb{P}(\boldsymbol{a};\boldsymbol{c})\mathbb{B}\boldsymbol{u} = \mathbb{P}\mathbb{B}f_2, &&\mathbb{P}(\boldsymbol{a};\boldsymbol{c})\mathbb{B}\boldsymbol{v} = \mathbb{P}\mathbb{B}f_1.\end{aligned} \tag{3.9}$$

Thus we have obtained the inversion formulas (3.7) and (3.6) for the block-Pick invertible matrix $\mathcal{P}(\boldsymbol{a};\boldsymbol{c})$, where $\boldsymbol{x},\boldsymbol{y},\boldsymbol{u},\boldsymbol{v}$ are defined as the solutions of the equations (3.9). □

4. Inversion formulas for block-Toeplitz integral operators in $L_n^2(0,a)$ $(a<\infty)$

4.1. Let

$$b(z) = \exp(-a(1-z)/(1+z))I_n, \qquad a>0.$$

Let $L^p_{n\times m}(\mathbb{R})$ denote the space of $n\times m$ m.f.s $H(\mu)$ which are measurable on the real axis with $\|H\|_p<\infty$ where

$$\begin{aligned}\|H\|_p &= \operatorname{ess\,sup}\{\|H(\mu)\| : \mu\in\mathbb{R}\} \quad \text{if } \ p=+\infty,\\ \|H\|_p^p &= \frac{1}{\pi}\int_{-\infty}^{\infty}(\operatorname{Sp}H^*(\mu)H(\mu))^{p/2}d\mu \quad \text{if } \ 1\le p<+\infty.\end{aligned}$$

$H^p_{n\times m}(\mathbb{C}_+)$ is the Hardy space of $n\times m$ m.f.s $H(\lambda)$ which are analytic in the open upper half-plane $\mathbb{C}_+=\{\lambda : \operatorname{Im}\lambda>0\}$ with $\|H\|_p<\infty$ where

$$\begin{aligned}\|H\|_p &= \sup\{\|H(\lambda)\| : \lambda\in\mathbb{C}_+\} \quad \text{if } \ p=+\infty,\\ \|H\|_p^p &= \frac{1}{\pi}\sup_{y>0}\int_{-\infty}^{\infty}(\operatorname{Sp}H^*(\mu+iy)H(\mu+iy))^{p/2}d\mu \quad \text{if } \ 1\le p<+\infty.\end{aligned}$$

By the identification of m.f.s $H(\lambda)$ $(\in H^p_{n\times m}(\mathbb{C}_+))$ with their boundary values $H(\mu)$ $(\in L^p_{n\times m}(\mathbb{R}))$ one can consider $H^p_{n\times m}(\mathbb{C}_+)$ as a subspace of $L^p_{n\times m}(\mathbb{R}))$; $L^2_{n\times m}(\mathbb{R})$ is a Hilbert space, $L^2_n(\mathbb{R}) = L^2_{n\times 1}(\mathbb{R})$, $H^2_n(\mathbb{C}_+) = H^2_{n\times 1}(\mathbb{C}_+)$.

We would like to consider the unitary operator $\mathcal{T}$ $(\in [L^2_n, L^2_n(\mathbb{R})])$ which is defined by the formula

$$(\mathcal{T}h)(\mu) = (i+\mu)^{-1}h((i-\mu)/(i+\mu)), \qquad h\in L^2_n.$$

(Recall that the symbols L^2_n and H^2_n without modifying brackets refer to spaces of functions defined on $\partial\mathcal{D}$ and $\mathcal{D}$, respectively.) It is known that $\mathcal{T}H^2_n = H^2_n(\mathbb{C}_+)$ and that for the specified choice of $b(z)$ we have

$$\mathcal{T}K_b = \{H(\lambda) : H(\lambda) = \frac{1}{\sqrt{2\pi}}\int_0^a e^{i\lambda t}\check{H}(t)dt,\ \check{H}\in L^2_n(0,a)\}.$$

Let $\mathcal{F}$ ($\in [L_n^2(\mathbb{R})]$) denote the Fourier transform

$$(\mathcal{F}\check{H})(\mu) = \frac{1}{\sqrt{2\pi}} \int_{-\infty}^{\infty} e^{i\mu t} \check{H}(t)dt, \qquad \check{H} \in L_n^2(\mathbb{R}).$$

Then the operator $\mathcal{L} = \mathcal{F}^{-1}\mathcal{T}$ is unitary from L_n^2 onto $L_n^2(\mathbb{R})$,

$$\mathcal{L}H_n^2 = L_n^2(0,\infty), \qquad \mathcal{L}K_b = L_n^2(0,a),$$

and its restriction $\mathcal{L}_0 = \mathcal{L}|K_b$ is unitary from K_b onto $L_n^2(0,a)$.

Let $F \in L_{n\times n}^{\infty}(\mathbb{R})$ and let M_F ($\in [L_n^2(\mathbb{R})]$) be the operator of multiplication by the m.f. $F(\mu)$. Then

$$\widetilde{K}(F) = \mathcal{F}^{-1} M_F \mathcal{F}$$

is the block-Toeplitz integral operator in $L_n^2(\mathbb{R})$ with symbol F, i.e.,

$$(4.1) \qquad (\widetilde{K}(F)\check{H})(t) = \int_{-\infty}^{+\infty} \check{F}(t-s)\check{H}(s)ds,$$

where the kernel $\check{F}(\cdot)$ of the considered integral operator is a generalized m.f.,

$$(4.2) \qquad \check{F}(t) = \frac{1}{2\pi} \int_{-\infty}^{+\infty} e^{-i\mu t} F(\mu)d\mu, \qquad F \in L_{n\times n}^{\infty}(\mathbb{R}).$$

If

$$(4.3) \qquad F \in L_{n\times n}^{\infty}(\mathbb{R}) \cap L_{n\times n}^{2}(\mathbb{R}),$$

then the m.f. $\check{F}(t)$ in (4.1) and (4.2) belongs to $L_{n\times n}^2(\mathbb{R})$. In this case one can consider the formula (4.1) in the ordinary sense.

We would like to consider the compression

$$K(F) = P_{L_n^2(0,a)} \widetilde{K}(f)|L_n^2(0,a),$$

i.e.,

$$(4.4) \qquad (K(F)\check{H})(t) = \int_{-\infty}^{\infty} \check{F}(t-s)\check{H}(s)ds, \qquad 0 < t < a, \ \ \check{H} \in L_n^2(0,a).$$

Here $\check{F}$ is a generalized m.f. on the interval $(-a,a)$. Under assumption (4.3), $\check{F}$ is a m.f. from $L_n^2(-a,a)$ and one can consider the formula (4.4) for $K(F)$ in the ordinary sense.

If $F \in L_{n\times n}^{\infty}(\mathbb{R})$, then one can write the formula for the compression $K(F)$ in the form

$$(4.5) \qquad (K(F)\check{H})(t) = \frac{d}{dt} \int_0^a \check{F}_0(t-s)\check{H}(s)ds, \qquad 0 < t < a, \ \ \check{H} \in L_n^2(0,a),$$

where $\check{F}_0 \in L^2_{n\times n}(-a,a)$,

$$\text{(4.6)}\qquad \begin{aligned} \check{F}_0(t) &= i\int_0^t \check{G}(\tau)d\tau - i\check{G}(t),\\ \check{G}(t) &= \frac{1}{2\pi}\int_{-\infty}^{\infty} e^{-i\mu t}(i-\mu)^{-1}F(\mu)d\mu, \end{aligned}$$

i.e.,

$$\check{F}_0(t) = \frac{-i}{2\pi}\int_{-\infty}^{\infty}\left(e^{-i\mu t} + \frac{e^{-i\mu t}-1}{i\mu}\right)(i-\mu)^{-1}F(\mu)d\mu.$$

Indeed, let

$$G(\mu) = (i-\mu)^{-1}F(\mu) \quad (\in L^\infty_{n\times n}(\mathbb{R}) \cap L^2_{n\times n}(\mathbb{R})).$$

Then if $H(\mu)$ belongs to $L^2_n(\mathbb{R})$ one has

$$G(\mu)H(\mu) \in L^2_n(\mathbb{R}), \qquad \mu G(\mu)H(\mu) \in L^2_n(\mathbb{R}),$$

and consequently

$$\check{\Phi}(t) := \int_{-\infty}^{\infty} G(t-s)\check{H}(s)ds \in L^2_n(\mathbb{R}) \quad \text{for } \check{H} \in L^2_n(\mathbb{R}).$$

Moreover, $\check{\Phi}(t)$ is an absolutely continuous function on $\mathbb{R}$ and $\frac{d}{dt}\check{\Phi}(t) \in L^2_n(\mathbb{R})$,

$$(\widetilde{K}(F)\check{H})(t) = i\check{\Phi}(t) - i\frac{d}{dt}\check{\Phi}(t),$$

i.e.,

$$(\widetilde{K}(F)\check{H})(t) = \frac{d}{dt}\int_0^a \check{F}_0(t-s)\check{H}(s)ds, \qquad \check{H} \in L^2_n(0,a),$$

where $\check{F}_0$ is defined by the formulas in (4.6).

The m.f. $\check{G}(t)$ in (4.6) is absolutely continuous and $\frac{d}{dt}\check{G}(t) \in L^2_n(\mathbb{R})$ iff the condition (4.3) holds. In this case

$$\text{(4.7)}\qquad \check{F}(t) = \check{F}_0'(t) = i\check{G}(t) - i\check{G}'(t).$$

Let $A \in H^\infty_{n\times n}(\mathbb{C}_+)$. Then

$$\text{(4.8)}\qquad \begin{aligned} A_r(\lambda) &:= (i+\lambda)^{-1}A(\lambda) \in H^\infty_{n\times n}(\mathbb{C}_+) \cap H^2_{n\times n}(\mathbb{C}_+),\\ A_r(\lambda) &= \int_0^\infty e^{i\lambda t}\check{A}_r(t)dt, \end{aligned}$$

and

$$\text{(4.9)}\qquad \begin{aligned} (K(A)\check{h})(t) &= \frac{d}{dt}\int_0^t \check{A}_0(t-s)\check{H}(s)ds, \qquad 0<t<a,\quad \check{H} \in L^2_n(0,a),\\ (K^*(A)\check{H})(t) &= -\frac{d}{dt}\int_t^a \check{A}_0^*(s-t)\check{H}(s)ds, \qquad 0<t<a,\quad \check{H} \in L^2_n(0,a), \end{aligned}$$

where

$$\check{A}_0(t) = i\int_0^t \check{A}_\tau(\tau)d\tau + i\check{A}_r(t). \tag{4.10}$$

The proof of the formulas (4.9) and (4.10) is the same as the proof of the formulas (4.5) and (4.6).

We would like to obtain the inversion formulas for an invertible compression $K(F)$ using Theorems 2.2 and 2.4. Let $F \in L^\infty_{n\times n}(\mathbb{R})$ and let

$$f(\zeta) = F(i(1-\zeta)/(1+\zeta)) \ \ (\in L^\infty_{n\times n}).$$

Then for the compressions $K(F)$ $(\in [L^2_n(0,a)])$ and $T(f)$ $(\in [K_b])$ we have

$$K(F) = \mathcal{L}_0 T(f)\mathcal{L}_0^{-1}. \tag{4.11}$$

Indeed,

$$\begin{aligned}\widetilde{K}(F) &= \mathcal{F}^{-1}M_F\mathcal{F} = \mathcal{F}^{-1}\mathcal{T}M_f\mathcal{T}^{-1}\mathcal{F} = \mathcal{L}M_f\mathcal{L}^{-1},\\ K(F) &= P_{L^2_n(0,a)}\widetilde{K}(F)|L^2_n(0,a) = P_{\mathcal{L}K_b}\widetilde{K}(F)|\mathcal{L}K_b\\ &= \mathcal{L}P_{K_b}\mathcal{L}^{-1}\widetilde{K}(F)|\mathcal{L}K_b = \mathcal{L}_0(P_{K_b}M_f|K_b)\mathcal{L}_0^{-1} = \mathcal{L}_0T(f)\mathcal{L}_0^{-1}.\end{aligned}$$

Consequently, $K(F)$ is invertible iff $T(f)$ is invertible,

$$K^{-1}(F) = \mathcal{L}_0T^{-1}(f)\mathcal{L}_0^{-1}. \tag{4.12}$$

For a m.f. $w(\zeta)$ from $L^2_{n\times n}$ we denote

$$\check{w}(t) = \mathcal{L}_0 w, \qquad \mathcal{W}(\mu) = (i+\mu)\mathcal{T}\check{w} = w((i-\mu)/(i+\mu)).$$

By computation we find for the m.f. considered in (2.1), (2.3),

$$\begin{aligned}\check{e}_1(t) &= -i\sqrt{2\pi}e^{-t}I_n, & \check{e}_2(t) &= -i\sqrt{2\pi}e^{-(a-t)}I_n,\\ \check{f}_1(t) &= \sqrt{2\pi}\check{G}(t), & \check{f}_2(t) &= -\sqrt{2\pi}\check{G}^*(a-t),\end{aligned} \tag{4.13}$$

where $\check{G}(t)$ is defined in (4.6).

The system (2.8) is equivalent to the following system of integral equations on the interval $(0,a)$:

$$-\frac{d}{dt}\int_0^a \check{F}_0^*(s-t)\check{x}(s)ds = \check{e}_1(t), \qquad \frac{d}{dt}\int_0^a \check{F}_0(t-s)\check{y}(s)ds = \check{e}_2(t),$$

$$-\frac{d}{dt}\int_0^a \check{F}_0^*(s-t)\check{u}(s)ds = \check{f}_2(t), \qquad \frac{d}{dt}\int_0^a \check{F}_0(s-t)\check{v}(s)ds = \check{f}_1(t).$$

If the solutions of these equations are absolutely continuous on $[0,a]$ and if their derivatives belong to $L^2_n(0,a)$, then the corresponding m.f.s x, y, u, v belong to

$H^\infty_{n\times n}$ and X, Y, U, V belong to $H^\infty_{n\times n}(\mathbb{C}_+)$. Thus we can apply Theorem 2.4. In this case, by the formulas (2.10), (4.11) and (4.12), we have the inversion formula

$$K^{-1}(F) = K^*(X) + K(Z)(K(Y)K^*(U) - K(V)K^*(X)), \tag{4.15}$$

where

$$Z(\mu) = (i-\mu)/(i+\mu)I_n, \qquad (K(Z)\check{H})(t) = -\check{H}(t) + 2\int_0^t e^{-(t-s)}\check{H}(s)ds,$$

$$X(\mu) = (i+\mu)\frac{1}{\sqrt{2\pi}}\int_0^a e^{i\mu t}\check{x}(t)dt, \quad Y(\mu) = (i+\mu)\frac{1}{\sqrt{2\pi}}\int_0^a e^{i\mu t}\check{y}(t)dt,$$

$$U(\mu) = (i+\mu)\frac{1}{\sqrt{2\pi}}\int_0^a e^{i\mu t}\check{u}(t)dt, \quad V(\mu) = (i+\mu)\frac{1}{\sqrt{2\pi}}\int_0^a e^{i\mu t}\check{v}(t)dt.$$

In the general case formula (2.9) gives the corresponding inversion formula for $K(F)$:

$$(K^{-1}(F)\check{H})(t)\ (= \mathcal{F}^{-1}\Phi) = \frac{1}{\sqrt{2\pi}}\int_{-\infty}^{\infty} e^{-i\mu t}\Phi(\mu)d\mu \tag{4.16}$$

where

$$\begin{aligned} \Phi(\lambda) &= \left(I_n - \frac{i-\lambda}{i+\lambda}V(\lambda)\right)\frac{1}{2\pi}\int_{-\infty}^{\infty}\frac{X^*(\mu)H(\mu)}{\mu-\lambda}d\mu \\ &\quad + \frac{i-\lambda}{i+\lambda}Y(\lambda)\frac{1}{2\pi}\int_{-\infty}^{\infty}\frac{U^*(\mu)H(\mu)}{\mu-\lambda}d\mu, \qquad H = \mathcal{F}\check{H}. \end{aligned} \tag{4.17}$$

Remark 4.6. The set $\mathcal{L}_0\mathbb{K}(b)\mathcal{L}_0^{-1}$ is the set of the operators K ($\in [L^2_n(0,a)]$) which are quasicommutative with the operator $\mathcal{I}$ of integration

$$\mathcal{I}\check{H}(t) = \int_0^t \check{H}(s)ds, \qquad 0 < t < a.$$

The operators $K(F)$ ($F \in L^\infty_{n\times n}$) belong to this set. One can obtain the inversion formulas for $K(F)$ using the commutator $K(F)\mathcal{I} - \mathcal{I}K(F)$ and the Cauchy formula for the m.f. $H(\lambda)$ in the upper half-plane. L. Sakhnovich obtained his inversion formula for $K(F)$ in another way, using this commutator [11].

Acknowledgements

The research described in this publication was made possible partly by Grants NR UCZ000 and NR UCZ200 from the International Science Foundation.
I wish to express my thanks to the Department of Theoretical Mathematics of the Weizmann Institute of Science for their hospitality. Thanks are also due to Professor Harry Dym for helping to smooth out the English and clarify the text, and to Mrs. Ruby Musrie for her expert typing of this manuscript.

References

[1] BERCOVICI, H., FOIAS, C., TANNENBAUM; A.: On skew Toeplitz operators, I, Operator Theory: Adv. Appl. 29 (1988), 21–43.

[2] FRIEDLANDER, B., MORF, M., KAILATH, T., LJUNG, L.: New inversion formulas for matrices classified in terms of their distance from Toeplitz matrices; Linear Algebra Appl. 27 (1979), 31–60.

[3] GOHBERG. I.C., SEMENCUL, A.: On inversion of finite-section Toeplitz matrices and their continuous analogues; Mat. Issled. 7:2 (1972), 201-224.

[4] GOMEZ, G., LERER, L.: Generalized bezoutians for analytic operator functions and inversion of structured operators; Systems and Networks: Mathematical Theory and Applications, Proceedings of the International Symposium MTNS 1993 held in Regensburg, Germany, August 2–6, 1993, Volume II, 691–696.

[5] HEINIG, G., ROST, K.: Algebraic method for Toeplitz like matrices and operators; Operator Theory: Adv. Appl. 13, Birkhäuser, Basel 1984.

[6] IOHVIDOV, I.S.: Hankel and Toeplitz Matrices and Forms, Algebraic Theory; Birkhäuser, Basel 1982.

[7] KAILATH, T., KUNG S., MORF, M.: Displacement rank matrices and linear equations; J. Math. Anal. Appl. 68 (1979), 395–407.

[8] KOLTRACHT, I., KON, B.A., LERER, L.: Inversion of structured operators; Integral Equations Operator Theory 20 (1994), 410–448.

[9] LERER, L., TISMENETSKY, M.: Generalized bezoutian and the inversion problem for block matrices, I. General Scheme; Integral Equations Operator Theory 9 (1986), 790–819.

[10] AROV, D.Z.: On strictly positive compressions of block-Toeplitz operators; to appear.

[11] SAKHNOVICH, L.A.: Equations with a difference kernel on a finite interval; Uspekhi Mat. Nauk 35:4 (1980), 69–129.

South-Ukrainian Pedagogical University
Prospekt Dimitrova 15, kv. 18
270104 Odessa
Ukraine

1991 Mathematics Subject Classification: Primary 47A62, 47A20; Secondary 45E10, 39B42

Submitted: April 22, 1996

Operator Theory:
Advances and Applications, Vol. 103

Contractive linear relations in Pontryagin spaces

T.YA. AZIZOV and A. DIJKSMA

Contractive linear relations in Hilbert spaces are (graphs of) operators, but in spaces with an indefinite metric this is not always true. The presence of a multivalued part gives difficulties in the characterization of maximal contractive linear relations and in the proof of the existence of invariant subspaces. These topics are considered in this note.

1. Introduction

Linear relations in Hilbert, Pontryagin or Kreĭn spaces play a role when dealing with adjoints of nondensely defined operators and inverses of noninvertible operators. They arise naturally in the study of boundary eigenvalue problems associated with differential operators (see, for example, the recent paper [DLS], where more references can be found) and in the study of the BALL–HELTON generalization of the commutant lifting theorem; see [BH]. In this generalization one no longer assumes that the intertwining operator A is a Hilbert space contraction, but instead that the negative part of the spectrum of the operator $I - A^*A$ is finite. For an alternative proof of the BALL–HELTON theorem we refer to [AADM] and [G]. If one tries to adapt AROCENA's proof of the classical commutant lifting theorem (see [A], and also [S]) to this situation one encounters an isometric linear relation in a Pontryagin space. We believe that the unitary extensions of this linear relation give rise to the liftings of the kind described by the BALL–HELTON theorem and can be used as parameters in the parametrization of all such liftings, in the same way as in [DDMS]. Rather than isometric linear relations we study the contractive ones. The reason is that for isometric linear relations we do not get any better results than for the more general contractive ones; for the exceptions, see Corollaries 4.6 and 5.4, Example 5.5, and Remark 5.8.

The study of linear relations in spaces with an indefinite metric goes back at least to SHMUL'YAN [Sh] who used them in connection with the theory of extensions of operators. RITZNER in [R] also studied algebraic, geometric and spectral properties of contractive and other linear relations in Kreĭn spaces; some of his results are reproven in Section 2.

We briefly outline the contents of this paper.

2. Contractive linear relations. After a review of linear relations we specialize to contractive and maximal contractive linear relations M in Kreĭn and Pontryagin spaces. We establish some relations between M and the linear manifold $\mathcal{N}(M) = \ker M \cap \ker M^{-1}$. We prove some criteria for a contractive closed linear relation to be maximal contractive (see Theorem 2.3).

3. Regularization. In this section we discuss eigenvalues, root spaces and chains associated with an eigenvalue of a linear relation M. In particular, we investigate the role of the linear manifold $\mathcal{R}(M) = \mathcal{R}_0(M) \cap \mathcal{R}_\infty(M)$, where $\mathcal{R}_\lambda(M)$ is the root space of M at the point λ (see Theorem 3.3). Note that the inclusion $\mathcal{N}(M) \subset \mathcal{R}(M)$ always holds. If $\mathcal{R}(M) = \mathcal{N}(M)$ then M is called regular. The regularization of a linear relation M in a Pontryagin space $\mathcal{P}$ is the linear relation (see (3.2))

$$M_r = \text{linear span}\,\{\mathcal{R}(M)^2, M \cap (\mathcal{R}(M)^2)^{\langle\perp\rangle}\},$$

where the orthogonal complement is taken with respect to a Kreĭn space inner product on the cartesian product $\mathcal{P}^2$ of $\mathcal{P}$ with itself (see (2.1)). Let M be a contractive closed linear relation in $\mathcal{P}$. We prove that M_r is regular (see Theorem 3.8) and that if $\mathcal{L} \subset \mathcal{P}$ is an invariant subspace for M then $\mathcal{L} + \mathcal{R}(M)$ is an invariant subspace for M_r (see Theorem 3.9). Finally, we establish a criterion for the maximal contractivity of M in terms of that of M_r (see Theorem 3.11).

4. Criteria for maximality. Here we study regular closed contractive linear relations M in a Pontryagin space $\mathcal{P}$ and investigate when M is maximal contractive (see Theorem 4.4). By going over to the quotient space $\mathcal{P}/\mathcal{R}(M)$ the investigation can be reduced to the case where $\mathcal{R}(M) = \{0\}$. In the proofs of the maximality criteria we make use of well known spectral properties of contractive operators, the Cayley transform, and the Riesz projection.

5. Properties of maximal contractive linear relations. We list some of the properties of the root spaces $\mathcal{R}_0(M)$, $\mathcal{R}_\infty(M)$ and their intersection $\mathcal{R}(M)$, where M is a maximal contractive linear relation. In various formulas these linear manifolds are related with those of the regularization and the adjoint of M.

6. Invariant subspaces. Here we prove some results on the existence of invariant subspaces for a collection $\mathbf{M} = \{M\}$ of maximal contractive linear relations M in a Pontryagin space whose Cayley transforms commute (see Theorems 6.3 and 6.4, and their Corollaries 6.6 and 6.7). The main assumption here is that $\mathcal{R}_\infty(M)$ and $\mathcal{R}(M)$ are independent of $M \in \mathbf{M}$ (see (6.1)).

We assume familiarity with operator theory on Pontryagin and Kreĭn spaces and the results in the books [AI], [B] and [IKL], and the paper [DR]; for additional information about contractive operators we refer to [AL]. For linear relations, see [R], [DS], and [AI], where also other references can be found. Some of the results in this note which are formulated in a Pontryagin space setting are also valid in Kreĭn spaces; we do not always point this out explicitly.

Let $(\mathcal{K}, [\cdot,\cdot])$, or $\mathcal{K}$ for short, be a Kreĭn space. Then $(\mathcal{K}, -[\cdot,\cdot])$ is also a Kreĭn space called the *anti-space* associated with $\mathcal{K}$. Let $\mathcal{K} = \mathcal{K}_+ \oplus \mathcal{K}_-$ be a fundamental decomposition of $\mathcal{K}$ into a positive subspace $\mathcal{K}_+$ and a negative subspace $\mathcal{K}_-$ of $\mathcal{K}$. We indicate the projections from $\mathcal{K}$ onto these subspaces by P_+ and P_-, respectively, and we set $J = P_+ - P_-$. We denote by $|\mathcal{K}|$ the Hilbert space $|\mathcal{K}| = \mathcal{K}_+ \oplus |\mathcal{K}_-|$, where $|\mathcal{K}_-|$ stands for the anti-space associated with $\mathcal{K}_-$, that is, $|\mathcal{K}|$ is the space $\mathcal{K}$ provided with the Hilbert space inner product $[J\cdot,\cdot]$. Whenever we use these notations we assume that a fundamental decomposition has been selected and fixed. The dimension (a natural number or ∞) of the negative subspace in a fundamental decomposition of a Kreĭn space $\mathcal{K}$ is independent of the fundamental decompositon. It is called the *negative index* of $\mathcal{K}$ and is denoted by $\operatorname{ind}_-\mathcal{K}$. A Pontryagin space in this note always is a Kreĭn space with finite negative index. By $\mathbf{L}(\mathcal{H},\mathcal{K})$, or $\mathbf{L}(\mathcal{H})$ if $\mathcal{H}=\mathcal{K}$, we denote the space of all bounded linear operators from the Kreĭn space $\mathcal{H}$ to the Kreĭn space $\mathcal{K}$. We use $\perp$ to denote the "orthogonal complement"; for example, $\langle\perp\rangle$ means that the orthogonal complement is taken with respect to the inner product $\langle\cdot,\cdot\rangle$. $\mathbb{R}$ and $\mathbb{C}$ stand for the sets of real and complex numbers, respectively, $\mathbb{D}$ is the unit disc in $\mathbb{C}$, $\overline{\mathbb{D}}$ its closure and $\partial\mathbb{D}$ its boundary.

2. Contractive linear relations

Let $\mathcal{P}$ be a linear space over the complex numbers $\mathbb{C}$. By definition a *linear relation* in $\mathcal{P}$ is a linear subset of the direct sum space $\mathcal{P}\oplus\mathcal{P}$. An element in $\mathcal{P}\oplus\mathcal{P}$ is written as a pair $\{x,y\}, x,y\in\mathcal{P}$. For example, the graph of an operator T in $\mathcal{P}$, $\{\{x,Tx\}\mid x\in\operatorname{dom}T\}$, is a linear relation in $\mathcal{P}$.

Linear relations are treated as operators; the latter are single valued, but the former may be multivalued. A linear relation M in $\mathcal{P}$ is (the graph of) an operator if and only if $\{0,y\}\in M\Rightarrow y=0$. The set $\{\{x,y\}\in M\mid x=0\}$ is called the *multivalued part* of the linear relation M. We identify an operator with its graph and use for linear relations the same notation as for operators. For example, $\operatorname{dom}M$, $\operatorname{ran}M$ and $\ker M$ stand for the domain, range and kernel of the linear relation M; M^{-1} is the inverse of M and consists of all elements $\{y,x\}$ for which $\{x,y\}$ belongs to M. Note that the range of the multivalued part of M coincides with $\ker M^{-1}$. For $\lambda\in\mathbb{C}$ we denote by $M-\lambda$ the linear relation

$$M-\lambda=\{\{x,y-\lambda x\}\mid\{x,y\}\in M\},$$

and we set $M-\infty=M^{-1}$, so that $(M-\infty)^{-1}$ stands for the linear relation M itself.

Let $\mathcal{P}$ be a Kreĭn space with inner product $[\cdot,\cdot]$. Then a linear relation M in $\mathcal{P}$ is called *closed* if it is a subspace (= closed linear subset) of $\mathcal{P}\oplus\mathcal{P}$. The *adjoint* M^* of M in $\mathcal{P}$ is the closed linear relation

$$M^*=\{\{u,v\}\mid[u,y]-[v,x]=0\text{ for all }\{x,y\}\in M\}.$$

A linear relation M in $\mathcal{P}$ is called *contractive* if for all elements $\{x,y\}\in M$,

$$[y,y]\leq[x,x].$$

It is called *isometric* if equality prevails, or equivalently, if $M^{-1}\subseteq M^*$, and it is called *unitary* if $M^{-1}=M^*$. M is called *maximal contractive* if it is contractive and not properly contained in any other contractive linear relation in $\mathcal{P}$. Clearly, a maximal contractive linear relation is closed. If $\mathcal{P}$ is a Hilbert space, then every contractive linear relation is an operator and bounded, but not necessarily everywhere defined. If the inner product on $\mathcal{P}$ is indefinite then $\ker M^{-1}$ need not be trivial. Evidently, it is a nonpositive subspace of $\mathcal{P}$ and hence $\dim\ker M^{-1} \leq \mathrm{ind}_-\mathcal{P}$.

With $\mathcal{P}$ we associate the Kreĭn space $\mathcal{P}^2 = \mathcal{P} \oplus \mathcal{P}$ provided with the inner product

$$(2.1) \qquad \langle\{x,y\},\{u,v\}\rangle = [x,u] - [y,v], \qquad \{x,y\},\{u,v\} \in \mathcal{P}^2.$$

If M is a linear relation in $\mathcal{P}$, then $M \subseteq \mathcal{P}^2$ and $M^* = (M^{\langle\perp\rangle})^{-1}$. A linear relation in $\mathcal{P}$ is contractive if and only if it is a nonnegative linear subset of $\mathcal{P}^2$.

We recall that a subspace M in a Kreĭn space $\mathcal{H}$ with fundamental decomposition $\mathcal{H} = \mathcal{H}_+ \oplus \mathcal{H}_-$ is nonnegative if and only if it is the "graph" of a contraction $K : \mathrm{dom}\, K \subseteq \mathcal{H}_+ \to |\mathcal{H}_-|$ ($|\mathcal{H}_-|$ is the anti-space of $\mathcal{H}_-$), that is,

$$M = \{x + Kx \mid x \in \mathrm{dom}\, K\}.$$

This connection between nonnegative subspaces and contractions is one–to–one and we indicate this by writing $K = K_M$. We call K_M the *graph contraction* for M. We also recall that a nonnegative subspace M is maximal nonnegative if and only if $\mathrm{dom}\, K_M = \mathcal{H}_+$.

For a linear relation M in $\mathcal{P}$ we set

$$\mathcal{N}(M) = \ker M \cap \ker M^{-1} \quad \text{and} \quad \mathcal{N}(M)^2 = \mathcal{N}(M) \oplus \mathcal{N}(M).$$

Lemma 2.1. *If M is a contractive closed linear relation in a Pontryagin space $\mathcal{P}$, then $\mathcal{N}(M)$ is a neutral subspace of $\mathcal{P}$, $\mathcal{N}(M)^2$ is a neutral subspace of $\mathcal{P}^2$, and the inclusions*

$$\mathcal{N}(M) \subseteq \mathrm{dom}\, M \cap (\mathrm{dom}\, M)^{[\perp]} \cap \mathrm{ran}\, M \cap (\mathrm{ran}\, M)^{[\perp]}, \quad \mathcal{N}(M)^2 \subseteq M \cap M^{\langle\perp\rangle}$$

are valid.

Proof. Assume that u belongs to $\mathcal{N}(M)$. Then $\{u,0\}$ and $\{0,u\}$ belong to M, and hence, since M is contractive, $[u,u]$ is nonpositive and nonnegative at the same time. Thus $[u,u] = 0$, and it follows that $\mathcal{N}(M)$ is a neutral subspace of $\mathcal{P}$. To prove the inclusion for $\mathcal{N}(M)$, we first show that

$$(2.2) \qquad \mathcal{N}(M) \subseteq \mathrm{dom}\, M \cap (\mathrm{dom}\, M)^{[\perp]}.$$

As $\ker M \subseteq \mathrm{dom}\, M$, we have that $\mathcal{N}(M) \subseteq \mathrm{dom}\, M$. M is a nonnegative subspace of $\mathcal{P}^2$. Hence the inner product $\langle\cdot,\cdot\rangle$ restricted to M is semidefinite there, and therefore we may apply the Cauchy-Schwarz-Bunyakowski inequality to the elements of M: We find that for $\{x,y\}$ and $\{u,0\} \in M$,

$$\begin{aligned} |[x,u]|^2 &= |\langle\{x,y\},\{u,0\}\rangle|^2 \\ &\leq \langle\{x,y\},\{x,y\}\rangle\langle\{u,0\},\{u,0\}\rangle \\ &= \langle\{x,y\},\{x,y\}\rangle[u,u]. \end{aligned}$$

Since for all $u \in \mathcal{N}(M)$, $[u,u]=0$, it follows that for all $\{x,y\}\in M$ and $u\in\mathcal{N}(M)$, $[x,u]=0$, that is, $\mathcal{N}(M)\subseteq(\operatorname{dom}M)^{[\perp]}$. Hence (2.2) is valid. Since M^{-1} is a nonpositive subspace of $\mathcal{P}^2$, it is a semidefinite inner product space with respect to the inner product $\langle\cdot,\cdot\rangle$, and the Cauchy-Schwarz-Bunyakowski inequality applied to elements in M^{-1} yields the inclusion

$$\mathcal{N}(M)\subseteq\operatorname{dom}M^{-1}\cap(\operatorname{dom}M^{-1})^{[\perp]}=\operatorname{ran}M\cap(\operatorname{ran}M)^{\perp},$$

which implies the inclusion for $\mathcal{N}(M)$ in the lemma.
The statements about $\mathcal{N}(M)^2$ can be proved in a similar way. □

Let M be a contractive closed linear relation in a Pontryagin space $\mathcal{P}$. Then $\mathcal{P}=(\mathcal{N}(M)+J\mathcal{N}(M))\oplus\mathcal{P}_1$, where $\mathcal{P}_1$ is a Pontryagin space with negative index

$$\operatorname{ind}_-\mathcal{P}_1=\operatorname{ind}_-\mathcal{P}-\dim\mathcal{N}(M).$$

Note that $\mathcal{N}(M)+J\mathcal{N}(M)$ is also a Pontryagin space with

$$\operatorname{ind}_-(\mathcal{N}(M)+J\mathcal{N}(M))=\frac{1}{2}\dim(\mathcal{N}(M)+J\mathcal{N}(M))=\dim\mathcal{N}(M).$$

We decompose $\mathcal{P}^2$ into

$$\mathcal{P}^2=\mathcal{P}_1^2\oplus(\mathcal{P}_1^2)^{\langle\perp\rangle}. \tag{2.3}$$

Lemma 2.2. *Let M be a contractive closed linear relation in a Pontryagin space $\mathcal{P}$ and consider M as a subspace of the Kreĭn space $\mathcal{P}^2$. Then with respect to the decomposition* (2.3) *of $\mathcal{P}^2$, M can be written as*

$$M=M_1\oplus\mathcal{N}(M)^2,$$

where $M_1=M\cap\mathcal{P}_1^2$; M_1 is a contractive closed linear relation in $\mathcal{P}_1$ and $\mathcal{N}(M_1)=\{0\}$.

Proof. By Lemma 2.1, we have that $\mathcal{N}(M)^2\subseteq M\cap(\mathcal{P}_1^2)^{\langle\perp\rangle}$, and this shows that $M_1\oplus\mathcal{N}(M)^2$ is an orthogonal sum decomposed in accordance with the decomposition of $\mathcal{P}^2$ and that $M_1\oplus\mathcal{N}(M)^2\subseteq M$. To prove the other inclusion we first show that $\operatorname{dom}M$ and $\operatorname{ran}M$ are contained in $\mathcal{P}_1\oplus\mathcal{N}(M)$. Let x belong to $\operatorname{dom}M$ or $\operatorname{ran}M$. Then $x=x_1+x_0+Jx_0'$ for some $x\in\mathcal{P}_1$ and $x_0,x_0'\in\mathcal{N}(M)$, and hence, since $\mathcal{N}(M)$ is orthogonal to $\operatorname{dom}M$ and $\operatorname{ran}M$,

$$0=[x,x_0']=[x_1+x_0+Jx_0',x_0']=[Jx_0',x_0'].$$

This shows that $x_0'=0$ and that $x=x_1+x_0$ belongs to $\mathcal{P}_1\oplus\mathcal{N}(M)$. Hence $\operatorname{dom}M$ and $\operatorname{ran}M$ are part of $\mathcal{P}_1\oplus\mathcal{N}(M)$. The inclusion $M\subseteq M_1\oplus\mathcal{N}(M)^2$ now easily follows: Let $\{x,y\}$ belong to M. Then, by what we just have proved, we may write $\{x,y\}$ as $\{x,y\}=\{x_1,y_1\}+\{x_0,y_0\}$, where x_1,y_1 belong to $\mathcal{P}_1$ and x_0,y_0 to $\mathcal{N}(M)$. Since $\mathcal{N}(M)^2$ is contained in M, we have that $\{x_0,y_0\}$ belongs to M and that $\{x_1, y_1\} = \{x, y\} - \{x_0, y_0\}$ also belongs to M and to $\mathcal{P}_1^2$ and hence to M_1. This proves the desired inclusion. Finally,

$$\ker M_1 \cap \ker M_1^{-1} \subseteq \mathcal{P}_1 \cap \ker M \cap \ker M^{-1} = \mathcal{P}_1 \cap \mathcal{N}(M) = \{0\},$$

which proves the last statement in the lemma. □

In the sequel we use the notation introduced above without each time recalling the relevant definitions.

Let $\mathcal{P} = \mathcal{P}_+ \oplus \mathcal{P}_-$ be a fundamental decomposition of $\mathcal{P}$ into a positive subspace $\mathcal{P}_+$ and a negative subspace $\mathcal{P}_-$ of $\mathcal{P}$. We fix a fundamental decomposition of $\mathcal{P}$. If M is a linear relation in $\mathcal{P}$ we denote by $\pi(M)$ the *Potapov-Ginzburg* transform of M. It is the linear relation in the Hilbert space $|\mathcal{P}|$ defined by

$$\pi(M) = \{\{P_+x + P_-y, P_+y + P_-x\} \mid \{x, y\} \in M\}.$$

Clearly, M is closed or contractive in $\mathcal{P}$ if and only if $\pi(M)$ is closed or contractive in $|\mathcal{P}|$, respectively. Some of the following equivalences already appear in [R].

Theorem 2.3. *For a contractive closed linear relation M in a Pontryagin space $\mathcal{P}$ the following statements are equivalent.*

(a) *M is a maximal contractive linear relation in $\mathcal{P}$.*
(b) *M is a maximal nonnegative subspace of $\mathcal{P}^2$.*
(c) *$\pi(M)$ belongs to $\mathbf{L}(\mathcal{P})$ and is a contraction on $|\mathcal{P}|$.*
(d) *The domain of the graph contraction K_M is all of $\mathcal{P}_+ \oplus |\mathcal{P}_-|$.*
(e) *M^* is a maximal contractive linear relation in $\mathcal{P}$.*
(f) *M is a bicontractive linear relation in $\mathcal{P}$, that is, M and M^* are both contractive.*
(g) *M_1 is a maximal contractive linear relation in $\mathcal{P}_1$ and if $\mathcal{N}'$ is a neutral subspace of $\mathcal{P}$ containing $\mathcal{N}(M)$ such that $M_1 + (\mathcal{N}')^2$, where $(\mathcal{N}')^2 = \mathcal{N}' \oplus \mathcal{N}'$, is contractive in $\mathcal{P}$, then $\mathcal{N}' = \mathcal{N}(M)$.*

Proof. The proof of the equivalence between (a) and (b) is an easy exercise and therefore omitted. That (a) and (b) are equivalent to (c) follows from the fact that a contractive linear relation in a Hilbert space (is a bounded linear operator and) is maximal contractive if and only if it is everywhere defined. (c) and (d) are equivalent because $\pi(M)$ and K_M coincide. If (d) holds then K_M^* is the graph contraction for $M^{\langle\perp\rangle} = (M^*)^{-1}$ which is a nonpositive subspace of $\mathcal{P}^2$. It is maximal nonpositive since the domain of K_M^* is all of $|\mathcal{P}_-| \oplus \mathcal{P}_+$. Hence M^* is maximal nonnegative in $\mathcal{P}^2$ and (e) is valid. Clearly, (e) implies (a), since $M^{**} = M$, and (e) implies (f). Now assume that (f) is valid and let $\widetilde{M}$ be a maximal contractive extension of M in $\mathcal{P}$. (A nonnegative subspace in a Kreĭn space is maximal or can be enlarged to a maximal one in the same space, for example, by extending its graph contraction.) Then $(\widetilde{M})^* \subseteq M^*$ and, since (a) and (e) are equivalent, $(\widetilde{M})^*$ is maximal contractive. Hence equality prevails and we see that $M = \widetilde{M}$, that is, (a) holds true. It remains to prove the equivalence between (a) and (g). From the definition of M_1 (see Lemma 2.2): $M_1 = M \cap \mathcal{P}_1^2$, it immediately follows that if M_1 is not maximal contractive in $\mathcal{P}_1$ then M is not maximal contractive in $\mathcal{P}$. Hence (a) implies the first part of (g) and it obviously implies the second part of (g). Now assume that (g) holds and let M' be a contractive closed linear relation in $\mathcal{P}$ which is an extension of M: $M \subseteq M'$. Then, by Lemma 2.1, $\mathcal{N}(M')$ is a neutral subspace of $\mathcal{P}$, $\mathcal{N}(M) \subseteq \mathcal{N}(M')$ and $M_1 + \mathcal{N}(M')^2$ is a contractive linear relation in $\mathcal{P}$, because it is contained in M'. The second part of (g) implies that $\mathcal{N}(M') = \mathcal{N}(M)$. It follows from Lemma 2.2 that

$$M' = M_1' \oplus \mathcal{N}(M)^2,$$

where $M_1' = M' \cap \mathcal{P}_1^2$. The linear relation M_1' is contractive in $\mathcal{P}_1$ and

$$M_1 = M \cap \mathcal{P}_1^2 \subseteq M' \cap \mathcal{P}_1^2 = M_1'.$$

Since, by (g), M_1 is maximal contractive in $\mathcal{P}_1$, the equality $M_1' = M_1$ prevails. The above decompostion of M' now yields that $M' = M$, that is, (a) is valid. □

As the proof shows the equivalence between the statements (a) through (f) of Theorem 2.3 remains true when $\mathcal{P}$ is a Kreĭn space. The equivalence between (a) and (f) can be formulated slightly better by saying (see [R]):

Corollary 2.4. *In a Kreĭn space a linear relation M is maximal contractive if and only if M is closed and bicontractive.*

Corollary 2.5. *Let M be a contractive closed linear relation in a Pontryagin space $\mathcal{P}$. Then M is a maximal contractive linear relation and an operator if and only if* $\operatorname{dom} M = \mathcal{P}$.

Proof. The proof is based on the equivalence between the statements (a) and (c) of Theorem 2.3. We begin with the "only if" part. Since M is an operator, we have that

$$\operatorname{dom} \pi(M) = (P_+ + P_- M) \operatorname{dom} M$$

and $\pi(M) = (P_- + P_+M)(P_+ + P_-M)^{-1}$. Assume that M is maximal. Then $\operatorname{dom} \pi(M) = \mathcal{P}$. We claim that for each $y_- \in \mathcal{P}_-$, there exists a unique element $x_- \in \mathcal{P}_-$ such that $P_-MP_-x_- = y_-$. Since $\mathcal{P}_-$ is finite dimensional, the claim implies that P_-MP_- is a bijection from $\mathcal{P}_-$ onto itself, and hence $\mathcal{P}_-$ is contained in $\operatorname{dom} M$. This in turn implies that $P_+\operatorname{dom} M \subseteq \operatorname{dom} M$ and from $\mathcal{P}_+ = P_+\operatorname{dom} \pi(M) = P_+\operatorname{dom} M$ it readily follows that $\mathcal{P} = \operatorname{dom} M$. We still have to prove the claim. Let y_- belong to $\mathcal{P}_-$. Then, since $\operatorname{dom} \pi(M) = \mathcal{P}$, there exists an $x \in \operatorname{dom} M$ such that $(P_+ + P_-M)x = y_-$. It follows that $P_+x = 0$ and $P_-Mx = y_-$. Writing $x = x_+ + x_-$ with $x_\pm \in \mathcal{P}_\pm$, we have that $x_+ = 0$ and

$P_-MP_-x_- = y_-$. This proves the existence of an element x_- with the desired property. If $y_- = 0$ then, since M is contractive, we have that

$$0 = [P_-MP_-x_-, P_-MP_-x_-] \leq [MP_-x_-, MP_-x_-] \leq [P_-x_-, P_-x_-] \leq 0.$$

and hence $x_- = P_-x_- = 0$. This implies the uniqueness part of the claim.
To prove the "if" part, assume that $\operatorname{dom} M = \mathcal{P}$. Then, as we show in a moment, M is an operator and, since M is contractive, P_-MP_- is a bijection from $\mathcal{P}_-$ onto itself. For each $y \in \mathcal{P}$ we have that

$$(P_+ + P_-M)x = y, \qquad x = P_+y + P_-(P_-MP_-)^{-1}(P_-y - P_-MP_+y).$$

Hence $\operatorname{dom} \pi(M) = \mathcal{P}$ and M is a maximal contractive linear relation. It remains to prove that the assumption that $\operatorname{dom} M = \mathcal{P}$ implies that M is an operator. Let $\operatorname{ind}_-\mathcal{P} = \kappa$ and let $x_1, x_2, \ldots, x_\kappa$ be a basis for $\mathcal{P}_-$. Then these elements belong to $\operatorname{dom} M$ and hence there exist elements $y_1, y_2, \ldots, y_\kappa$ in $\mathcal{P}$ such that the pairs $\{x_i, y_i\}$ belong to M, $i = 1, 2, \ldots, \kappa$. Since M is contractive, the y_i's are negative and linearly independent. Thus there exists a fundamental decomposition $\mathcal{P} = \mathcal{P}'_+ \oplus \mathcal{P}'_-$ where the negative subspace $\mathcal{P}'_-$ is spanned by the y_i's. Assume that M contains an element of the form $\{0, y\}$. We prove that M is an operator by showing that $y = 0$. Write y as $y = y_+ + \sum_{i=1}^{\kappa} \lambda_i y_i$ for some $y_+ \in \mathcal{P}'_+$ and $\lambda_i \in \mathbb{C}$. Then also $\{\sum_{i=1}^{\kappa} \lambda_i x_i, \sum_{i=1}^{\kappa} \lambda_i y_i\}$ belongs to M and so does the pair $\{\sum_{i=1}^{\kappa} \lambda_i x_i, -y_+\}$. Since M is a contraction,

$$0 \leq [y_+, y_+] \leq \Big[\sum_{i=1}^{\kappa} \lambda_i x_i, \sum_{i=1}^{\kappa} \lambda_i x_i\Big] \leq 0.$$

Hence $y_+ = 0$ and $\lambda_1 = \lambda_2 = \cdots = \lambda_\kappa = 0$, that is, $y = 0$. □

If M is a contractive operator in a Pontryagin space $\mathcal{P}$ with inner product $[\cdot, \cdot]$, then for all $x \in \operatorname{dom} M$, $[Mx, Mx] \leq [x, x]$ and hence

$$[P_+Mx, P_+Mx] \leq [P_+x, P_+x] - [P_-x, P_-x] - [P_-Mx, P_-Mx],$$

so that

$$[JMx, Mx] = [P_+Mx, P_+Mx] - [P_-Mx, P_-Mx] \leq [Jx, x] + 2[JP_-Mx, P_-Mx].$$

Since $\dim \mathcal{P}_- < \infty$, this last inequality readily implies the following special case of [AI, Corollary 4.7].

Corollary 2.6. *A closed contractive operator in a Pontryagin space is bounded on its domain.*

3. Regularization

In this section we prove some lemmas to prepare for Section 4 in which we provide more statements which are equivalent to the ones in Theorem 2.3. Some of these statements involve conditions on the *resolvent set* $\rho(M)$ or the *spectrum* $\sigma(M)$ of a linear relation M in a Pontryagin space $\mathcal{P}$. By definition,

$$\rho(M) = \{\lambda \in \mathbb{C} \cup \{\infty\} \mid (M-\lambda)^{-1} \in \mathbf{L}(\mathcal{P})\} \quad \text{and} \quad \sigma(M) = (\mathbb{C} \cup \{\infty\}) \setminus \rho(M).$$

If for $\lambda \in \mathbb{C} \cup \{\infty\}$, $\ker(M-\lambda)$ contains more than just the zero element then λ is called an *eigenvalue* of M; the nonzero elements in $\ker(M-\lambda)$ are called *eigenvectors* associated with λ. The set of all eigenvalues of M is called the *point spectrum* of M and is denoted by $\sigma_p(M)$; it is a subset of the spectrum of M. The definitions of $\rho(M)$ and $\sigma(M)$ involve the topology of $\mathcal{P}$, the definiton of $\sigma_p(M)$ does not and makes sense in any linear space. In the following paragraphs we assume that $\mathcal{P}$ is a linear space. We define the *root space* $\mathcal{R}_\lambda(M)$ of M at the point $\lambda \in \mathbb{C} \cup \{\infty\}$ by

$$\begin{aligned}\mathcal{R}_\lambda(M) = \big\{x \in \operatorname{dom} M \mid \ &\text{there exist } x_1, x_2, \dots, x_{k-1}, x_k \in \operatorname{dom} M \text{ such that} \\ &\{x, \lambda x + x_1\}, \{x_1, \lambda x_1 + x_2\}, \dots, \{x_{k-1}, \lambda x_{k-1} + x_k\}, \{x_k, \lambda x_k\} \in M\big\}\end{aligned}$$

if $\lambda \neq \infty$, and by

$$\mathcal{R}_\infty(M) = \mathcal{R}_0(M^{-1})$$

if $\lambda = \infty$. $\mathcal{R}_\lambda(M)$ is a linear subset of $\mathcal{P}$ and it is nontrivial if and only if λ belongs to the point spectrum of M. Moreover, if λ belongs to $\mathbb{C}$ and

$$\{x_0, \lambda x_0 + x_1\}, \{x_1, \lambda x_1 + x_2\}, \dots, \{x_{k-1}, \lambda x_{k-1} + x_k\}, \{x_k, \lambda x_k\} \in M$$

then all elements $x_0, x_1, \dots, x_k$ belong to $\mathcal{R}_\lambda(M)$; if $x_k \neq 0$ then x_k is an eigenvector and the ordered set $\{x_0, x_1, \dots, x_k\}$ is called a *Jordan chain of length* $k+1$ of M at λ. Similar definitions are used in connection with the eigenvalue ∞: If

$$\{x_1, x_0\}, \{x_2, x_1\}, \dots, \{x_k, x_{k-1}\}, \{0, x_k\} \in M$$

then $x_0, x_1, \dots, x_k$ belong to $\mathcal{R}_\infty(M)$, and if $x_k \neq 0$ then x_k is an eigenvector and the ordered set $\{x_0, x_1, \dots, x_k\}$ is called a *Jordan chain of length* $k+1$ of M at ∞. For a linear relation M in $\mathcal{P}$ we set

$$\mathcal{R}(M) = \mathcal{R}_0(M) \cap \mathcal{R}_\infty(M) \quad \text{and} \quad \mathcal{R}(M)^2 = \mathcal{R}(M) \oplus \mathcal{R}(M).$$

Clearly, $\mathcal{R}(M)$ is contained in $\operatorname{dom} M \cap \operatorname{ran} M$.

Lemma 3.1. *Let M be a linear relation in a complex linear space $\mathcal{P}$. If $\mathcal{R}(M) \neq \{0\}$ then $\sigma_p(M) = \mathbb{C} \cup \{\infty\}$ and there exists a finite dimensional linear subset $\mathcal{H}$ of $\mathcal{P}$ such that for all $\lambda \in \mathbb{C} \cup \{\infty\}$, $\mathcal{H} \cap \ker(M-\lambda) \neq \{0\}$.*

Proof. Let $\mathcal{R}(M)$ contain the nonzero element x_0, say. Then there exist numbers $m, n \in \mathbb{N}$ and elements $x_j \in \operatorname{dom} M \cap \operatorname{ran} M$, $j = -m, \ldots, -1, 0, 1, \ldots, n$, such that x_{-m} and x_n are nonzero and for all j's between $-m-1$ and n, the pair $\{x_j, x_{j+1}\}$ belongs to M, where we have set $x_j = 0$ if $j < -m$ and if $j > n$. Denote by L the linear span of these pairs. Then L is a linear relation in the finite dimensional linear subset $\mathcal{H}$ of $\mathcal{P}$ spanned by the x_j's. Clearly, the points 0 and ∞ belong to $\sigma_p(L)$ and $\sigma_p(L)$ is contained in $\sigma_p(M)$. Hence to prove the first part of the lemma, it suffices to show that $\mathbb{C} \setminus \{0\}$ is contained in $\sigma_p(L)$. Assume that this is not the case. Then there exists a complex number $\lambda \neq 0, \infty$ such that $T = (L - \lambda)^{-1}$ is an everywhere defined linear operator on $\mathcal{H}$. The points $-1/\lambda$ and 0 belong to $\sigma_p(T)$. In fact, $Tx_{-m} = 0$ and $Tx_n = (-1/\lambda)x_n$. We have that for all j's between $-m-1$ and n, the pair $\{x_{j+1} - \lambda x_j, x_j\}$ belongs to T. This pair can be written as

$$\{x_{j+1} - \lambda x_j, x_j\} = \{x_{j+1} - \lambda x_j, 0(x_{j+1} - \lambda x_j) + x_j\},$$

and using an induction argument starting from $j = -m-1$ and going upwards we find that

$$\mathcal{R}_0(T) = \text{l.s.}\, \{x_j \mid j = -m, \ldots, -1, 0, 1, \ldots, n\} = \mathcal{H},$$

where l.s. stands for "linear span". On the other hand this pair can also be written as

$$\{x_{j+1} - \lambda x_j, x_j\} = \{x_{j+1} - \lambda x_j, (-1/\lambda)(x_{j+1} - \lambda x_j) + (1/\lambda)x_{j+1}\},$$

and using induction beginning with $j = n$ and going downwards we also find that

$$\mathcal{R}_{-1/\lambda}(T) = \text{l.s.}\, \{x_j \mid j = -m, \ldots, -1, 0, 1, \ldots, n\} = \mathcal{H}.$$

Hence $\mathcal{R}_0(T) = \mathcal{R}_{-1/\lambda}(T)$. It follows that if k is the maximum of the lengths of the Jordan chains of T at 0,

$$(-1/\lambda)^k x_n \in T^k \mathcal{R}_{-1/\lambda}(T) = T^k \mathcal{R}_0(T) = \{0\}.$$

This contradicts the fact that x_n is nonzero, and we conclude that $\sigma_p(M) = \sigma_p(L) = \mathbb{C} \cup \{\infty\}$. Evidently, the space $\mathcal{H}$ has the properties stated in the last part of the lemma. □

Lemma 3.2. *Let M be a linear relation in a complex linear space of dimension $n < \infty$. If M has more than n different eigenvalues in $\mathbb{C} \setminus \{0\}$, then $\mathcal{R}(M) \neq \{0\}$.*

Proof. Let $\lambda_1, \lambda_2, \ldots, \lambda_{n+1}$ be $n+1$ different eigenvalues in $\mathbb{C} \setminus \{0\}$ with (nonzero) eigenvectors $x_1, x_2, \ldots, x_{n+1}$, respectively. The space has dimension n and since there are $n+1$ of them, the eigenvectors are linearly dependent: there exist complex numbers $\alpha_1, \alpha_2, \ldots, \alpha_{n+1}$, not all equal to zero, such that

$$\sum_{j=1}^{n+1} \alpha_j x_j = 0.$$

We claim that there is a natural number k between 0 and n such that

$$\sum_{j=1}^{n+1} \lambda_j^k \alpha_j x_j \neq 0. \tag{3.1}$$

Indeed, if these sums are equal to zero for all $k = 0, 1, \ldots, n$, then, as the Vandermonde matrix $(\lambda_j^{k-1})_{j,k=1}^{n+1}$ is invertible, $\alpha_j x_j = 0$ for all $j = 1, 2, \ldots, n$ and hence all α_j's are zero. This contradiction proves the claim. Let k be the smallest integer such that (3.1) holds; then $k \geq 1$. We set $\beta_j = \lambda_j^{k-1}\alpha_j$; since the eigenvalues are different from zero, not all β_j's are equal to zero and

$$\sum_{j=1}^{n+1} \beta_j x_j = 0, \qquad x = \sum_{j=1}^{n+1} \lambda_j \beta_j x_j \neq 0.$$

Since for every $j = 1, 2, \ldots, n+1$, $\{x_j, \lambda_j x_j\}$ belongs to M, we have that $\{0, x\}$ belongs to M and also that for every polynomial $p(\lambda)$, the pair

$$\Big\{\sum_{j=1}^{n+1} \lambda_j \beta_j p(\lambda_j) x_j, \sum_{j=1}^{n+1} \lambda_j^2 \beta_j p(\lambda_j) x_j\Big\}$$

belongs to M. Consider the polynomial

$$p(\lambda) = (\lambda - \lambda_1)(\lambda - \lambda_2)\cdots(\lambda - \lambda_{n+1}) = \lambda^{n+1} + a_n\lambda^n + \cdots + a_1\lambda + a_0,$$

and write $p_{n+1}(\lambda) = p(\lambda)$ and for $k = 0, 1, \ldots, n$,

$$p_k(\lambda) = (p_{n+1}(\lambda) - (a_{n-k}\lambda^{n-k} + \cdots + a_0))/\lambda^{n+1-k} = (p_{k+1}(\lambda) - a_{n-k})/\lambda.$$

(Note $p_0(\lambda) = 1$.) Then for $k = 0, 1, \ldots, n$, the pair

$$\Big\{\sum_{j=1}^{n+1} \lambda_j \beta_j p_k(\lambda_j) x_j, \sum_{j=1}^{n+1} \lambda_j \beta_j p_{k+1}(\lambda_j) x_j\Big\}$$

$$= \Big\{0, a_{n-k}\sum_{j=1}^{n+1} \lambda_j \beta_j x_j\Big\} + \Big\{\sum_{j=1}^{n+1} \lambda_j \beta_j p_k(\lambda_j) x_j, \sum_{j=1}^{n+1} \lambda_j^2 \beta_j p_k(\lambda_j) x_j\Big\}$$

belongs to M. Since $\sum_{j=1}^{n+1} \lambda_j \beta_j p_{n+1}(\lambda_j) x_j = 0$, the elements

$$\sum_{j=1}^{n+1} \lambda_j \beta_j p_k(\lambda_j) x_j, \qquad k = 0, 1, \ldots, n+1,$$

form a Jordan chain of M at 0, which shows that the element

$$\sum_{j=1}^{n+1} \lambda_j \beta_j p_0(\lambda_j) x_j = \sum_{j=1}^{n+1} \lambda_j \beta_j x_j = x$$

belongs to $\mathcal{R}_0(M)$. We have already seen and used that x belongs to $\ker M^{-1}$, and hence x belongs to $\mathcal{R}(M)$, that is, $\mathcal{R}(M) \neq \{0\}$. □

Theorem 3.3. *Let M be a contractive closed linear relation in a Pontryagin space with negative index κ. Then $\mathcal{R}(M) \neq \{0\}$ if and only if M has more than κ eigenvalues in $\mathbb{C} \setminus \overline{\mathbb{D}}$.*

Proof. The "only if" part follows from Lemma 3.1 To prove the "if" part, we assume that $\lambda_1, \lambda_2, \ldots, \lambda_{\kappa+1}$ are $\kappa+1$ different eigenvalues of M in $\mathbb{C} \setminus \overline{\mathbb{D}}$ and denote the corresponding eigenvectors by $x_1, x_2, \ldots, x_{\kappa+1}$. Then for arbitrary complex numbers $a_1, a_2, \ldots, a_{\kappa+1}$ and $n \in \mathbb{N}$, the pair

$$\left\{\sum_{j=1}^{\kappa+1} \frac{a_j}{\lambda_j^n} x_j, \sum_{j=1}^{\kappa+1} \frac{a_j}{\lambda_j^{n-1}} x_j\right\}$$

belongs to M. Since M is contractive, we obtain via induction on n that for all $n \in \mathbb{N}$,

$$\left[\sum_{j=1}^{\kappa+1} a_j x_j, \sum_{j=1}^{\kappa+1} a_j x_j\right] \leq \left[\sum_{j=1}^{\kappa+1} \frac{a_j}{\lambda_j^n} x_j, \sum_{j=1}^{\kappa+1} \frac{a_j}{\lambda_j^n} x_j\right].$$

The righthand side tends to zero as n tends to infinity, and hence the linear span $\mathcal{L}$ of all x_j's is a nonpositive subspace of the Pontryagin space with negative index κ. It follows that $\dim \mathcal{L} \leq \kappa$. Now the linear relation $M_1 = M \cap (\mathcal{L} \oplus \mathcal{L})$ in $\mathcal{L}$ has at least $\kappa+1$ different eigenvalues in $\mathbb{C} \setminus \overline{\mathbb{D}}$, namely $\lambda_1, \lambda_2, \ldots, \lambda_{\kappa+1}$, and hence, according to Lemma 3.2, $\mathcal{R}(M_1) \neq \{0\}$. Since $\mathcal{R}_0(M_1) \subseteq \mathcal{R}_0(M)$ and $\mathcal{R}_\infty(M_1) \subseteq \mathcal{R}_\infty(M)$, we have that $\mathcal{R}(M_1) \subseteq \mathcal{R}(M)$ and hence $\mathcal{R}(M) \neq \{0\}$. □

Let M be a linear relation in $\mathcal{P}$. A subspace $\mathcal{L}$ of $\mathcal{P}$ is called *M–invariant* if for every $x \in \mathcal{L}$ there exists an element $y \in \mathcal{L} \cap \operatorname{ran} M$ such that the pair $\{x, y\}$ belongs to M, or equivalently, if

$$\operatorname{dom}(\mathcal{L}^2 \cap M) = \mathcal{L}$$

(in particular, $\mathcal{L} \subseteq \operatorname{dom} M$). For example, $\mathcal{R}(M)$ is M and M^{-1}–invariant. Indeed, if $\mathcal{R}(M)$ contains the nonzero element x_0, say, there exist numbers $m, n \in \mathbb{N}$ and elements $x_j \in \mathcal{P}$, $j = -m, \ldots, -1, 0, 1, \ldots, n$, such that x_{-m} and x_n are nonzero and all pairs

$$\{0, x_{-m}\}, \ldots, \{x_{-2}, x_{-1}\}, \{x_{-1}, x_0\}, \{x_0, x_1\}, \ldots, \{x_n, 0\}$$

belong to M and then all points x_j belong to $\mathcal{R}(M)$. More can be said about $\mathcal{R}(M)$ when M is contractive.

Lemma 3.4. *Let M be a contractive closed linear relation in a Pontryagin space $\mathcal{P}$. Then $\mathcal{R}(M)$ is a neutral subspace (hence $\dim \mathcal{R}(M) \leq \operatorname{ind}_- \mathcal{P}$), and*

$$(\mathcal{R}(M) \oplus \mathcal{P}) \cap M \subseteq \mathcal{R}(M) \oplus \mathcal{R}(M)^{[\perp]}, \quad (\mathcal{R}(M)^{[\perp]} \oplus \mathcal{P}) \cap M \subseteq \mathcal{R}(M)^{[\perp]} \oplus \mathcal{R}(M)^{[\perp]}.$$

Proof. Let x_0 be an element in $\mathcal{R}_0(M)$ and $\{x_0, x_1, \ldots, x_{m-1}, x_m\}$ be a Jordan chain. Then x_m is an eigenvector of M at zero, and, since M is contractive,

$$0 \leq [x_0, x_0] \leq [x_1, x_1] \leq \cdots \leq [x_{m-1}, x_{m-1}] \leq [x_m, x_m].$$

It follows that $\mathcal{R}_0(M)$ is a nonnegative subspace of $\mathcal{P}$. In a similar way it can be proved that $\mathcal{R}_\infty(M)$ is a nonpositive subspace of $\mathcal{P}$, and hence $\mathcal{R}(M)$ is neutral. This proves the first statement. To prove the first inclusion we decompose the space $\mathcal{P}$ into

$$\mathcal{P} = (\mathcal{R}(M) + J\mathcal{R}(M)) \oplus \mathcal{P}_1,$$

where $\mathcal{P}_1$ is also a Pontryagin space. Let $\{x, y\}$ belong to the intersection

$$(\mathcal{R}(M) \oplus \mathcal{P}) \cap M$$

and write $y = v_0 + Jv_0' + y_1$ for some $v_0, v_0' \in \mathcal{R}(M)$ and $y_1 \in \mathcal{P}_1$. We have to prove that $v_0' = 0$, because then y belongs to $\mathcal{R}(M) \oplus \mathcal{P}_1 = \mathcal{R}(M)^{[\perp]}$. Since $\mathcal{R}(M)$ is M^{-1}–invariant, there exists an element $u_0' \in \mathcal{R}(M)$ such that $\{u_0', v_0'\}$ belongs to M. Since x also belongs to $\mathcal{R}(M)$, we have $[u_0', x] = 0$. M is a nonnegative subspace of the Kreĭn space $\mathcal{P}^2$, and thus we may apply the Cauchy-Schwarz-Bunyakowski inequality to elements of M. We find that

$$\begin{aligned} \left|[v_0', Jv_0']\right|^2 &= \left|[v_0', y]\right|^2 = \left|\langle\{u_0', v_0'\}, \{x, y\}\rangle\right|^2 \\ &\leq \langle\{u_0', v_0'\}, \{u_0', v_0'\}\rangle\langle\{x, y\}, \{x, y\}\rangle \\ &= 0. \end{aligned}$$

Here we again used that the space $\mathcal{R}(M)$ is neutral and hence that $[u_0', u_0'] = [v_0', v_0'] = 0$. So indeed, $v_0' = 0$. The proof of the other inclusion is almost the same; now we have that $[u_0', x] = 0$, because x belongs to $\mathcal{R}(M)^{[\perp]}$ and u_0' to $\mathcal{R}(M)$. □

For any linear relation M the inclusion $\mathcal{N}(M) \subseteq \mathcal{R}(M)$ holds. We call M *regular* if equality prevails. Thus M is regular if and only if $\mathcal{R}(M)^2$ is contained in M. If M is an operator, then $\mathcal{R}_\infty(M) = \mathcal{N}_\infty(M) = \{0\}$ and M is regular. We give two other examples.

Example 3.5. In a Pontryagin space with negative index ≤ 1 every contractive linear relation is regular.

Proof. If the space is a Hilbert space, then the linear relation is an operator and hence regular. We consider the remaining case where M is a contractive linear relation in the Pontryagin space $\mathcal{P}$ with $\operatorname{ind}_-\mathcal{P} = 1$ and we assume that $\mathcal{R}(M) \neq \{0\}$. (Evidently, if equality holds then M is regular.) From the proof of Lemma 3.4 we know that $\mathcal{R}_0(M)$ is a nonnegative and $\mathcal{R}_\infty(M)$ is a nonpositive subspace of $\mathcal{P}$. Since $\operatorname{ind}_-\mathcal{P} = 1$, we have that $\dim \mathcal{R}_\infty(M) = 1$ and hence

$\ker M^{-1} = \mathcal{R}_\infty(M)$. This implies that $\mathcal{R}(M)$ is contained in $\ker M^{-1}$, and so to show that $\mathcal{R}(M) = \mathcal{N}(M)$ it suffices to prove that $\mathcal{R}(M)$ is also contained in $\ker M$. Let x_0 belong to $\mathcal{R}(M)$; then x_0 belongs to $\mathcal{R}_0(M)$, and let $\{x_0, x_1, \ldots, x_k\}$ be the corresponding Jordan chain at 0, where $x_k \neq 0$ is an eigenvector . Since M is a contraction and $\mathcal{R}(M)$ is a neutral subspace,

$$0 \leq [x_k, x_k] \leq \cdots \leq [x_1, x_1] \leq [x_0, x_0] = 0,$$

which implies that all x_j's are neutral. Since they belong to the nonnegative subspace $\mathcal{R}_0(M)$, an argument based on the Cauchy-Schwarz-Bunyakowski inequality readily yields that these elements also belong to $\mathcal{R}_0(M)^{[\perp]}$ and therefore they span a neutral subspace (contained in $\mathcal{R}_0(M) \cap \mathcal{R}_0(M)^{[\perp]}$). Since $\operatorname{ind}_-\mathcal{P} = 1$, this subspace is one dimensional and hence x_0 is equal to αx_k for some scalar α, that is, x_0 belongs to $\ker M$. □

If M is a contractive linear relation in a Pontryagin space $(\mathcal{P}, [\cdot,\cdot])$ with $\dim \mathcal{P}_+ = 1$ then M is regular. Indeed, M^{-1} is a contractive linear relation in the Pontragin space $(\mathcal{P}, -[\cdot,\cdot])$ with negative index 1. Example 3.5 shows that M^{-1} is regular, and since $\mathcal{R}(M) = \mathcal{R}(M^{-1})$ and $\mathcal{N}(M) = \mathcal{N}(M^{-1})$, M is also regular.

Example 3.6. In every Pontryagin space $\mathcal{P}$ with $\dim \mathcal{P}_+ \geq 2$ and $\dim \mathcal{P}_- \geq 2$ there is a contractive linear relation which is not regular.

Proof. It suffices to consider the four dimensional Pontryagin space $\mathcal{P} = \mathcal{P}_+ \oplus \mathcal{P}_-$ in which e_1^+, e_2^+ and e_1^-, e_2^- are orthonormal bases for $\mathcal{P}_+$ and $|\mathcal{P}_-|$, respectively. Set $e_1 = e_1^+ + e_1^-$ and $e_2 = e_2^+ + e_2^-$ and consider the linear relation

$$M = \text{l.s.}\,\{\{0, e_1\}, \{e_1, e_2\}, \{e_2, 0\}\}.$$

Then M is contractive, $\mathcal{R}(M) = \mathcal{R}_0(M) = \mathcal{R}_\infty(M) = \text{l.s.}\,\{e_1, e_2\}$ and $\mathcal{N}(M) = \{0\}$. □

For a linear relation M in $\mathcal{P}$ we set

$$M_r = \text{l.s.}\left\{\mathcal{R}(M)^2, M \cap (\mathcal{R}(M)^2)^{\langle\perp\rangle}\right\} \tag{3.2}$$

and call M_r the *regularization* of M. Note that

$$(M^{-1})_r = (M_r)^{-1}. \tag{3.3}$$

We begin with the following simple result.

Lemma 3.7. *Let M be a contractive closed linear relation in a Pontryagin space. Then $M_r = M$ if and only if M is regular.*

Proof. Since the inclusion $\mathcal{N}(M) \subseteq \mathcal{R}(M)$ always holds, the three "statements"

$$\mathcal{R}(M)^2 \subseteq M, \quad \mathcal{R}(M) \subseteq \mathcal{N}(M) \quad \text{and} \quad \mathcal{R}(M) = \mathcal{N}(M)$$

are equivalent. If $M = M_r$, then $\mathcal{R}(M)^2 \subseteq M$ and $\mathcal{R}(M) = \mathcal{N}(M)$. Conversely, assume that $\mathcal{R}(M) = \mathcal{N}(M)$. Then $\mathcal{R}(M)^2 \subseteq M$, and using the Cauchy-Schwarz-Bunyakowski inequality and the fact that, by Lemma 3.4, $\mathcal{R}(M)$ is neutral, we find that $M \subseteq (\mathcal{R}(M)^2)^{\langle\perp\rangle}$. These inclusions readily imply $M = M_r$. □

Theorem 3.8. *Let M be a contractive closed linear relation in a Pontryagin space $\mathcal{P}$. Then its regularization M_r is a regular contractive closed linear relation in $\mathcal{P}$ and*

$$\mathcal{R}_0(M_r) = \mathcal{R}_0(M), \qquad \mathcal{R}_\infty(M_r) = \mathcal{R}_\infty(M),$$

so that also $\mathcal{R}(M_r) = \mathcal{R}(M)$.

Proof. Consider pairs $\{u, v\} \in \mathcal{R}(M)^2$ and $\{x, y\} \in M \cap (\mathcal{R}(M)^2)^{\langle\perp\rangle}$ and complex numbers a and b. Then $[u, x] = [v, y]$, $[y, y] \leq [x, x]$ and $[u, u] = [v, v] = 0$ (since, by Lemma 3.4, $\mathcal{R}(M)$ is neutral), and hence

$$[av + by, av + by] \leq [au + bx, au + bx],$$

that is, M_r is contractive. M_r is closed as it is the sum in $\mathcal{P}^2$ of $M \cap (\mathcal{R}(M)^2)^{\langle\perp\rangle}$ which is a subspace and $\mathcal{R}(M)^2$ which is a finite dimensional subspace of $\mathcal{P}^2$. Since $\mathcal{R}(M)^2 \subseteq M_r$,

$$\mathcal{R}(M) \subseteq \mathcal{N}(M_r) \subseteq \mathcal{R}(M_r).$$

That these sets are actually equal, and hence that M_r is regular, follows from the third equality in the theorem. Since this equality follows from the first two it remains to prove these.

To show the inclusion $\mathcal{R}_0(M_r) \subseteq \mathcal{R}_0(M)$, assume that x_0 belongs to $\mathcal{R}_0(M_r)$ and let $\{x_0, x_1, \ldots, x_k\}$ be a Jordan chain of M_r beginning with x_0 and ending with an eigenvector $x_k \neq 0$ of M_r at zero. Then $\{x_k, 0\}$ belongs to M_r and can be written as

$$\{x_k, 0\} = \{u, v\} + \{x, y\}$$

for some elements $u, v \in \mathcal{R}(M)$ and a pair $\{x, y\} \in M \cap (\mathcal{R}(M)^2)^{\langle\perp\rangle}$. Then $y = -v \in \mathcal{R}_0(M)$ and there is a Jordan chain $\{y, y_1, \ldots, y_r\}$ of M at zero. Since $\{x, y\}$ belongs to M this Jordan chain can be extended by adjoining x on the lefthand side to get what is again a Jordan chain of M at zero. Hence x belongs to $\mathcal{R}_0(M)$. From the equality $x_k = x + u$, where u is in $\mathcal{R}_0(M)$, we see that also x_k belongs to $\mathcal{R}_0(M)$. We now show that the same is true for x_{k-1}. As above we may write the pair $\{x_{k-1}, x_k\}$, since it belongs to M_r, as

$$\{x_{k-1}, x_k\} = \{u, v\} + \{x, y\}$$

with $u, v \in \mathcal{R}(M)$ and $\{x, y\} \in M \cap (\mathcal{R}(M)^2)^{\langle\perp\rangle}$. Then $y = x_k - v \in \mathcal{R}_0(M)$ and proceeding as before we find that x_{k-1} belongs to $\mathcal{R}_0(M)$. A simple mathematical induction shows that also x_0 belongs to this space. So the desired inclusion has been established. We now show that equality holds by proving the reversed inclusion. By Lemma 3.4, $\mathcal{R}(M)$ is neutral. Since it is contained in the nonnegative subspace $\mathcal{R}_0(M)$, an argument involving the Cauchy-Schwarz-Bunyakowksi inequality yields the inclusion $\mathcal{R}_0(M) \subseteq \mathcal{R}(M)^{[\perp]}$. Let x_0 belong to the root subspace $\mathcal{R}_0(M)$ and let $\{x_0, x_1, \ldots, x_k\}$ be a corresponding Jordan chain of M at 0. Then all elements x_j belong to the root subspace and hence to the space $\mathcal{R}(M)^{[\perp]}$. Therefore all pairs $\{x_0, x_1\}, \{x_1, x_2\}, \ldots, \{x_k, 0\}$ are elements of $M \cap (\mathcal{R}(M)^2)^{\langle\perp\rangle}$. Since this set is a subset of M_r, the chain is also a Jordan chain of M_r at 0, that is, x_0 also belongs to $\mathcal{R}_0(M_r)$. This completes the proof of the first equality of the lemma. The arguments in this proof almost verbally apply to M^{-1}; the only difference is that $\mathcal{R}_\infty(M)$ is nonpositive rather than nonnegative. Hence the first equality also holds if M is replaced by M^{-1}, and the second equality then follows from (3.3). □

Theorem 3.9. *Let M be a contractive closed linear relation in a Pontryagin space $\mathcal{P}$. If $\mathcal{L}$ is an M_r–invariant subspace of $\mathcal{P}$, then the subspace $\mathcal{R}(M) + \mathcal{L}$ is M–invariant.*

Proof. From the definition of M_r it follows that

$$(\mathcal{R}(M) + \mathcal{L})^2 \cap M_r = \mathcal{R}(M)^2 + \mathcal{F}$$

for some subspace $\mathcal{F} \subseteq (\mathcal{R}(M)^2)^{\langle\perp\rangle} \cap M$. Since $\mathcal{L}$ is M_r–invariant, $\mathcal{R}(M) + \mathcal{L}$ is also M_r–invariant and hence

$$\mathcal{R}(M) + \operatorname{dom} \mathcal{F} = \mathcal{R}(M) + \mathcal{L}.$$

Moreover, there exists a subspace $\mathcal{F}_1 \supseteq \mathcal{F}$ such that

$$(\mathcal{R}(M) + \mathcal{L})^2 = \mathcal{R}(M)^2 + \mathcal{F}_1.$$

It follows that

$$(\mathcal{R}(M) + \mathcal{L})^2 \cap M \supseteq \mathcal{R}(M)^2 \cap M + \mathcal{F}_1 \cap M \supseteq \mathcal{R}(M)^2 \cap M + \mathcal{F}.$$

Hence

$$\operatorname{dom} (\mathcal{R}(M) + \mathcal{L})^2 \cap M \supseteq \mathcal{R}(M) + \operatorname{dom} \mathcal{F} = \mathcal{R}(M) + \mathcal{L},$$

that is, $\mathcal{R}(M) + \mathcal{L}$ is M–invariant. □

To prove Theorem 3.11 below we use a general result in Kreĭn spaces.

Lemma 3.10. *Let M and L be nonnegative subspaces in a Kreĭn space with direct sum decompositions*

$$M = N + Q \quad \textit{and} \quad L = N + P,$$

where $N = M \cap L$, and P and Q are finite dimensional subspaces.

(a) *If M is maximal nonnegative, then* $\dim P \leq \dim Q$.
(b) *If* $\dim P = \dim Q$, *then M is maximal nonnegative if and only if L is maximal nonnegative.*

Proof. In terms of graph contractions, the hypotheses of the lemma imply that

$$\operatorname{dom} K_M = \operatorname{dom} K_N + \operatorname{dom} K_Q \quad \text{and} \quad \operatorname{dom} K_L = \operatorname{dom} K_N + \operatorname{dom} K_P$$

and that these sums are direct, since K_N and K_Q are both restrictions of K_M and K_N and K_P are both restrictions of K_L.
We now prove (a). Since M is maximal nonnegative,

$$\mathcal{H}_+ = \operatorname{dom} K_N + \operatorname{dom} K_Q, \text{ direct sum.}$$

Hence

$$\begin{aligned}\dim P &= \dim(\operatorname{dom} K_P) = \dim(\operatorname{dom} K_L/\operatorname{dom} K_N)\\ &\leq \dim(\mathcal{H}_+/\operatorname{dom} K_N) = \dim(\operatorname{dom} K_Q)\\ &= \dim Q.\end{aligned}$$

The proof of (b) is similar and therefore omitted. □

Theorem 3.11. *Let M be a contractive closed linear relation in a Pontryagin space $\mathcal{P}$. Then M is maximal contractive if and only if M_r is maximal contractive and*

$$\dim M/(M \cap (\mathcal{R}(M)^2)^{\langle\perp\rangle}) = \dim \mathcal{R}(M)^2/(M \cap \mathcal{R}(M)^2).$$

Proof. There exist finite dimensional subspaces P and Q of $\mathcal{P}^2$ such that

$$\mathcal{R}(M)^2 = M \cap \mathcal{R}(M)^2 + P \quad \text{and} \quad M = M \cap (\mathcal{R}(M)^2)^{\langle\perp\rangle} + Q$$

are direct sums in $\mathcal{P}^2$. (For example, one can arrange it so that these sums are orthogonal in the Hilbert space $|\mathcal{P}^2|$.) Then also

$$M_r = M \cap (\mathcal{R}(M)^2)^{\langle\perp\rangle} + P$$

and the sum is direct. Note that

$$M \cap M_r = M \cap (\mathcal{R}(M)^2)^{\langle\perp\rangle}.$$

Since $\dim \mathcal{R}(M)^2 = 2\dim \mathcal{R}(M)$, $\dim(M \cap \mathcal{R}(M)^2) \geq \dim \mathcal{R}(M)$ (as $\mathcal{R}(M)$ is M-invariant) and $\dim \mathcal{R}(M) \leq \operatorname{ind}_- \mathcal{P}$ (as $\mathcal{R}(M)$ is neutral), we have that

$$\dim P \leq \operatorname{ind}_- \mathcal{P}.$$

We claim that

$$\dim Q \leq \dim P.$$

To prove the claim we assume the contrary. Then $Q \cap P^{\langle\perp\rangle}$ is a nontrivial subspace of $\mathcal{P}^2$. Indeed, if $\{x_1, y_1\}, \{x_2, y_2\}, \dots, \{x_p, y_p\}$ is a basis for P and

$$\{\widetilde{x}_1, \widetilde{y}_1\}, \{\widetilde{x}_2, \widetilde{y}_2\}, \dots, \{\widetilde{x}_p, \widetilde{y}_p\}, \{\widetilde{x}_{p+1}, \widetilde{y}_{p+1}\}$$

form a collection of linearly independent elements from Q, then the $p \times (p+1)$ matrix with entries $\langle\{x_i, y_i\}, \{\widetilde{x}_j, \widetilde{y}_j\}\rangle$ is singular, and hence there is a nonzero linear combination $\{\widetilde{x}, \widetilde{y}\}$ of the elements $\{\widetilde{x}_j, \widetilde{y}_j\}$ which is orthogonal to all of P and hence belongs to $Q \cap P^{\langle\perp\rangle}$. Put $\widetilde{M} = \text{l.s.}\,\{M_r, \{\widetilde{x}, \widetilde{y}\}\}$; then every element in $\widetilde{M}$ can be written in the form

$$\{x, y\} + \{u, v\} + a\{\widetilde{x}, \widetilde{y}\}$$

for some $\{x, y\} \in M$, $\{u, v\} \in P$ and $a \in \mathbb{C}$. Since $\{x, y\} + a\{\widetilde{x}, \widetilde{y}\}$ belongs to M, $[x, u] = [y, v] = 0$ (as u and v belong to $\mathcal{R}(M)$) and $\langle\{\widetilde{x}, \widetilde{y}\}, \{u, v\}\rangle = 0$, we find that

$$\langle\{x, y\} + \{u, v\} + a\{\widetilde{x}, \widetilde{y}\}, \{x, y\} + \{u, v\} + a\{\widetilde{x}, \widetilde{y}\}\rangle \geq 0,$$

that is, $\widetilde{M}$ is a nonnegative subspace of $\mathcal{P}^2$. Because $\mathcal{R}(M)^2 \subseteq \widetilde{M}$ and by the Cauchy-Schwarz-Bunyakowski inequality, we get that $\widetilde{M} \subseteq (\mathcal{R}(M)^2)^{\langle\perp\rangle}$. In particular, this implies for the nonzero element $\{\widetilde{x}, \widetilde{y}\}$ that

$$\{\widetilde{x}, \widetilde{y}\} \in Q \cap (\mathcal{R}(M)^2)^{\langle\perp\rangle} \subseteq Q \cap (M \cap (\mathcal{R}(M)^2)^{\langle\perp\rangle}) = \{\{0, 0\}\}.$$

This contradiction proves the claim.
We can now apply Lemma 3.10 in the Kreĭn space $\mathcal{P}^2$. If M is maximal contractive, then, by (a) and what we have just proved, $\dim P = \dim Q$, and, by (b) with $L = M_r$, this in turn implies that M_r is maximal nonnegative. This proves the "only if" part of the lemma. The "if" part follows directly from part (b) of Lemma 3.10. □

4. Criteria for maximality

Let M be a contractive closed linear relation in the Pontryagin space $\mathcal{P}$, and assume that M is regular, that is, $\mathcal{R}(M) = \mathcal{N}(M)$. Denote by $\widehat{M}$ the linear relation generated by M in the quotient space $\widehat{\mathcal{P}} = \mathcal{R}(M)^{[\perp]}/\mathcal{R}(M)$. This quotient space consists of equivalence classes of the form $\widehat{x} = x + \mathcal{R}(M)$, where x is an element of $\mathcal{R}(M)^{[\perp]}$. Equipped with the inner product

$$[\widehat{x}, \widehat{y}] = [x, y], \qquad x, y \in \mathcal{R}(M)^{[\perp]},$$

it is a Pontryagin space with $\text{ind}_-\widehat{\mathcal{P}} = \text{ind}_-\mathcal{P} - \dim \mathcal{R}(M)$. Write $\mathcal{P}$ as

$$\mathcal{P} = (\mathcal{R}(M) + J\mathcal{R}(M)) \oplus \mathcal{P}_1$$

where $\mathcal{P}_1$ is a Pontryagin space with $\text{ind}_-\mathcal{P}_1 = \text{ind}_-\mathcal{P} - \dim \mathcal{R}(M)$. Then

$$\mathcal{R}(M)^{[\perp]} = \mathcal{R}(M) \oplus \mathcal{P}_1,$$

and we can identify the space $\mathcal{P}_1$ with the quotient space $\widehat{\mathcal{P}}$ via the unitary map $x \mapsto \widehat{x}$. Since $\mathcal{R}(M)^2 \subseteq M \subseteq (\mathcal{R}(M)^2)^{\langle\perp\rangle}$,

$$\widehat{M} = \{\{\widehat{x}, \widehat{y}\} \mid \{x, y\} \in M\}.$$

We may identify $\widehat{\mathcal{P}}$ with $\mathcal{P}_1$, and then we have that $\widehat{M} = M_1$, where M_1 is defined in Section 2; see Lemma 2.2.

Lemma 4.1. *If M is a regular contractive closed linear relation in a Pontryagin space $\mathcal{P}$, then $\widehat{M}$ is a contractive closed linear relation in $\widehat{\mathcal{P}}$ and*

$$\mathcal{R}_0(M)\widehat{\ } = \mathcal{R}_0(\widehat{M}), \quad \mathcal{R}_\infty(M)\widehat{\ } = \mathcal{R}_\infty(\widehat{M}) \text{ and } \mathcal{R}(\widehat{M}) = \mathcal{N}(\widehat{M}) = \{0\}.$$

In particular, $\widehat{M}$ is regular.

Proof. By Lemma 2.2, $\widehat{M}$ is a contractive closed linear relation in $\widehat{\mathcal{P}}$. The equalities follow from the inclusion $\mathcal{R}(M)^2 \subseteq M$. We only prove that $\mathcal{R}(\widehat{M}) = \mathcal{R}_0(\widehat{M}) \cap \mathcal{R}_\infty(\widehat{M})$ is trivial. If $\widehat{x}$ belongs to this intersection then M contains pairs of the form

$$\{\widehat{x}, \widehat{x}_1\}, \{\widehat{x}_1, \widehat{x}_2\}, \ldots, \{\widehat{x}_{k-1}, \widehat{x}_k\}, \{\widehat{x}_k, 0\} \in \widehat{M}$$

and pairs of the form

$$\{\widehat{y}_1, \widehat{x}\}, \{\widehat{y}_2, \widehat{y}_1\}, \ldots, \{\widehat{y}_k, \widehat{y}_{k-1}\}, \{0, \widehat{y}_k\} \in \widehat{M}.$$

Since $\mathcal{R}(M)^2 \subseteq M$, we can choose the representatives in the corresponding equivalence classes, that is, $x \in \widehat{x}, x_j \in \widehat{x}_j$ and $y_j \in \widehat{y}_j$, such that the above inclusions hold without the "$\widehat{\ }$". It follows that x belongs to $\mathcal{R}(M)$ and hence $\widehat{x} = 0$. □

For a more general version of the following lemma, see [AI, Theorem 2.2.21] and also [IKL, Theorem 11.1].

Lemma 4.2. *If M is a regular contractive closed linear relation in a Pontryagin space with negative index κ, then $\widehat{M}$ has at most $\kappa - \dim \mathcal{R}(M)$ eigenvalues in $\mathbb{C} \setminus \overline{\mathbb{D}}$.*

Proof. Apply Theorem 3.3 and Lemma 4.1. □

In the proof of the following lemma we use the Cayley transform $C_\mu(M)$ of a linear relation M with respect to a real parameter $\mu \neq \pm 1$:

$$C_\mu(M) = \{\{y - \mu x, \mu y - x\} \mid \{x, y\} \in M\}.$$

It plays an important role in the remainder of this paper. We shall mention its properties when we make use of them.

Lemma 4.3. *Let M be a regular contractive closed linear relation in a Pontryagin space $\mathcal{P}$. Then there exists a regular maximal contractive linear relation $\widetilde{M}$ in $\mathcal{P}$ with $M \subseteq \widetilde{M}$ and $\mathcal{R}(\widetilde{M}) = \mathcal{R}(M)$.*

Proof. We first consider the special case that $\mathcal{R}(M) = \{0\}$. Assume that M is not maximal. Then, by Theorem 2.3, the subspace $\operatorname{dom}\pi(M)$ does not coincide with $\mathcal{P}$. Hence its orthogonal complement in the Hilbert space $|\mathcal{P}|$ denoted by $\operatorname{dom}\pi(M)^{|\perp|}$ is a nontrivial subspace. Define the linear relation $\widetilde{M}$ in $\mathcal{P}$ by

$$\widetilde{M} = M + \left\{\{P_+x, P_-x\} \mid x \in \operatorname{dom}\pi(M)^{|\perp|}\right\}.$$

Then $\pi(\widetilde{M})$ is the trivial contractive extension of the contraction $\pi(M)$, and since the domain of this extension is the whole space $\mathcal{P}$, $\widetilde{M}$ is a maximal contractive linear relation which contains M. We show that $\mathcal{R}(\widetilde{M})$ contains only the zero element. Assume that this is not true and apply Lemma 3.1: Let $\widetilde{\mathcal{H}}$ be the finite dimensional subspace of $\mathcal{P}$ such that for all $\zeta \in \mathbb{C}$, $\widetilde{\mathcal{H}} \cap \ker(\widetilde{M} - \zeta) \neq \{0\}$.
Since $\mathcal{R}(M) = \{0\}$, there exists a real number μ, say, with $\mu > 1$ and $\mu \notin \sigma_p(M)$; see Theorem 3.3. The Cayley transform $C_\mu(M)$ of M with respect to this μ is a closed contractive operator, which, on account of Corollary 2.6, is also bounded on its domain. Moreover for all $\lambda \in \mathbb{C}$,

$$\ker(C_\mu(M) - \lambda) = \ker(M - \zeta(\lambda)), \qquad \zeta(\lambda) = \frac{1 - \lambda\mu}{\mu - \lambda}.$$

Note that the mapping $\lambda \mapsto \zeta(\lambda)$ is a bijection on the unit circle $\partial\mathbb{D}$. As $C_\mu(M)$ is an operator, the spaces $\ker(C_\mu(M) - \lambda)$ for finitely many mutually different values of $\lambda \in \partial\mathbb{D}$ are linearly independent. Hence, since $\widetilde{\mathcal{H}}$ is finite dimensional, there exists a $\zeta \in \partial\mathbb{D}$ such that $\widetilde{\mathcal{H}} \cap \ker(M - \zeta) = \{0\}$.
For this ζ we select a nonzero element in the intersection $\widetilde{\mathcal{H}} \cap \ker(\widetilde{M} - \zeta)$ and denote it by y_ζ. Then $\{y_\zeta, \zeta y_\zeta\}$ belongs to $\widetilde{M}$ and there exist a pair $\{u, v\} \in M$ and an element $x \in \operatorname{dom}\pi(M)^{|\perp|}$ such that

$$\{y_\zeta, \zeta y_\zeta\} = \{u, v\} + \{P_+x, P_-x\}.$$

It follows that

$$v = \zeta u + \zeta P_+x - P_-x,$$

and, since $P_+u + P_-v$ belongs to $\operatorname{dom}\pi(M)$, $[P_+u + P_-v, Jx] = 0$ which implies

$$[u, P_+x] = [v, P_-x] = \zeta[u, P_-x] - [P_-x, P_-x].$$

Combining this with the inequality $[u, u] \geq [v, v]$, which holds as M is contractive, we obtain that

$$\begin{aligned}
[u, u] &\geq [\zeta u + \zeta P_+x - P_-x, \zeta u + \zeta P_+x - P_-x] \\
&= [u, u] + ([u, P_+x] - \zeta[u, P_-x]) + [P_+x, P_+x] \\
&\quad + ([P_+x, u] - \overline{\zeta}[P_-x, u]) + [P_-x, P_-x] \\
&= [u, u] + [P_+x, P_+x] - [P_-x, P_-x]
\end{aligned}$$

and hence that

$$0 \leq [P_+x, P_+x] \leq [P_-x, P_-x] \leq 0.$$

This implies that $x = P_+x + P_-x = 0$ and $\{y_\zeta, \zeta y_\zeta\}$ belongs to M, that is, y_ζ belongs to $\widetilde{\mathcal{H}} \cap \ker(M - \zeta) = \{0\}$, contradicting the fact that y_ζ is nonzero. Hence $\mathcal{R}(\widetilde{M}) = \{0\}$.

We still have to consider the general case where $\mathcal{R}(M) = \mathcal{N}(M) \neq \{0\}$. Then, by Lemma 4.1, $\mathcal{R}(\widehat{M}) = \{0\}$ and, as we have shown in the foregoing special case, there exists a maximal contractive linear relation $\widehat{L}$ in $\widehat{\mathcal{P}}$ with $\widehat{M} \subseteq \widehat{L}$ and $\mathcal{R}(\widehat{L}) = \{0\}$. Define the linear relation $\widetilde{M}$ in $\mathcal{P}$ by

$$\widetilde{M} = \left\{ \{x, y\} \mid \{\widehat{x}, \widehat{y}\} \in \widehat{L} \right\}.$$

It is not difficult to verify that $\widetilde{M}$ has the properties mentioned in the lemma. □

Theorem 4.4. *For a regular contractive closed linear relation M in a Pontryagin space $\mathcal{P}$ the following statements and formulas are equivalent.*

(1) *M is a maximal contractive linear relation.*
(2) *$\widehat{M}$ is a maximal contractive linear relation.*
(3) $\rho(\widehat{M}) \cap \mathbb{R} \cap (\mathbb{C} \setminus \overline{\mathbb{D}}) \neq \emptyset$.
(4) *The set $(\mathbb{C} \setminus \overline{\mathbb{D}}) \setminus \rho(\widehat{M})$ consists of at most $(\mathrm{ind}_-\mathcal{P} - \dim \mathcal{R}(M))$ points and these points are normal eigenvalues of $\widehat{M}$.*
(5) $\mathrm{dom}\, \widehat{M} + \mathcal{R}_\infty(\widehat{M}) = \widehat{\mathcal{P}}$.
(6) $\mathrm{dom}\, M + \mathcal{R}_\infty(M) = \mathcal{R}(M)^{[\perp]}$.
(7) $\dim(\widehat{\mathcal{P}}/\mathrm{dom}\, \widehat{M}) = \dim \ker \widehat{M}^{-1}$.
(8) $\dim(\mathcal{P}/\mathrm{dom}\, M) = \dim \ker M^{-1}$.

Proof. The equivalences (1) ⇔ (2), (5) ⇔ (6) and (7) ⇔ (8) follow directly from the fact that $\mathcal{R}(M)^2$ is contained in M. So to prove the equivalence between (1) through (8) we may (and do) assume that $\mathcal{R}(M) = \{0\}$, so that $\widehat{\mathcal{P}} = \mathcal{P}$ and $\widehat{M} = M$. By Lemma 4.2, there is a real point μ, say, with $|\mu| > 1$ and $\mu \notin \sigma_p(M)$. Then, as in the proof of Lemma 4.3, the Cayley transform $C_\mu(M)$ is a bounded closed contractive operator in $\mathcal{P}$ with closed domain. Moreover,

$$\ker(C_\mu(M) - (1/\mu)) = \ker M \quad \text{and} \quad \ker(C_\mu(M) - \mu) = \ker M^{-1},$$

$$\mathcal{R}_{1/\mu}(C_\mu(M)) = \mathcal{R}_0(M) \quad \text{and} \quad \mathcal{R}_\mu(C_\mu(M)) = \mathcal{R}_\infty(M),$$

and $C_\mu(M)$ is a maximal contractive linear relation in $\mathcal{P}$ if and only if M is a maximal contractive linear relation there.

We are now ready to prove the implication (2) ⇒ (3): If (2) holds then, according to Corollary 2.5, $\mathrm{ran}\,(M - \mu) = \mathrm{dom}\, C_\mu(M) = \mathcal{P}$, and hence μ belongs to $\rho(M)$, that is, (3) is valid.

For the proof of the implication (3) $\Rightarrow$ (4) we use the following theorem (see [Br] and [IKL, Section 11]): If $T \in \mathbf{L}(\mathcal{H})$ is a contraction on a Pontryagin space $\mathcal{H}$, then the number of points in the resolvent set $(\mathbb{C} \setminus \overline{\mathbb{D}}) \setminus \rho(T)$ is at most equal to $\operatorname{ind}_{-}\mathcal{H}$ and these points are normal eigenvalues of T. So if (3) holds and μ, say, belongs to $\rho(M) \cap \mathbb{R} \cap (\mathbb{C} \setminus \overline{\mathbb{D}})$, then $C_\mu(M)$ is a bounded contraction defined on the whole space $\mathcal{P}$ and hence we may apply the theorem with $T = C_\mu(M)$ and $\mathcal{H} = \mathcal{P}$. Now (4) easily follows; see also [AI, Theorem 2.6.4].
We now show the implication (4) $\Rightarrow$ (5). Assume that (4) is valid. Then, as before, for some real $\mu > 1$, $C_\mu(M)$ is a bounded contraction defined on the whole space $\mathcal{P}$. If μ belongs to $\rho(C_\mu(M))$, then $\mathcal{P} = \operatorname{ran}(C_\mu(M) - \mu) = \operatorname{dom} M$, and (5) holds. If μ does not belong to $\rho(C_\mu(M))$, there exists a real number $\varepsilon > 0$ such that the intersection

$$\sigma(C_\mu(M)) \cap \{\lambda \in \mathbb{C} \mid |\lambda - \mu| \leq \varepsilon\}$$

only contains the point μ. Denote by $\Gamma_{\mu,\varepsilon}$ the circle $\{\lambda \in \mathbb{C} \mid |\lambda - \mu| = \varepsilon\}$ and by $P_\mu = P_\mu(M)$ the Riesz projection onto the space $\mathcal{R}_\mu(C_\mu(M))$, that is, the operator defined by the contour integral

$$P_\mu = -\frac{1}{2\pi i} \int_{\Gamma_{\mu,\varepsilon}} (C_\mu(M) - \lambda)^{-1}\, d\lambda.$$

Then

$$\mathcal{P} = \operatorname{ran} P_\mu + \operatorname{ran}(1 - P_\mu), \quad \text{direct sum},$$

$\operatorname{ran} P_\mu$ and $\operatorname{ran}(1 - P_\mu)$ are $C_\mu(M)$–invariant subspaces of $\mathcal{P}$,

$$\operatorname{ran} P_\mu = \mathcal{R}_\mu(C_\mu(M)) = \mathcal{R}_\infty(M)$$

and μ belongs to the set $\rho(C_\mu(M)\,|_{\operatorname{ran}(1-P_\mu)})$. The latter implies that $\operatorname{ran}(1 - P_\mu)$ is contained in $\operatorname{dom} M$, and hence (5) is valid.
We prove (5) $\Rightarrow$ (7). The operator $C_\mu(M)$ is a bounded operator, so we can construct a basis of Jordan chains

$$\{x_{10}, x_{11}, \ldots, x_{1k_1}\}, \{x_{20}, x_{21}, \ldots, x_{2k_2}\}, \ldots, \{x_{m0}, x_{m1}, \ldots, x_{mk_m}\}$$

of M at ∞. (A (maximal) Jordan chain $\{y_0, y_1, \ldots, y_k\}$ of $C_\mu(M)$ at μ corresponds to a (maximal) Jordan chain $\{x_0, x_1, \ldots, x_k\}$ of M at ∞ of the same length, where each x_i can be expressed as a linear combination with nonzero coefficients of $y_i, \ldots, y_k$, $i = 0, 1, \ldots, k$. It follows that a basis of Jordan chains of $C_\mu(M)$ at μ corresponds to a basis of Jordan chains of M at ∞.) Here a Jordan chain can have length 1, but since together they form a basis, the elements $x_{1k_1}, x_{2k_2}, \ldots, x_{mk_m}$ span the subspace $\ker M^{-1}$ and the elements $x_{10}, x_{20}, \ldots, x_{m0}$ are linear independent modulo $\operatorname{dom} M$. Hence, on account of (5),

$$\operatorname{dom} M + \text{l.s.}\{x_{10}, x_{20}, \ldots, x_{m0}\} = \mathcal{P}$$

and the sum on the lefthand side is direct. Now (7) follows.
It remains to prove the implication (8) $\Rightarrow$ (1). Suppose that M is not maximal.

Then, by Lemma 4.3, there exists a maximal contractive linear relation $\widetilde{M}$ in $\mathcal{P}$ such that $M \subseteq \widetilde{M}$ and $\mathcal{R}(\widetilde{M}) = \{0\}$. It follows that $\ker M^{-1}$ is contained in $\ker \widetilde{M}^{-1}$ and that $\operatorname{dom} M$ is contained in $\operatorname{dom} \widetilde{M}$. Since $\widetilde{M}$ is maximal and, as we have shown above, (1) implies (8), we have that (8) holds with M replaced by $\widetilde{M}$. Thus

$$\dim \ker M^{-1} \leq \dim \ker \widetilde{M}^{-1} = \dim (\mathcal{P}/\operatorname{dom} \widetilde{M}) \leq \dim (\mathcal{P}/\operatorname{dom} M).$$

From (8) it follows that equality prevails everywhere, and hence

$$\ker M^{-1} = \ker \widetilde{M}^{-1} \quad \text{and} \quad \operatorname{dom} M = \operatorname{dom} \widetilde{M}.$$

This implies that $M = \widetilde{M}$, and (1) is valid. □

As in Theorem 3.11, the equality

$$\dim M/(M \cap (\mathcal{R}(M)^2)^{\langle\perp\rangle}) = \dim \mathcal{R}(M)^2/(M \cap \mathcal{R}(M)^2) \tag{4.1}$$

plays an important role in the following corollary to Theorem 4.4.

Corollary 4.5. *Let M be a contractive closed linear relation in a Pontryagin space $\mathcal{P}$ and denote by M_r its regularization. Then the seven statements* (a)–(g) *in Theorem* 2.3 *and the following statements are equivalent.*

(h) *M_r is maximal contractive and* (4.1) *is valid.*

(i) *$\widehat{M_r}$ is maximal contractive and* (4.1) *is valid.*

(j) *The set $\rho(\widehat{M_r}) \cap \mathbb{R} \cap (\mathbb{C} \setminus \overline{\mathbb{D}})$ is not empty and* (4.1) *is valid.*

(k) *The set $(\mathbb{C} \setminus \overline{\mathbb{D}}) \setminus \rho(\widehat{M_r})$ consists of at most $(\operatorname{ind}_{-}\mathcal{P} - \dim \mathcal{R}(M))$ points and these points are normal eigenvalues of $\widehat{M_r}$ and* (4.1) *is valid.*

(l) *The equalities $\operatorname{dom} \widehat{M_r} + \mathcal{R}_\infty(\widehat{M_r}) = \widehat{\mathcal{P}}$ and* (4.1) *are valid.*

(m) *The equalities $\operatorname{dom} M_r + \mathcal{R}_\infty(M_r) = \mathcal{R}(M)^{[\perp]}$ and* (4.1) *are valid.*

(o) *The equalities $\dim (\widehat{\mathcal{P}}/\operatorname{dom} \widehat{M_r}) = \dim \ker (\widehat{M_r}^{-1})$ and* (4.1) *are valid.*

(p) *The equalities $\dim (\mathcal{P}/\operatorname{dom} M_r) = \dim \ker (M_r^{-1})$ and* (4.1) *are valid.*

Proof. That (h) is equivalent to (a) of Theorem 2.3 follows from Theorem 3.11. The rest follows from Theorem 4.4. □

Corollary 4.6. *A linear relation M in a Pontryagin space is unitary if and only if M_r is unitary and* (4.1) *is valid.*

Proof. The equivalence follows from the equivalence between the statements Theorem 2.3(a) and Corollary 4.5(h), the equality (3.3) and the following two observations:

(1) M is unitary if and only if M and M^{-1} are maximal contractive linear relations.

(2) The equality (4.1) holds true if and only if the equality (4.1) with M replaced by M^{-1} holds true. □

In Example 3.6 we have that $M_r = \mathcal{R}(M)^2$ and hence M_r is unitary. However, M itself is not unitary and so (4.1) does not hold for M.

5. Properties of maximal contractive linear relations

In this section we list some properties of maximal contractive linear relations.

Theorem 5.1. *If M is a maximal contractive linear relation in a Pontryagin space $\mathcal{P}$ then $\mathcal{R}(M) = \mathcal{R}(M^*)$, $(M^*)_r = (M_r)^*$ and hence $(\widehat{M_r})^* = (M_r^*)\widehat{\ }$, where the adjoint on the lefthand side is taken in the quotient space $\widehat{\mathcal{P}}$.*

Proof. The first equality follows from the inclusion $\mathcal{R}(M) \subseteq \mathcal{R}(M^*)$. Indeed, since $M^{**} = M$ and, by Theorem 2.3, M^* is a maximal contractive linear relation in $\mathcal{P}$ also, we may replace M by M^* in this inclusion and obtain the converse inclusion and the desired equality is proved. To prove the inclusion, consider an element $x_0 \in \mathcal{R}(M)$. Then there exist numbers $m, n \in \mathbb{N}$ and elements $x_j \in \mathcal{R}(M)$, $j = -m, \ldots, -1, 1, \ldots, n$, such that all pairs

$$\{0, x_{-m}\}, \ldots, \{x_{-2}, x_{-1}\}, \{x_{-1}, x_0\}, \{x_0, x_1\}, \ldots, \{x_n, 0\}$$

belong to M. By Lemma 3.4, $\mathcal{R}(M)$ is a neutral subspace of $\mathcal{P}$ and hence these pairs are neutral elements in the Kreĭn space $\mathcal{P}^2$. Since M is a nonnegative subspace of this Kreĭn space, the Cauchy-Schwarz-Bunyakowski inequality implies that these pairs belong to $M^{\langle\perp\rangle} = (M^*)^{-1}$. Hence the pairs

$$\{0, x_n\}, \ldots, \{x_1, x_0\}, \{x_0, x_{-1}\}, \{x_{-1}, x_{-2}\}, \ldots, \{x_{-m}, 0\}$$

belong to M^*. In particular, x_0 belongs to $\mathcal{R}_0(M^*) \cap \mathcal{R}_\infty(M^*) = \mathcal{R}(M^*)$, and the desired inclusion holds true. The second equality in the theorem is obtained as follows:

$$\begin{aligned}(M_r)^* &= \left(\mathcal{R}(M)^2 + M \cap (\mathcal{R}(M)^2)^{\langle\perp\rangle}\right)^* = \left(\mathcal{R}(M)^2 + M^{-1} \cap (\mathcal{R}(M)^2)^{\langle\perp\rangle}\right)^{\langle\perp\rangle} \\ &= (\mathcal{R}(M)^2)^{\langle\perp\rangle} \cap (M^* + \mathcal{R}(M)^2) = \mathcal{R}(M)^2 + M^* \cap (\mathcal{R}(M)^2)^{\langle\perp\rangle} \\ &= \mathcal{R}(M^*)^2 + M^* \cap (\mathcal{R}(M^*)^2)^{\langle\perp\rangle} = (M^*)_r.\end{aligned}$$

The last equality in the third line holds since $\mathcal{R}(M)^2 \subseteq (\mathcal{R}(M)^2)^{\langle\perp\rangle}$. □

The second equality in Theorem 5.1 implies the following result.

Corollary 5.2. *A maximal contractive linear relation M in a Pontryagin space is regular if and only if M^* is regular.*

Theorem 5.3. *If M is a maximal contractive linear relation in a Pontryagin space, then*

$$\dim \mathcal{R}_\infty(M) = \dim \mathcal{R}_\infty(M^*).$$

Proof. By Theorem 2.3, M^* also is a maximal contractive linear relation. On account of Theorem 3.8 and Theorem 5.1,

$$\mathcal{R}_\infty(M) = \mathcal{R}_\infty(M_r), \quad \mathcal{R}_\infty(M^*) = \mathcal{R}_\infty(M_r^*), \quad \mathcal{R}(M_r) \subseteq \mathcal{R}_\infty(M_r) \cap \mathcal{R}_\infty(M_r^*).$$

Hence the theorem is proved once it is established that

$$\dim \mathcal{R}_\infty(\widehat{M_r}) = \dim \mathcal{R}_\infty((\widehat{M_r})^*).$$

To prove this equality we may and do assume that M is a regular maximal contractive linear relation in $\mathcal{P}$ with $\mathcal{R}(M) = \{0\}$. Let P_μ be the Riesz projection onto $\mathcal{R}_\mu(C_\mu(M))$ considered in the proof of Theorem 4.4. Then P_μ^* is the Riesz projection onto $\mathcal{R}_\mu(C_\mu(M)^*)$. Since $C_\mu(M)^* = C_\mu(M^*)$, we obtain

$$\begin{aligned}\dim \mathcal{R}_\infty(M) &= \dim \mathcal{R}_\mu(C_\mu(M)) = \dim \operatorname{ran} P_\mu \\ &= \dim \operatorname{ran} P_\mu^* = \dim \mathcal{R}_\mu(C_\mu(M^*)) \\ &= \dim \mathcal{R}_\infty(M^*).\end{aligned}$$

This is what we had to show. □

Corollary 5.4. *If M is a maximal contractive isometric linear relation in a Pontryagin space then* $\dim \mathcal{R}_0(M) \le \dim \mathcal{R}_\infty(M)$*; equality holds when M is unitary.*

Example 5.5. There exists a maximal contractive isometric linear relation M with $\mathcal{R}_\infty(M) \neq \mathcal{R}_\infty(M^*)$ and $\dim \mathcal{R}_0(M) < \dim \mathcal{R}_\infty(M)$.

Proof. Let $\mathcal{P} = \mathcal{P}_+ \oplus \mathcal{P}_-$ be a Pontryagin space where $\mathcal{P}_+$ has an orthonormal basis $(e_i^+)_{i=1}^\infty$ and $\mathcal{P}_-$ is one dimensional and spanned by e^- with $[e^-, e^-] = -1$ and consider the linear relation

$$M = \text{l.s.}\,\left\{\{0, e^- + e_1^+\}, \{e_i^+, e_{i+1}^+\}_{i=1}^\infty\right\}.$$

Then

$$M^* = \text{l.s.}\,\left\{\{e^- + e_1^+, 0\}, \{0, e^-\}, \{e_{i+1}^+, e_i^+\}_{i=1}^\infty\right\}.$$

Then $M^{-1} \subseteq M^*$, M is isometric, $\operatorname{dom} M = \mathcal{P}_+$, $\mathcal{R}_\infty(M) = \ker(M^{-1}) = \text{l.s.}\,\{e^- + e_1^+\}$, $\mathcal{R}_0(M) = \ker M = \{0\}$, $\mathcal{R}(M) = \mathcal{N}(M) = \{0\}$, and M is regular. In particular, $\mathcal{R}_0(M) \neq \mathcal{R}_\infty(M)$. From Corollary 4.5 (p) it follows that M is maximal contractive. Hence M^* is also maximal contractive and regular. We have $\mathcal{R}_0(M^*) = \ker M^* = \text{l.s.}\,\{e^- + e_1^+\}$, $\mathcal{R}_\infty(M^*) = \ker(M^*)^{-1} = \text{l.s.}\,\{e^-\}$. In particular, $\mathcal{R}_\infty(M) \neq \mathcal{R}_\infty(M^*)$. □

Theorem 5.6. *If M is a maximal contractive linear relation in a Pontryagin space $\mathcal{P}$, then*

(5.1) $$\mathcal{R}(M)^{[\perp]} = \mathcal{R}_\infty(M) + \mathcal{R}_\infty(M^*)^{[\perp]}, \quad \mathcal{R}_\infty(M) \cap \mathcal{R}_\infty(M^*)^{[\perp]} = \mathcal{R}(M),$$

or equivalently,

$$\widehat{\mathcal{P}} = \mathcal{R}_\infty(\widehat{M}_r) + \mathcal{R}_\infty((\widehat{M}_r)^*)^{[\perp]}, \quad \textit{direct sum.} \tag{5.2}$$

Moreover, every maximal nonnegative subspace of $\mathcal{R}_\infty(M^*)^{[\perp]}$ *is a maximal nonnegative subspace of* $\mathcal{P}$ *and the dimension of every maximal nonpositive subspace of* $\mathcal{R}_\infty(M^*)^{[\perp]}$ *is equal to*

$$\tau\left(\mathcal{R}_\infty(M^*)^{[\perp]}\right) := \operatorname{ind}_-\mathcal{P} - \dim \mathcal{R}_\infty(M) + \dim \mathcal{R}_\infty(M^*)^0,$$

where $\mathcal{R}_\infty(M^*)^0$ *is the isotropic part of* $\mathcal{R}_\infty(M^*)$.

Proof. The equivalence of (5.1) and (5.2) follows from considerations as in the proof of Theorem 5.3. To prove the equality (5.2), we again may and do assume that M is a regular maximal contractive linear relation in $\mathcal{P}$ with $\mathcal{R}(M) = \{0\}$. Let P_μ be the Riesz projection onto $\mathcal{R}_\mu(C_\mu(M))$ considered in the proofs of Theorem 4.4 and Theorem 5.3. Then the equalities $\mathcal{P} = \operatorname{ran} P_\mu + \operatorname{ran}(I - P_\mu)$, direct sum, and $\operatorname{ran}(I - P_\mu) = (\operatorname{ran} P_\mu^*)^{[\perp]}$ readily imply

$$\mathcal{P} = \mathcal{R}_\infty(M) + \mathcal{R}_\infty(M^*)^{[\perp]}, \quad \text{direct sum},$$

which we had to show. To prove the last statements we set $\mathcal{H} = \mathcal{R}_\infty(M^*)^{[\perp]}$. Since M^* is contractive, $\mathcal{H}^{[\perp]}$ is nonpositive and hence can be decomposed as $\mathcal{H}^{[\perp]} = \mathcal{H}_- \oplus \mathcal{H}^0$, where $\mathcal{H}_-$ is a maximal negative subspace of $\mathcal{H}^{[\perp]}$ and $\mathcal{H}^0$ is the isotropic part of $\mathcal{H}$. In terms of these spaces $\mathcal{P}$ and $\mathcal{H}$ can be decomposed as

$$\mathcal{P} = \mathcal{H}_- \oplus (\mathcal{H}^0 + J\mathcal{H}^0) \oplus \mathcal{P}' \quad \text{and} \quad \mathcal{H} = \mathcal{H}^0 \oplus \mathcal{P}',$$

where $\mathcal{P}'$ is a Pontryagin space. Let $\mathcal{L}$ be a maximal nonnegative subspace of $\mathcal{H}$. Then $\mathcal{H}^0$ is contained in $\mathcal{L}$. Let $\mathcal{L}'$ be a nonnegative subspace of $\mathcal{P}$ containing $\mathcal{L}$. We show that the two subspaces coincide. Let x belong to $\mathcal{L}'$ and write $x = x_- + x_0 + Jy_0 + x'$ with $x_- \in \mathcal{H}_-$, $x_0, y_0 \in \mathcal{H}^0$ and $x' \in \mathcal{P}'$. Then for all $\lambda \in \mathbb{R}$,

$$[x + \lambda y_0, x + \lambda y_0] = [x_-, x_-] + 2\lambda[Jy_0, y_0] + 2\operatorname{Re}[x_0, Jy_0][x', x'] \geq 0,$$

which implies that $y_0 = 0$. The linear span of $\mathcal{L}$ and the element $x_0 + x'$ is a nonnegative subspace of $\mathcal{H}$. Since $\mathcal{L}$ is maximal nonnegative, $x_0 + x'$ belongs to $\mathcal{L}$. It follows that $x_- = x - (x_0 + x')$ belongs to $\mathcal{L}'$ which yields $x_- = 0$. Hence $x = x_0 + x'$ and belongs to $\mathcal{L}$, and we have proved that $\mathcal{L}' = \mathcal{L}$, that is, $\mathcal{L}$ is also a maximal nonnegative subspace of $\mathcal{P}$. As to the dimension of a maximal nonpositive subspace of $\mathcal{H}$: it is equal to $\tau(\mathcal{H}) = \dim \mathcal{H}^0 + \operatorname{ind}_-\mathcal{P}'$. The desired formula now follows from the identities

$$\begin{aligned}
\operatorname{ind}_-\mathcal{P} &= \dim \mathcal{H}_- + \dim \mathcal{H}^0 + \operatorname{ind}_-\mathcal{P}' \\
&= \dim \mathcal{R}_\infty(M^*) - \dim \mathcal{H}^0 + \tau(\mathcal{H}) \\
&= \dim \mathcal{R}_\infty(M) - \dim \mathcal{R}_\infty(M^*)^0 + \tau\left(\mathcal{R}_\infty(M^*)^{[\perp]}\right),
\end{aligned}$$

where we have used Theorem 5.3. □

Corollary 5.7. *Let M be a maximal contractive linear relation in a Pontryagin space $\mathcal{P}$ and assume that $\mathcal{R}_\infty(M)$ is a neutral subspace of $\mathcal{P}$. Then*

$$\mathcal{R}_\infty(M) + \mathcal{R}_\infty(M^*) = \mathcal{P}_2 \oplus \mathcal{R}(M),$$

where $\mathcal{P}_2$ is a finite dimensional Pontryagin space with

$$\dim \mathcal{P}_2 = 2\operatorname{ind}_-\mathcal{P}_2, \qquad \operatorname{ind}_-\mathcal{P}_2 = \dim \mathcal{R}_\infty(M) - \dim \mathcal{R}(M).$$

Proof. Set $\mathcal{L} = \mathcal{R}_\infty(M) + \mathcal{R}_\infty(M^*)$. We show $\mathcal{L} \cap \mathcal{L}^{[\perp]} = \mathcal{R}(M)$. Since M and M^* are contractive, $\mathcal{R}_\infty(M)$ and $\mathcal{R}_\infty(M^*)$ are nonpositive and contain the neutral subspace $\mathcal{R}(M)$. Hence $\mathcal{L}$ is finite dimensional, $\dim \mathcal{L} \le 2\operatorname{ind}_-\mathcal{P}$, and the Cauchy-Schwarz-Bunyakowski inequality implies that $\mathcal{R}(M) \subseteq \mathcal{L} \cap \mathcal{L}^{[\perp]}$. To prove the converse inclusion, consider an element $x \in \mathcal{L} \cap \mathcal{L}^{[\perp]}$. Then $x = y + y^*$ with $y \in \mathcal{R}_\infty(M)$ and $y^* \in \mathcal{R}_\infty(M^*)$. Since x belongs to $\mathcal{R}_\infty(M)^{[\perp]}$ and $\mathcal{R}_\infty(M)$ is neutral or equivalently, $\mathcal{R}_\infty(M) \subseteq \mathcal{R}_\infty(M)^{[\perp]}$, we have

$$y^* = x - y \in \mathcal{R}_\infty(M^*) \cap \mathcal{R}_\infty(M)^{[\perp]} = \mathcal{R}(M).$$

Here we used the first equality in Theorem 5.6. Similarly, since x and $\mathcal{R}(M)$ are contained in $\mathcal{R}_\infty(M^*)^{[\perp]}$, we also have

$$y = x - y^* \in \mathcal{R}_\infty(M) \cap \mathcal{R}_\infty(M^*)^{[\perp]} = \mathcal{R}(M).$$

Now we used the second equality in Theorem 5.6. It follows that x belongs to $\mathcal{R}(M)$. Hence $\mathcal{L} \cap \mathcal{L}^{[\perp]} = \mathcal{R}(M)$. This yields that $\mathcal{L}$ admits a decomposition described in the corollary. By going over to equivalence classes we obtain

$$\widehat{\mathcal{P}_2} = \mathcal{R}_\infty(\widehat{M_r}) + \mathcal{R}_\infty((\widehat{M_r})^*), \quad \text{direct sum.}$$

On account of Theorem 5.3, the two summands on the righthand side have the same dimension, κ, say. Since the first summand is neutral, $\kappa \le \dim(\widehat{\mathcal{P}_2})_+$, and since the second summand is nonpositive, $\kappa \le \dim(\widehat{\mathcal{P}_2})_-$, where $\widehat{\mathcal{P}_2} = (\widehat{\mathcal{P}_2})_+ \oplus (\widehat{\mathcal{P}_2})_-$ is a fundamental decomposition of $\widehat{\mathcal{P}_2}$. Finally, since $\dim \widehat{\mathcal{P}_2} \le 2\kappa$, the three last inequalities are equalities. Hence

$$\dim \widehat{\mathcal{P}_2} = 2\operatorname{ind}_-\widehat{\mathcal{P}_2}, \qquad \operatorname{ind}_-\widehat{\mathcal{P}_2} = \dim \mathcal{R}_\infty(\widehat{M_r}) = \dim \mathcal{R}_\infty(M) - \dim \mathcal{R}(M).$$

These equalities and the equalities $\dim \mathcal{P}_2 = \dim \widehat{\mathcal{P}_2}$ and $\operatorname{ind}_-\mathcal{P}_2 = \operatorname{ind}_-\widehat{\mathcal{P}_2}$ now imply the ones in the corollary. □

Remark 5.8. Let M be a linear relation in a Pontryagin space.

(1) If M is isometric, then $\mathcal{R}_\infty(M)$ is neutral.

(2) If M is a contraction and $\mathcal{R}_\infty(M)$ is neutral, then $\mathcal{R}_\infty(M) \subseteq \mathcal{R}_0(M^*)$.

(3) If M is maximal contractive and $\mathcal{R}_\infty(M)$ is neutral, then

$$\mathcal{R}_\infty(M) \subseteq \mathcal{R}_0(M^*) \cap \mathcal{R}_0(M^*)^{[\perp]}.$$

Equality holds if M is unitary.

Proof. Let x belong to $\mathcal{R}_\infty(M)$. Then there exist elements $x_1,\dots,x_k$ in the Pontryagin space $\mathcal{P}$, say, such that the pairs

$$\{x_1,x\},\{x_2,x_1\},\dots,\{x_k,x_{k-1}\},\{0,x_k\}$$

belong to M. If M is isometric, then

$$0=[x_k,x_k]=[x_{k-1},x_{k-1}]=\cdots=[x_1,x_1]=[x,x].$$

This proves (1). If M is contractive and $\mathcal{R}_\infty(M)$ is neutral then M is nonnegative and $\mathcal{R}_\infty(M)^2$ is neutral in the Kreĭn space $\mathcal{P}^2$. Using the Cauchy-Schwarz-Bunyakowski inequality we find that the pairs

$$\{x,x_1\},\{x_1,x_2\},\dots,\{x_{k-1},x_k\},\{x_k,0\}$$

belong to M^*. Hence x belongs to $\mathcal{R}_0(M^*)$. This proves (2). The inclusion in (3) follows from (2) and the fact that, since now M^* also is a contraction, $\mathcal{R}_0(M^*)$ is nonnegative in $\mathcal{P}$. If M is unitary equality holds since $M^*=M^{-1}$ and, on account of (1), $\mathcal{R}_\infty(M)$ is neutral. □

6. Invariant subspaces

NAĬMARK's theorem [N] asserts that for every commuting family of unitary operators U in a Pontryagin space with negative index κ there exists a κ-dimensional, nonpositive subspace which is invariant under all $U's$; for a new and short proof, see [Sa]. For a study of the invariant subspaces of contractions or commuting families of contractions, we refer to [AI], [IKL] and [AL]. In this section we prove some results on the existence of invariant subspaces for a collection $\mathbf{M}=\{M\}$ of maximal contractive linear relations M in a Pontryagin space $\mathcal{P}$. We assume throughout that the subspaces

(6.1) $\quad \mathcal{R}_\infty:=\mathcal{R}_\infty(M)$ and $\mathcal{R}:=\mathcal{R}(M)$ are independent of $M\in\mathbf{M}$.

Then all linear relations $\widehat{M_r}$ act in the same space $\widehat{\mathcal{P}}=\mathcal{R}^{[\perp]}/\mathcal{R}$ and are maximal contractive there. Note that, on account of Theorem 3.8, we also have that for all $M\in\mathbf{M}$, $\mathcal{R}_\infty=\mathcal{R}_\infty(M_r)$ and $\mathcal{R}=\mathcal{R}(M_r)$. By Corollary 4.5, the set $\rho(\widehat{M_r})\cap\mathbb{R}\cap(\mathbb{C}\setminus\overline{\mathbb{D}})$ is not empty. For each $M\in\mathbf{M}$, we select and fix a point $\mu(M)$ from this set and consider the corresponding Cayley transform $C_{\mu(M)}(\widehat{M_r})$; we often suppress the dependence on M and simply write μ and C_μ for $\mu(M)$ and $C_{\mu(M)}(\widehat{M_r})$, respectively. Each C_μ belongs to $\mathbf{L}(\widehat{\mathcal{P}})$ and is a contraction. We say that the collection $\mathbf{M}=\{M\}$ is *commutative* if the family of Cayley transforms $\{C_\mu\}$ is commutative in the usual sense. This definition is independent of the choices of the points $\mu(M)\in\rho(\widehat{M_r})\cap\mathbb{R}\cap(\mathbb{C}\setminus\overline{\mathbb{D}})$, $M\in\mathbf{M}$; this follows from the equality

$$C_{\mu'}=((\mu-\mu')-(1-\mu\mu')C_\mu)((\mu-\mu')C_\mu-(1-\mu\mu'))^{-1}.$$

Lemma 6.1. *Let $\mathbf{M} = \{M\}$ be a commutative collection of maximal contractive linear relations in a Pontryagin space $\mathcal{P}$ with property* (6.1). *Then $\mathcal{R}_\infty(M^*)$ is independent of $M \in \mathbf{M}$.*

Proof. $\mathcal{R}_\infty(M^*) = \mathcal{R}_\infty(M_r^*)$ is independent of $M \in \mathbf{M}$ if and only if

$$\mathcal{R}_\infty((\widehat{M_r})^*) = \mathcal{R}_\infty(M_r^*)/\mathcal{R}$$

is independent of $\widehat{M_r}$, $M \in \mathbf{M}$. Therefore it suffices to prove the lemma under the additional assumption that all M's are regular and that $\mathcal{R} = \{0\}$. As in the proof of Theorem 4.4 we consider the Riesz projection

$$P_\mu = -\frac{1}{2\pi i}\int_{\Gamma_{\mu,\varepsilon}} (C_\mu - \lambda)^{-1}\, d\lambda,$$

where in the notation we have suppressed the dependence of C_μ, μ and ε and hence of P_μ on M. The Riesz projections belong to $\mathbf{L}(\mathcal{P})$, and from property (6.1) it follows that their ranges are independent of $M \in \mathbf{M}$:

$$\operatorname{ran} P_{\mu(M)} = \operatorname{ran} \mathcal{R}_{\mu(M)}(C_{\mu(M)}(M)) = \mathcal{R}_\infty.$$

Since the C_μ's commute, the projections P_μ commute, and since these projections have the same range, they coincide. The lemma now is a consequence of the equality

$$\mathcal{R}_\infty(M^*) = \mathcal{R}_\mu(C_\mu^*) = \operatorname{ran} P_\mu^*,$$

which follows from the argument in the last part of the proof of Theorem 5.3. □

Under the conditions of Lemma 6.1, we set

$$\mathcal{R}_\infty^* := \mathcal{R}_\infty(M^*),$$

and we note that the root subspace $(\mathcal{R}_\infty^*)^0 := \mathcal{R}_\infty^* \cap (\mathcal{R}_\infty^*)^{[\perp]}$ is also independent of $M \in \mathbf{M}$.

To prove the next theorem we recall [IKL, Lemma 11.5].

Lemma 6.2. *Let $\mathcal{P}$ be a Pontryagin space and let $T \in \mathbf{L}(\mathcal{P})$ be a contraction. Assume $\mu \in \mathbb{C} \setminus \overline{\mathbb{D}}$ is an eigenvalue of T and let $\mathcal{R}^0 = \mathcal{R}_\mu(T) \cap \mathcal{R}_\mu(T)^\perp$ be the isotropic part of the root space of T at μ. Then*

$$T\mathcal{R}^0 = \mathcal{R}^0, \qquad T^*\mathcal{R}^0 = \mathcal{R}^0,$$

and $T|_{\mathcal{R}^0}$ is invertible with inverse $T^*|_{\mathcal{R}^0}$.

Theorem 6.3. *Let $\mathbf{M} = \{M\}$ be a commutative collection of maximal contractive linear relations in a Pontryagin space $\mathcal{P}$ with property* (6.1). *Then there exist a maximal nonnegative subspace $\mathcal{H}_+$ in $\mathcal{P}$ and a nonpositive subspace $\mathcal{H}_-$ with*

$$\dim \mathcal{H}_- = \operatorname{ind}_-\mathcal{P} - \dim \mathcal{R}_\infty + \dim (\mathcal{R}_\infty^*)^0$$

which are M–invariant for all $M \in \mathbf{M}$.

Proof. We first assume that all M's are regular and that $\mathcal{R} = \{0\}$. Then the decomposition

$$\mathcal{P} = \operatorname{ran} P_\mu \dotplus \operatorname{ran}(I - P_\mu), \quad \text{direct sum},$$

considered in the proof of Theorem 5.6 is independent of $M \in \mathbf{M}$. We use the same notation and set

$$\mathcal{H} = \operatorname{ran}(I - P_\mu) = \operatorname{ran} \mathcal{R}_\mu(C_\mu^*)^{[\perp]} = \mathcal{R}_\infty(M^*)^{[\perp]}.$$

As in the proof of the implication (4) $\Rightarrow$ (5) of Theorem 4.4, the space $\mathcal{H}$ is invariant under all C_μ's and μ belongs to the resolvent set $\rho(C_\mu|_\mathcal{H})$ of the restriction of C_μ to $\mathcal{H}$. Hence for each $M \in \mathbf{M}$, $\mathcal{H}$ is contained in $\operatorname{dom} M$ and is M–invariant. It follows that $L_M := M \cap \mathcal{H}^2$ is (the graph of) a contractive operator on the whole space $\mathcal{H}$; in fact,

$$L_M = (\mu C_\mu|_\mathcal{H} - 1)(C_\mu|_\mathcal{H} - \mu)^{-1}.$$

Write $\mathcal{H} = \mathcal{H}^0 \oplus \mathcal{P}'$, where $\mathcal{H}^0$ is the isotropic part of $\mathcal{H}$ and $\mathcal{P}'$ is a Pontryagin space with $\operatorname{ind}_-\mathcal{P}' = \tau(\mathcal{H}) - \dim \mathcal{H}^0 = \operatorname{ind}_-\mathcal{P} - \dim \mathcal{R}_\infty$; see Theorem 5.6. Since C_μ^* is a contraction on $\mathcal{P}$, we may apply Lemma 6.2 with $T = C_\mu^*$ and $\mathcal{R}_\mu(T) = \mathcal{R}_\mu(C_\mu^*) = \mathcal{H}^\perp$. We conclude that $\mathcal{H}^0$ is invariant under C_μ and hence also under L_M.
Denote by P' the projection from $\mathcal{H}$ onto $\mathcal{P}'$ parallel to $\mathcal{H}^0$. Then the operator $L'_M = P' L_M|_{\mathcal{P}'}$ belongs to $\mathbf{L}(\mathcal{P}')$ and is a contraction. The collection of contractions $\{L'_M\}_{M \in \mathbf{M}}$ is commutative in the usual sense and hence we may invoke [AI, Theorem 3.4.11], which implies that there exist a maximal nonnegative subspace $\mathcal{H}'_+$ in $\mathcal{P}'$ and a nonpositive subspace $\mathcal{H}'_-$ in $\mathcal{P}'$ with $\dim \mathcal{H}'_- = \operatorname{ind}_-\mathcal{P}'$ which are L'_M–invariant for all $M \in \mathbf{M}$. It follows that the spaces $\mathcal{H}_+ = \mathcal{H}'_+ \oplus \mathcal{H}^0$ and $\mathcal{H}_- = \mathcal{H}'_- \oplus \mathcal{H}^0$ are contained in $\operatorname{dom} M$ and are L_M–invariant and hence M–invariant for all $M \in \mathbf{M}$; the first one is a maximal nonnegative subspace and the second one is a maximal nonpositive subspace of $\mathcal{H}$. By Theorem 5.6, $\mathcal{H}_+$ is a maximal nonnegative subspace of $\mathcal{P}$ and $\dim \mathcal{H}_- = \tau(\mathcal{H})$.
For the general case we note that the above applies to the family of linear relations $\{\widehat{M_r}\}$ which act in $\widehat{\mathcal{P}}$. That is, there exist a maximal nonnegative subspace $\widehat{\mathcal{H}}_+$ in $\widehat{\mathcal{P}}$ and a nonpositive subspace $\widehat{\mathcal{H}}_-$ with $\dim \widehat{\mathcal{H}}_- = \tau(\mathcal{H})$ which are contained in $\operatorname{dom} \widehat{M_r}$ and are $\widehat{M_r}$–invariant for all $M \in \mathbf{M}$. The spaces $\mathcal{H}_\pm = \{x \mid \widehat{x} \in \widehat{\mathcal{H}}_\pm\}$ are M_r–invariant, and by Theorem 3.9, also M–invariant. These spaces also have the other properties mentioned in the theorem. □

For the following theorem we fix a Pontryagin subspace $\mathcal{P}_1$ of $\mathcal{P}$ such that

$$\mathcal{P} = (\mathcal{R} \dotplus J\mathcal{R}) \oplus \mathcal{P}_1, \qquad \mathcal{R}^{[\perp]} = \mathcal{R} \oplus \mathcal{P}_1.$$

Theorem 6.4. *Let $\mathbf{M} = \{M\}$ be a commutative collection of maximal contractive linear relations in a Pontryagin space $\mathcal{P}$ with property* (6.1) *and assume that for all $M \in \mathbf{M}$,*

$$\mathcal{R}_\infty = \mathcal{R}_0(M^*) \cap \mathcal{R}_0(M^*)^{[\perp]}.$$

Then the subspace

$$\mathcal{P}_3 = \mathcal{P}_1 \cap \left(\mathcal{R}_\infty(M) + \mathcal{R}_\infty(M^*)^{[\perp]}\right)$$

is a Pontryagin space which does not depend on $M \in \mathbf{M}$ and

$$\operatorname{ind}_-\mathcal{P}_3 = \operatorname{ind}_-\mathcal{P} - \dim \mathcal{R}_\infty,$$

and for all $M \in \mathbf{M}$, the space $\mathcal{P}_3$ is M_r–invariant and the space $\mathcal{P}_3 + \mathcal{R}$ is M–invariant.

Proof. By Lemma 6.1 the space $\mathcal{P}_3$ is independent of $M \in \mathbf{M}$ (even without the assumption that $\mathcal{R}_\infty$ is neutral). On account of Corollary 5.7, $\mathcal{P}_3 = \mathcal{P}_1 \cap \mathcal{P}_2^{[\perp]}$, $\mathcal{P}_3$ is a Pontryagin space, and

$$\operatorname{ind}_-\mathcal{P}_3 = \operatorname{ind}_-\mathcal{P}_1 - \operatorname{ind}_-\mathcal{P}_2 = \operatorname{ind}_-\mathcal{P} - \dim \mathcal{R}_\infty.$$

To prove that $\mathcal{P}_3$ is M–invariant, we first consider the special case that all $M \in \mathbf{M}$ are regular and that $\mathcal{R} = \{0\}$. Then $\mathcal{P}_1 = \mathcal{P}$ and $\mathcal{P}_3 = \mathcal{R}_\infty^{[\perp]} \cap \mathcal{H}$, where $\mathcal{H} = \mathcal{R}_\infty(M^*)^{[\perp]} = \mathcal{R}_\mu(C_\mu^*)^{[\perp]}$. Evidently, $\mathcal{H}$ is C_μ–invariant. Since $\mathcal{R}_\infty$ is the isotropic part of $\mathcal{R}_0(M^*) = \mathcal{R}_{1/\mu}(C_\mu(M^*))$ and $C_\mu(M^*) = C_\mu(M)^*$ is a contraction, we may apply Lemma 6.2. We conclude that $\mathcal{R}_\infty$ is $C_\mu(M)^*$–invariant and hence $\mathcal{R}_\infty^{[\perp]}$ is C_μ–invariant. It follows that $\mathcal{P}_3$ is C_μ–invariant. The point μ belongs to the resolvent set of the restriction of C_μ to $\mathcal{P}_3$, and therefore this space is invariant under all $M \in \mathbf{M}$. As to the general case note that, because of Theorem 3.8 and Theorem 5.1,

$$\mathcal{P}_3 = \mathcal{P}_1 \cap \left(\mathcal{R}_\infty(\widehat{M_r}) + \mathcal{R}_\infty(\widehat{M_r})^*\right)^{[\perp]},$$

where elements in the space $\widehat{\mathcal{P}}$ are identified with their representatives in $\mathcal{P}_1$. The special case just proved implies that $\mathcal{P}_3$ is invariant under $\widehat{M_r}$, and hence also under M_r. From Theorem 3.9 it follows that $\mathcal{P}_3 + \mathcal{R}$ is M–invariant. □

Let $\mathcal{G}$ be a Hilbert space with inner product $(\cdot,\cdot)$ and let G be a bounded selfadjoint operator with zero kernel such that $\mathcal{G}$ admits the $(\cdot,\cdot)$–orthogonal decomposition $\mathcal{G} = \mathcal{G}^+ \oplus \mathcal{G}^-$ into G–invariant subspaces $\mathcal{G}^\pm$ with the property that the restrictions $\pm G\,|_{\mathcal{G}^\pm}$ are positive operators and $\kappa = \dim \mathcal{G}^- < \infty$. Let $\mathcal{P}_G$ be the completion of $\mathcal{G}$ relative to the norm $\|x\| = (|G|x,x)^{1/2}$, $x \in \mathcal{G}$. Then the inner product $[x,y] = (Gx,y)$ can be continuously extended to all of $\mathcal{P}_G$, and $\mathcal{P}_G$ provided with this extended form, which we also denote by $[\cdot,\cdot]$, is a Pontryagin space with negative index $\mathcal{P}_G = \kappa$. The space $\mathcal{G}$ provided with the inner product $[x,y] = (Gx,y)$ will be denoted by $\mathcal{G}^\kappa$. For more details we refer to [AI, Proposition 1.6.14]. There one can also find the proof of the following lemma; see [AI, Proposition 1.8.18].

Lemma 6.5. *If $\mathcal{H}_+$ is a maximal nonnegative subspace in $\mathcal{P}_G$, then $\mathcal{H}_+ \cap \mathcal{G}$ is a maximal nonnegative subspace in the space $\mathcal{G}^\kappa$.*

Corollary 6.6. *Every commutative collection of maximal contractive linear relations in the space $\mathcal{G}^\kappa$ which satisfy* (6.1) *has a maximal nonnegative subspace which is invariant under all these linear relations.*

Proof. For each M in the collection, let $\widetilde{M}$ be its maximal contractive extension in $\mathcal{P}_G$. By Theorem 6.3, there exists a maximal nonnegative subspace $\widetilde{\mathcal{H}}_+$ of $\mathcal{P}_G$ which is invariant under all these $\widetilde{M}$'s. It follows from Lemma 6.5 that the space $\widetilde{\mathcal{H}}_+ \cap \mathcal{G}$ has the desired properties. □

Corollary 6.7. *Let $\{V\}$ be a commutative collection of contractive operators V in the space $\mathcal{G}^\kappa$ with* $\operatorname{dom} V = \mathcal{G}$. *Then the V's are bounded operators on the Hilbert space $\mathcal{G}$ and they have a common invariant subspace which is maximal nonnegative in the space $\mathcal{G}^\kappa$.*

Proof. The closure of (the graph of) the operator V in $\mathcal{P}_G$ is a contractive linear relation $\overline{V}$ in $\mathcal{P}_G$ with domain equal to $\mathcal{P}_G$. By Corollary 2.5, $\overline{V}$ is (the graph of) an operator in $\mathbf{L}(\mathcal{P}_G)$. Hence each V is bounded on $\mathcal{G}$. Since the collection $\{\overline{V}\}$ is commutative, the corollary follows from Corollary 6.6. □

Acknowledgements

The work of Tomas Ya. Azizov was supported by the Netherlands Organization for Scientific Research, NWO, the International Science Foundation (grant NZP000), and INTAS (project 93–0249).

References

[A] Arocena, R.: Unitary extensions of isometries and contractive intertwining dilations; Operator Theory: Adv. Appl. 41 (1989), 13–23.

[AADM] Arocena, R., Azizov, T. Ya., Dijksma, A., Marcantognini, S.A.M.: On commutant lifting with finite defect II; J. Functional Anal., to appear.

[AI] Azizov, T.Ya., Iokhvidov, I.S.: Foundations of the theory of linear operators in spaces with indefinite metric; Nauka, Moscow 1986 (in Russian); English translation: Linear operators in spaces with an indefinite metric; John Wiley & Sons, New York 1989.

[AL] Azizov, T.Ya., Langer, H.: Some spectral properties of contractive and expansive operators in indefinite inner product spaces; Math. Nachr. 162 (1993), 247–259.

[BH] Ball, J.A., Helton, J.W.: A Beurling–Lax theorem for the Lie group $U(m,n)$ which contains most classical interpolation theory; J. Operator Theory 9 (1983), 107–142.

[B] Bognar, J.: Indefinite inner product spaces; Springer Verlag, Berlin 1974.

[Br] BRODSKII, M.L.: On the properties of an operator mapping the nonnegative part of a space with an indefinite metric into itself; Uspekhi Mat. Nauk 14:1 (1959), 147–152 (in Russian).

[DDMS] DIJKSMA, A., DRITSCHEL, M., MARCANTOGNINI, S.A.M., DE SNOO, H.S.V.: The commutant lifting theorem for contractions on Kreĭn spaces; Operator Theory: Adv. Appl. 61 (1993), 65–83.

[DLS] DIJKSMA, A., LANGER, H., DE SNOO, H.S.V.: Eigenvalues and pole functions of Hamiltonian systems with eigenvalue depending boundary conditions; Math. Nachr. 161 (1993), 107–154.

[DS] DIJKSMA, A., DE SNOO, H.S.V.: Symmetric and selfadjoint linear relations in Kreĭn spaces I; Operator Theory: Adv. Appl. 24 (1987), 145–166.

[DR] DRITSCHEL, M., ROVNYAK, J.: Extension theorems for contractive operators on Kreĭn spaces; Operator Theory: Adv. Appl. 47 (1990), 221–305.

[G] GHEONDEA, A.: Contractive intertwining dilations for quasi–contractions; Z. Anal. Anwendungen 16 (1996), 31–44.

[IKL] IOKHVIDOV, I.S., KREĬN, M.G., LANGER, H.: Introduction to the spectral theory of operators in spaces with an indefinite metric; Mathematical Research 9, Akademie Verlag, Berlin 1982.

[N] NAĬMARK, M.A.: On commuting unitary operators in spaces with indefinite metric; Acta Sci. Math. Szeged 24 (1963), 177–189.

[R] RITZNER, V.S.: Linear relations and indefinite geometry; Thesis, Voronezh State University, Voronezh 1982 (in Russian).

[S] SARASON, D.: New Hilbert spaces from old; in: Paul Halmos, Celebrating 50 years of Mathematics, EWING, J., GEHRING F.W. (eds.), Springer Verlag, Berlin 1991, 195–204.

[Sa] SASVÁRI, Z.: New proof of Naĭmark's theorem on the existence of nonpositive invariant subspaces for commuting families of unitary operators in Pontryagin spaces; Monatsh. Math. 109 (1990), 153–156.

[Sh] SHMUL'YAN, YU. L.: Theory of linear relations and indefinite metric spaces; Funktsional Anal. i Prilozhen. 10:1 (1976), 68–72 (in Russian); English translation: Funct. Anal. Appl. 10:1 (1976), 56–60.

Department of Mathematics
Voronezh State University
Universitetskaja pl., 1
394693 Voronezh
Russia
azizov@tom.vsu.ru

Department of Mathematics
University of Groningen
P.O. Box 800
9700 AV Groningen
The Netherlands
a.dijksma@math.rug.nl

1991 Mathematics Subject Classification: Primary 46C20, 47A15, 47A20

Submitted: September 16, 1995

Operator Theory:
Advances and Applications, Vol. 103

On a new algorithm for almost periodic factorization

M.A. BASTOS, YU.I. KARLOVICH, I.M. SPITKOVSKY and P.M. TISHIN

An algorithm is proposed allowing to find necessary and sufficient conditions for existence of an almost periodic factorization for several new classes of triangular and block triangular matrix functions.

1. Introduction

Let AP denote the *Bohr algebra* of almost periodic functions, that is, the smallest C^*-subalgebra of $L^\infty(\mathbb{R})$ containing all the monomials

$$e_\lambda(x) = e^{i\lambda x}, \qquad \lambda \in \mathbb{R}.$$

It is well known (see [2, 7, 8] for these and other properties of AP) that for every $f \in AP$ there exists the *Bohr mean value*

$$\mathbf{M}(f) = \lim_{T\to\infty} \frac{1}{2T} \int_{-T}^{T} f(x)\,dx,$$

and that the set

$$\Omega(f) = \{\lambda \in \mathbb{R} : \widehat{f}(\lambda) \neq 0\},$$

where $\widehat{f}(\lambda) = \mathbf{M}(e_{-\lambda} f)$, is at most countable. This set is called the *Fourier spectrum* of f, and the (formal) series

$$\sum_{\lambda\in\Omega(f)} \widehat{f}(\lambda) e_\lambda \tag{1.1}$$

is the *Fourier series* of f. We say that $f \in AP_W$ if $\sum_{\lambda\in\Omega(f)} |\widehat{f}(\lambda)| < \infty$; in the latter case, of course, the series (1.1) converges absolutely and uniformly on $\mathbb{R}$.

Set

$$AP^\pm = \{f \in AP : \widehat{f}(\lambda) = 0 \text{ for } \pm\lambda < 0\}, \qquad AP_W^\pm = AP^\pm \cap AP_W.$$

We say that an $n \times n$ matrix function X belongs to AP (AP_W, $AP^\pm$, etc.) if each entry of X has this property. We also define $\mathbf{M}(X)$ elementwise and $\Omega(X)$ as the union of the Fourier spectra of the entries.

By an AP-factorization of an $n \times n$ matrix function G we understand its representation in the form

$$G(x) = G_+(x)\Lambda(x)G_-(x), \qquad x \in \mathbb{R}, \tag{1.2}$$

where

$$G_+^{\pm 1} \in AP^+, \qquad G_-^{\pm 1} \in AP^-, \tag{1.3}$$

and

$$\Lambda(x) = \operatorname{diag}[e^{i\lambda_1 x}, \dots, e^{i\lambda_n x}], \qquad \lambda_j \in \mathbb{R}.$$

We say that (1.2) is an AP_W-factorization of G if, instead of (1.3), the more restrictive condition

$$G_+^{\pm 1} \in AP_W^+, \qquad G_-^{\pm 1} \in AP_W^-,$$

holds. Of course, the AP- (AP_W-) factorization can possibly exist only for matrix functions invertible in AP (respectively, AP_W).

As observed in [3], the numbers $\lambda_1, \dots, \lambda_n$ are defined by the matrix G uniquely, provided that its factorization (1.2) exists; we call them the *partial AP-indices* of G.

Our interest in the AP-factorization and partial AP-indices is based on the role they play in the theory of Toeplitz operators, singular integral and convolution type equations. A partial understanding of this role can be obtained from the following result, see [1].

Theorem 1.1. *Let $G \in AP_W$ and $P_\pm = \frac{1}{2}(I \pm S)$, where the operator S is defined by*

$$(S\phi)(x) = \frac{1}{\pi i} \int_{-\infty}^{+\infty} \phi(y) \frac{dy}{y - x}.$$

Then the following statements are equivalent.
(1) *the operator $R_G = P_+ + GP_-$ is Fredholm on $L^2(\mathbb{R})$,*
(2) *the operator R_G is invertible on all the weighted spaces $L^p(\mathbb{R}, \rho)$, where $1 < p < \infty$ and $\rho(x)$ is a Muckenhoupt weight:*

$$\sup_{-\infty < a < b < \infty} \frac{1}{b-a} \left(\int_a^b \rho(x)^p \, dx \right)^{1/p} \left(\int_a^b \rho(x)^{-p/(p-1)} \, dx \right)^{1-1/p} < \infty,$$

(3) *G admits an AP-factorization* (1.2) *with zero partial AP-indices,*
(4) *G admits an AP_W-factorization* (1.2) *with zero partial AP-indices.*

Due to its distinguished role, the AP- (or AP_W-) factorization with zero partial AP-indices is called a *canonical* AP- (respectively, AP_W-) factorization. When a canonical AP-factorization exists, the matrix

$$\mathbf{d}(G) = \mathbf{M}(G_+)\mathbf{M}(G_-)$$

does not depend on the choice of factorization (1.2) and is defined by G uniquely. The value of $\mathbf{d}(G)$ is important in the consideration of singular integral and Wiener-Hopf operators with semi almost periodic symbols, see [3, 4].

Of particular interest is the special case

$$G(x) = \begin{bmatrix} e^{i\lambda x} & 0 \\ f(x) & e^{-i\lambda x} \end{bmatrix}, \tag{1.4}$$

related with convolution type equations on the interval $(0, \lambda)$. However, even in this special case the existence of factorization (1.2), computation of the partial AP-indices and of $\mathbf{d}(G)$ are very non-trivial.

2. Known results

The following simple observation regarding the AP-factorization of matrices (1.4) was made in [3] and has been used repeatedly thereafter.

For any $c \in \mathbb{C}$, $\lambda' \in \mathbb{R} \setminus (-\lambda, \lambda)$ the matrices (1.4) and

$$G'(x) = \begin{bmatrix} e^{i\lambda x} & 0 \\ f(x) + ce^{i\lambda' x} & e^{-i\lambda x} \end{bmatrix}$$

are AP-factorable only simultaneously, the sets of their partial AP-indices are the same, and, if the latter equal zero,

$$\mathbf{d}(G') = \begin{cases} \begin{bmatrix} 1 & 0 \\ c & 1 \end{bmatrix} \mathbf{d}(G) & \text{if } \lambda' = \lambda, \\ \mathbf{d}(G) \begin{bmatrix} 1 & 0 \\ c & 1 \end{bmatrix} & \text{if } \lambda' = -\lambda, \\ \mathbf{d}(G) & \text{otherwise.} \end{cases}$$

Therefore, without loss of generality it can always be assumed that

$$\Omega(f) \subset (-\lambda, \lambda). \tag{2.1}$$

For several particular classes of f, the factorability properties of G are known.

CASE 1. The distances between all the points of $\Omega(f)$ are commensurable, that is,

$$\Omega(f) \subset \mu + \alpha\mathbb{Z}$$

for certain $\mu, \alpha \in \mathbb{R}$.

It was shown in [5] that an AP-factorization in such a setting always exists, the case of zero partial indices was singled out, and $\mathbf{d}(G)$ was computed in this case.

CASE 2. One-sided f, that is,

$$\Omega(f) \subset (-\infty, 0] \text{ or } \Omega(f) \subset [0, +\infty).$$

This is another case in which an AP-factorization exists and can be found explicitly (including computation of the partial AP-indices and $\mathbf{d}(G)$), see [3, Theorem 2.4] or [6].

CASE 3. f is a trinomial,

$$f(x) = c_{-1}e^{-i\nu x} - c_0 + c_1 e^{i\alpha x} \tag{2.2}$$

$(0 < \alpha, \nu;\ \alpha + \nu = \lambda)$.

If $\beta = \nu/\alpha$ is rational, this becomes a subcase of Case 1, so that only the case $\beta \notin \mathbb{Q}$ is of interest. In the latter case, an AP-factorization of G exists if and only if

$$|c_1^\beta c_{-1}| \neq |c_0|^{\beta+1} \tag{2.3}$$

or

$$c_1 c_{-1} = c_0 = 0. \tag{2.4}$$

Under condition (2.3), the partial AP-indices of G equal zero, and

$$\mathbf{d}(G) = \begin{cases} \begin{bmatrix} 0 & c_0^{-1} \\ -c_0 & 0 \end{bmatrix} & \text{if } |c_1^\beta c_{-1}| < |c_0|^{\beta+1}, \\ \begin{bmatrix} -c_1^{-1}c_{-1} & 0 \\ -2c_0 & -c_1 c_{-1}^{-1} \end{bmatrix} & \text{if } |c_1^\beta c_{-1}| > |c_0|^{\beta+1}. \end{cases}$$

Under condition (2.4), the partial AP-indices of G equal $\pm\alpha$ if $c_1 \neq 0$, $\pm\nu$ if $c_{-1} \neq 0$, and $\pm\lambda$ if $c_1 = c_{-1} = 0$.

These results were obtained in [6][1]. The main tool used there was a certain transformation of matrices (1.4) with f given by (2.2) which preserved both AP-factorability and the algebraic structure of G as well as the trinomial structure of its block f. In this paper we present another transformation of (1.4). It still preserves the AP-factorability and structure of (1.4), but allows us to consider more complicated blocks f. As we will see, in certain cases an application of this procedure (maybe, repeated several times) reduces the matrix G at hand to one covered by Case 3 and therefore solves the problem of its AP-factorization completely.

3. The reduction procedure

From now on we suppose that f in (1.4) is an AP polynomial, that is, $\Omega(f)$ is a finite set. The situation when $\Omega(f) \cap (-\lambda, 0) = \emptyset$ is covered by Case 2 (if

[1]Formally, it was supposed there that $c_{-1}c_1c_0 \neq 0$. In the remaining cases, however, f becomes a binomial, and the corresponding results can be extracted from Case 1 or even from earlier considerations of [3].

one takes (2.1) into consideration). Hence, we may suppose that the intersection $\Omega(f) \cap (-\lambda, 0)$ is not empty; denote its leftmost point by $-\nu$ (of course, $a := \widehat{f}(-\nu) \neq 0$). We will write f in the form

$$f(x) = ae^{-i\nu x}\left(1 - \sum_{k=1}^{m} b_k e^{i\gamma_k x}\right), \tag{3.1}$$

where

$$\gamma_1, \dots, \gamma_m \in (0, \lambda + \nu).$$

Denote the m-tuple $(\gamma_1, \dots, \gamma_m)$ by γ, and for any $n_1, \dots, n_m \in \mathbb{Z}_+$ let $n = (n_1, \dots, n_m)$, $\langle n, \gamma \rangle = \sum_{k=1}^{m} n_k \gamma_k$. For $\xi \in \mathbb{R}$, let also $N(\gamma, \xi)$ ($\widetilde{N}(\gamma, \xi)$, $N^0(\gamma, \xi)$) be the set of all $n \in \mathbb{Z}_+^m$ such that $\langle n, \gamma \rangle < \xi$ (respectively, $\leq \xi$, $= \xi$). Finally, let

$$f_1(x) = \sum_{n \in N(\gamma, \lambda+\nu) \setminus \widetilde{N}(\gamma, \lambda-\nu)} d_n a^{-1} e^{i(\langle n, \gamma \rangle - \lambda)x}, \tag{3.2}$$

where

$$d_n = \frac{(n_1 + n_2 + \cdots + n_m)!}{n_1! n_2! \cdots n_m!} b_1^{n_1} b_2^{n_2} \cdots b_m^{n_m}.$$

Theorem 3.1. *The matrix* (1.4) *with f given by* (3.1) *and the matrix*

$$G_1(x) = \begin{bmatrix} e^{i\nu x} & 0 \\ f_1(x) & e^{-i\nu x} \end{bmatrix} \tag{3.3}$$

with f_1 given by (3.2) *are AP-factorable only simultaneously and their partial AP-indices coincide. In the case of canonical AP-factorization,*

$$\mathbf{d}(G_1) = \begin{bmatrix} -a & 0 \\ X & a^{-1} \end{bmatrix} \mathbf{d}(G) \begin{bmatrix} -Y & 1 \\ 1 & 0 \end{bmatrix}, \tag{3.4}$$

where

$$X = \sum_{k=1}^{m} \sum_{n \in N^0(\gamma, \lambda+\nu-\gamma_k)} d_n b_k, \qquad Y = \sum_{n \in N^0(\gamma, \lambda-\nu)} d_n a^{-1}.$$

Proof. Put

$$M_1^+(x) = \sum_{n \in N(\gamma, \lambda+\nu)} d_n a^{-1} e^{i\langle n, \gamma \rangle x},$$

$$M_2^+(x) = \sum_{k=1}^{m} \sum_{n \in N_k(\gamma, \lambda+\nu)} d_n b_k e^{i(\langle n, \gamma \rangle + \gamma_k - \lambda - \nu)x}$$

and

$$M^-(x) = \sum_{n \in \widetilde{N}(\gamma, \lambda-\nu)} d_n a^{-1} e^{i(\langle n, \gamma \rangle - \lambda + \nu)x},$$

where, for $\xi \in \mathbb{R}$,

$$\begin{aligned} N_k(\gamma,\xi) = \{ \ n = (n_1, \dots, n_{k-1}, n_k, n_{k+1}, \dots, n_m) \in \mathbb{Z}^m : \\ (n_1, \dots, n_{k-1}, 0, n_{k+1}, \dots, n_m) \in N(\gamma,\xi), \\ n_k = f_k(n_1, \dots, n_{k-1}, n_{k+1}, \dots, n_m)\}, \end{aligned}$$

and, in its turn,

$$\begin{aligned} & f_k(n_1, \dots, n_{k-1}, n_{k+1}, \dots, n_m) \\ & = \max\{z \in \mathbb{Z} : (n_1, \dots, n_{k-1}, z, n_{k+1}, \dots, n_m) \in N(\gamma,\xi)\}. \end{aligned}$$

Direct computations show that

$$M_1^+(x)e^{-i\lambda x} - M^-(x)e^{-i\nu x} = f_1(x), \qquad M_1^+(x)f(x)e^{i\nu x} + M_2^+(x)e^{i(\lambda+\nu)x} = 1,$$

which in a matrix form can be rewritten as

$$\begin{bmatrix} -f(x)e^{i\nu x} & e^{i(\lambda+\nu)x} \\ M_2^+(x) & M_1^+(x) \end{bmatrix} G(x) \begin{bmatrix} -M^-(x) & 1 \\ 1 & 0 \end{bmatrix} = G_1(x). \tag{3.5}$$

The factor

$$A^+ = \begin{bmatrix} -fe_\nu & e_{\lambda+\nu} \\ M_2^+ & M_1^+ \end{bmatrix}$$

is invertible in AP_W^+, the factor

$$A^- = \begin{bmatrix} -M^- & 1 \\ 1 & 0 \end{bmatrix}$$

is invertible in AP_W^-, so that indeed G and G_1 are AP-factorable only simultaneously. Their partial AP-indices are obviously the same, and (3.4) follows from (3.5) and the simple observation that $\mathbf{M}(fe_\nu) = a$, $\mathbf{M}(M_1^+) = a^{-1}$, $\mathbf{M}(M_2^+) = X$, $\mathbf{M}(M^-) = Y$. □

4. Matrices with regular Fourier spectra

In this section we consider matrix functions (1.4) for which $\Omega(f)$ lies in the union of two shifted grids $\nu\mathbb{Z}_+$:

$$\Omega(f) \subset (-\nu + \nu\mathbb{Z}_+) \cup (\alpha + \nu\mathbb{Z}_+), \qquad \alpha, \nu > 0. \tag{4.1}$$

We suppose that α/ν is irrational (otherwise we would be in Case 1 of commensurable distances between points of $\Omega(f)$). For the applicability of Theorem 3.1 it is then important that λ lies on the same grid as $\Omega(f)$, that is,

$\lambda \in (-\nu + \nu\mathbb{Z}_+) \cup (\alpha + \nu\mathbb{Z}_+)$. Due to condition (2.1), this means that either $\lambda \in \nu + \nu\mathbb{N}$ or $\lambda \in \alpha + \nu\mathbb{N}$. We set

$$p = \left\lfloor \frac{\lambda - \alpha}{\nu} \right\rfloor, \qquad k = \left\lfloor \frac{\alpha}{\nu} \right\rfloor$$

(where, as usual, $\lfloor \xi \rfloor$ stands for the integral part of ξ) and consider the two possibilities separately.

4.1. The case $\lambda \in \alpha + \nu\mathbb{N}$

In this case $\lambda = \alpha + p\nu$, $p \in \mathbb{N}$. Suppose, in addition, that $\dfrac{\alpha}{\nu} > p - 1$. From (4.1) and (2.1) it follows that

$$f(x) = ae^{-i\nu x}\left(1 - \sum_{s=1}^{k+p+1} b_s^{(1)} e^{is\nu x} - \sum_{s=1}^{p} b_s^{(2)} e^{i(\alpha+s\nu)x}\right). \tag{4.2}$$

The m-tuple γ can in this case be written as

$$\gamma = (\nu, \dots, (k+p+1)\nu, \alpha + \nu, \dots, \alpha + p\nu),$$

and the set $N(\gamma, \lambda + \nu) \setminus \widetilde{N}(\gamma, \lambda - \nu)$ consists of all $n \in \mathbb{Z}_+^m$ such that

$$\alpha + (p-1)\nu < l_1\nu + l_2\alpha < \alpha + (p+1)\nu, \tag{4.3}$$

where

$$l_1 = \sum_{j=1}^{k+p+1} jn_j + \sum_{j=k+p+2}^{k+2p+1} (j - (k+p+1))n_j, \qquad l_2 = \sum_{j=k+p+2}^{k+2p+1} n_j.$$

Condition (4.3) holds only for the following combinations of l_1, l_2: $l_1 = p$, $l_2 = 1$; $l_1 = k + p$ or $l_1 = k + p + 1$, $l_2 = 0$.

Indeed, it is clear that $l_1 \geq l_2$. Thus, $l_2 \geq 2$ implies $l_1\nu + l_2\alpha \geq 2(\alpha + \nu) \geq \alpha + (k+2)\nu \geq \alpha + (p+1)\nu$. If $l_2 = 0$, then $n_j = 0$ for all $j = k+p+2, \dots, k+2p+1$, which implies that $l_1 = \sum_{j=1}^{k+p+1} jn_j$. In its turn, $l_2 = 1$ if and only if exactly one of n_j ($j = k+p+2, \dots, k+2p+1$) equals 1 and the others all equal zero. From formula (3.2) we conclude that $\Omega(f_1) \subseteq \{k\nu - \alpha, 0, (k+1)\nu - \alpha\}$. Since

$$k\nu - \alpha < 0 < (k+1)\nu - \alpha, \qquad (\alpha - k\nu) + (k+1)\nu - \alpha = \nu,$$

and $\dfrac{(k+1)\nu - \alpha}{\alpha - k\nu}$ is irrational along with $\dfrac{\alpha}{\nu}$, the matrix (3.3) with f_1 given by (4.4) satisfies the conditions of Case 3. To apply Theorem 3.1, we need the explicit values of the coefficients of f_1 as well as X and Y.

Somewhat lengthy but straightforward computations show that

$$f_1(x) = \left(x_{k+p}e^{i(k\nu-\alpha)x} + y_p + x_{k+p+1}e^{i((k+1)\nu-\alpha)x}\right)a^{-1}, \tag{4.4}$$

$$X = y_{p+1}, \qquad Y = y_{p-1},$$

where

$$x_s = \sum_{n_1+2n_2+\cdots+\sigma n_\sigma = s} \frac{(n_1+n_2+\cdots+n_\sigma)!}{n_1!n_2!\cdots n_\sigma!} b_1^{(1)^{n_1}} b_2^{(1)^{n_2}} \cdots b_\sigma^{(1)^{n_\sigma}} \tag{4.5}$$

and

(4.6)

$$y_s = \sum_{m=1}^{\min\{s,p\}} \sum_{\substack{n_1+2n_2+\cdots \\ +\sigma n_\sigma+m=s}} \frac{(n_1+n_2+\cdots+n_\sigma+1)!}{n_1!n_2!\cdots n_\sigma!} b_1^{(1)^{n_1}} b_2^{(1)^{n_2}} \cdots b_\sigma^{(1)^{n_\sigma}} b_m^{(2)}$$

($\sigma = \min\{s, k+p+1\}$).

From here and Theorem 3.1 we conclude the following.

Theorem 4.1. *Let the block f of the matrix (1.4) be given by (4.2) with $\frac{\alpha}{\nu}$ being irrational, $p = \frac{\lambda-\alpha}{\nu} \in \mathbb{N}$ and $k = \left\lfloor \frac{\alpha}{\nu} \right\rfloor \geq p-1$. Define $x_{k+p}, x_{k+p+1}, y_p, y_{p\pm1}$ by (4.5), (4.6) and set $\xi = \frac{\alpha - k\nu}{(k+1)\nu-\alpha}$. Then:*

(1) *G is AP-factorable if and only if $y_p = x_{k+p}x_{k+p+1} = 0$ or*

$$|x_{k+p+1}^{\xi}x_{k+p}| \neq |y_p^{1+\xi}|. \tag{4.7}$$

(2) *The partial AP-indices of G equal*

$$\begin{aligned} \pm\nu \quad &\text{if} \quad y_p = x_{k+p} = x_{k+p+1} = 0, \\ \pm(k\nu-\alpha) \quad &\text{if} \quad y_p = x_{k+p+1} = 0, x_{k+p} \neq 0, \\ \pm((k+1)\nu-\alpha) \quad &\text{if} \quad y_p = x_{k+p} = 0, x_{k+p+1} \neq 0, \\ 0 \quad &\text{if} \quad (4.7) \text{ holds.} \end{aligned}$$

(3) *In the case of zero partial AP-indices, $\mathbf{d}(G)$ is the matrix*

$$\begin{bmatrix} y_p^{-1} & y_p^{-1}y_{p-1}a^{-1} \\ -ay_{p+1}y_p^{-1} & y_p - y_{p+1}y_p^{-1}y_{p-1} \end{bmatrix}$$

if $|x_{k+p+1}^{\xi}x_{k+p}| < |y_p^{1+\xi}|$ and the matrix

$$\begin{bmatrix} 0 & x_{k+p+1}^{-1}x_{k+p}a^{-1} \\ -x_{k+p+1}x_{k+p}^{-1}a & -y_{p+1}x_{k+p+1}^{-1}x_{k+p} + 2y_p - y_{p-1}x_{k+p+1}x_{k+p}^{-1} \end{bmatrix}$$

if $|x_{k+p+1}^{\xi}x_{k+p}| > |y_p^{1+\xi}|$.

4.2. The case $\lambda \in \nu\mathbb{N}$

We now have

$$\lambda = \nu \left\lfloor \frac{\lambda}{\nu} \right\rfloor = \nu \left(\left\lfloor \frac{\lambda - \alpha}{\nu} \right\rfloor + \left\lfloor \frac{\alpha}{\nu} \right\rfloor + 1 \right) = \nu(k + p + 1).$$

Suppose in addition that $\alpha + \nu < \lambda < 2\alpha + \nu$. Condition (2.1) implies that

$$f(x) = ae^{-i\nu x} \left(1 - \sum_{s=1}^{k+p+1} b_s^{(1)} e^{is\nu x} - \sum_{s=1}^{p+1} b_s^{(2)} e^{i(\alpha + s\nu)x} \right) \tag{4.8}$$

(the only difference from (4.2) being in the upper bound of summation in the second sum), and in the case at hand the tuple γ equals

$$\gamma = (\nu, \dots, (k+p+1)\nu, \alpha + \nu, \dots, \alpha + (p+1)\nu).$$

The set $N(\gamma, \lambda + \nu) \setminus \widetilde{N}(\gamma, \lambda - \nu)$ is now described by the formula

$$\nu(k+p) < l_1\nu + l_2\alpha < \nu(k+p+2), \tag{4.9}$$

where

$$l_1 = \sum_{j=1}^{k+p+1} jn_j + \sum_{j=k+p+2}^{k+2(p+1)} (j - (k+p+1))n_j, \qquad l_2 = \sum_{j=k+p+2}^{k+2(p+1)} n_j.$$

If $l_2 \geq 2$, then $l_1\nu + l_2\alpha \geq 2(\nu + \alpha) > \lambda + \nu = \nu(k+p+2)$, so that (4.9) is possible only if $l_2 = 0$ or $l_2 = 1$. If $l_2 = 0$, then (4.9) is equivalent to $l_1 = k+p+1$. Finally, if $l_2 = 1$, then (4.9) holds for $l_1 = p$ and $l_1 = p + 1$.

Formula (3.2) takes the form

$$f_1(x) = \left(y_p e^{i(\alpha - (k+1)\nu)x} + x_{k+p+1} + y_{p+1} e^{i(\alpha - k\nu)x} \right) a^{-1}$$

with x_{k+p+1} defined by (4.5) and

$$y_s = \sum_{m=1}^{s} \sum_{\substack{n_1 + 2n_2 + \cdots \\ + (s-1)n_{s-1} + m \\ = s}} \frac{(n_1 + n_2 + \cdots + n_{s-1} + 1)!}{n_1! n_2! \cdots n_{s-1}!} b_1^{(1)n_1} b_2^{(1)n_2} \cdots b_{s-1}^{(1)\, n_{s-1}} b_m^{(2)} \tag{4.10}$$

$(s = p, p+1)$.

As in the previous subsection, the resulting matrix (3.3) falls into Case 3. Theorem 3.1 yields the following.

Theorem 4.2. *Let the matrix G be given by formula (1.4) with f satisfying (4.8), $1 \leq p = \left\lfloor \frac{\lambda - \alpha}{\nu} \right\rfloor$, $k = \left\lfloor \frac{\alpha}{\nu} \right\rfloor$, and let $\frac{\lambda}{\nu} \in \mathbb{N}$, $\lambda < 2\alpha + \nu$. Define x_s*

$(s = k+p, k+p+1, k+p+2)$ *and* y_s $(s = p, p+1)$ *by* (4.5) *and* (4.10), *respectively, and set* $\xi = \dfrac{\alpha - k\nu}{(k+1)\nu - \alpha}$. *Then:*

(1) *G is AP-factorable if and only if* $x_{k+p+1} = y_p y_{p+1} = 0$ *or*

$$|y_{p+1} y_p^{\xi}| \neq |x_{k+p+1}^{1+\xi}|. \tag{4.11}$$

(2) *The partial AP-indices of G equal*

$$\begin{aligned} \pm\nu \quad &\text{if} \quad y_p = y_{p+1} = x_{k+p+1} = 0, \\ \pm(k\nu - \alpha) \quad &\text{if} \quad y_p = x_{k+p+1} = 0, \; y_{p+1} \neq 0, \\ \pm((k+1)\nu - \alpha) \quad &\text{if} \quad y_{p+1} = x_{k+p+1} = 0, \; y_p \neq 0, \\ 0 \quad &\text{if} \quad (4.11) \text{ holds.} \end{aligned}$$

(3) *In the case of zero partial AP-indices,* $\mathbf{d}(G)$ *equals*

$$\begin{bmatrix} x_{k+p+1}^{-1} & x_{k+p+1}^{-1} x_{k+p} a^{-1} \\ -x_{k+p+2} x_{k+p+1}^{-1} a & x_{k+p+1} - x_{k+p+2} x_{k+p+1}^{-1} x_{k+p} \end{bmatrix}$$

if $|y_{p+1} y_p^{\xi}| < |x_{k+p+1}^{1+\xi}|$ *and*

$$\begin{bmatrix} 0 & a^{-1} y_{p+1}^{-1} y_p \\ -a y_{p+1} y_p^{-1} & 2x_{k+p+1} - x_{k+p+2} y_{p+1}^{-1} y_p - x_{k+p} y_{p+1} y_p^{-1} \end{bmatrix}$$

if $|y_{p+1} y_p^{\xi}| > |x_{k+p+1}^{1+\xi}|$.

5. Repeated use of the reduction procedure

In the previous section, the use of Theorem 3.1 allowed us to transform a given matrix G into a matrix G_1 with known factorability properties. Of course, for that to happen we had to impose rather restrictive conditions on our matrix G. The applicability of Theorem 3.1, however, may be extended by applying it repeatedly. We illustrate this point by considering the following case.

Let G be given by (1.4) with

$$f(x) = ae^{-i\nu x}\left(1 - \sum_{n=1}^{k+p+1} b_n^{(1)} e^{in\nu x} - \sum_{n=1}^{p} b_n^{(2)} e^{i(\alpha+n\nu)x} - b^{(3)} e^{2i(\alpha+\nu)x}\right), \tag{5.1}$$

where $p = \dfrac{\lambda - \alpha}{\nu}$ is integer, $\dfrac{\alpha}{\nu}$ is irrational, and $k = \left\lfloor \dfrac{\alpha}{\nu} \right\rfloor = p - 2$.

When applied in this setting, Theorem 3.1 leads to the following conclusion.

Lemma 5.1. *The matrix G with f given by (5.1) is AP-factorable simultaneously with and has the same partial AP-indices as the matrix G_1 for which*

$$f_1(x) = \left(x_{k+p}e^{i(k\nu-\alpha)x} + y_p + x_{k+p+1}e^{i((k+1)\nu-\alpha)x} + ze^{i(\alpha-k\nu)x}\right)a^{-1},$$

where $z = b^{(3)} + b_1^{(2)^2}$. In the case of zero AP-partial indices,

$$\mathbf{d}(G) = \begin{bmatrix} -a^{-1} & 0 \\ y_{p+1} & a \end{bmatrix} \mathbf{d}(G_1) \begin{bmatrix} 0 & 1 \\ 1 & y_{p-1}a^{-1} \end{bmatrix}. \tag{5.2}$$

Here is a remarkable consequence of the preceding lemma.

Theorem 5.2. *Let G be given by the formula (1.4) with f as in (5.1), where $p = \dfrac{\lambda-\alpha}{\nu} \in \mathbb{N}$, $k = \left\lfloor \dfrac{\alpha}{\nu} \right\rfloor = p-2$, and $\dfrac{\alpha}{\nu}$ itself is irrational. Define x_{k+p}, x_{k+p+1} by (4.5), y_{p-1}, y_p, y_{p+1} by (4.6), let $z = b^{(3)} + b_1^{(2)^2}$, and for $x_{k+p} \neq 0$ set*

$$\widetilde{\nu} = \alpha - k\nu, \qquad \widetilde{\alpha} = (k+1)\nu - \alpha, \qquad \widetilde{k} = \left\lfloor \frac{\widetilde{\alpha}}{\widetilde{\nu}} \right\rfloor, \qquad \widetilde{\xi} = \frac{\widetilde{\alpha} - \widetilde{k}\widetilde{\nu}}{(\widetilde{k}+1)\widetilde{\nu} - \widetilde{\alpha}},$$

and

$$\widetilde{x}_s = \sum_{n_1+2n_2=s} \frac{(n_1+n_2)!}{n_1!n_2!}(-1)^{n_1+n_2} x_{k+p}^{-(n_1+n_2)} y_p^{n_1} z^{n_2} \qquad (s \le \widetilde{k}+2).$$

Then:

(1) G is AP-factorable if and only if $x_{k+p}x_{k+p+1} = 0$ or

$$x_{k+p} \neq 0 \text{ and } |\widetilde{x}^{\widetilde{\xi}}_{\widetilde{k}+2}\widetilde{x}_{\widetilde{k}+1}| \neq |x^{-1}_{k+p}x_{k+p+1}|^{1+\widetilde{\xi}}. \tag{5.3}$$

(2) The partial AP-indices of G equal zero if and only if either $x_{k+p} = 0$, $y_p \neq 0$, or (5.3) holds.

(3) In the case of zero partial AP-indices, $\mathbf{d}(G)$ is the matrix

$$y_p^{-1}\begin{bmatrix} 1 & y_{p-1}a^{-1} \\ a(-y_{p+1} + 2y_p^{-1}x_{k+p+1}z) & y_p^2 - y_{p+1}y_{p-1} + 2y_p^{-1}y_{p-1}x_{k+p+1}z \end{bmatrix} \tag{5.4}$$

if $x_{k+p} = 0$, $y_p \neq 0$, the matrix

$$x_{k+p}^{-1}\begin{bmatrix} -\widetilde{x}^{-1}_{\widetilde{k}+2}\widetilde{x}_{\widetilde{k}+1} & -\widetilde{x}^{-1}_{\widetilde{k}+2}\widetilde{x}_{\widetilde{k}+1}y_{p-1}a^{-1} \\ a\psi & \psi y_{p-1} - x^2_{k+p}\widetilde{x}_{\widetilde{k}+2}\widetilde{x}^{-1}_{\widetilde{k}+1} \end{bmatrix},$$

where

$$\psi = (x^{-1}_{k+p}y_p x_{k+p+1} - y_{p+1})\widetilde{x}^{-1}_{\widetilde{k}+2}\widetilde{x}_{\widetilde{k}+1} + 2x_{k+p+1},$$

if $|\widetilde{x}^{\widetilde{\xi}}_{\widetilde{k+2}}\widetilde{x}_{\widetilde{k+1}}| > |x^{-1}_{k+p}x_{k+p+1}|^{1+\widetilde{\xi}}$, *and the matrix*

$$\begin{bmatrix} 0 & x_{k+p}x^{-1}_{k+p+1}a^{-1} \\ -x^{-1}_{k+p}x_{k+p+1}a & y_p - x^{-1}_{k+p}x_{k+p+1}y_{p-1} - x_{k+p}x^{-1}_{k+p+1}y_{p+1} \end{bmatrix}$$

if $|\widetilde{x}^{\widetilde{\xi}}_{\widetilde{k+2}}\widetilde{x}_{\widetilde{k+1}}| < |x^{-1}_{k+p}x_{k+p+1}|^{1+\widetilde{\xi}}$.

Proof. If $x_{k+p} = 0$, the matrix (3.3) falls into Case 2. Thus, it is AP-factorable, its partial AP-indices equal zero if and only if $y_p \neq 0$, and in the latter case

$$\mathbf{d}(G_1) = \begin{bmatrix} 0 & -y_p^{-1}a \\ y_p a^{-1} & \sum_{\langle n,\gamma\rangle=\nu} \widehat{c}_n \end{bmatrix}, \tag{5.5}$$

where $n = (n_1, n_2)$, $\gamma = (\gamma_1, \gamma_2)$, $\gamma_1 = (k+1)\nu - \alpha$, $\gamma_2 = \alpha - k\nu$,

$$\widehat{c}_n = \frac{(n_1+n_2)!}{n_1!n_2!}\widehat{b}_1^{n_1}\widehat{b}_2^{n_2}, \qquad \widehat{b}_1 = -y_p^{-1}x_{k+p+1}, \ \widehat{b}_2 = -y_p^{-1}z.$$

Since $\langle n,\gamma\rangle = \nu$ if and only if $n_1((k+1)\nu-\alpha) + n_2(\alpha - k\nu) = \nu$, or, equivalently, $((n_1-n_2)k + n_1 - 1)\nu + (n_2 - n_1)\alpha = 0$, the only possible choice for n_1, n_2 is $n_1 = n_2 = 1$. Hence,

$$\sum_{\langle n,\gamma\rangle=\nu} \widehat{c}_n = 2\widehat{b}_1\widehat{b}_2 = 2y_p^{-2}x_{k+p+1}z. \tag{5.6}$$

From (5.2), (5.5) and (5.6) we deduce that

$$\begin{aligned}
&\mathbf{d}(G) \\
&= \begin{bmatrix} -a^{-1} & 0 \\ y_{p+1} & a \end{bmatrix} \begin{bmatrix} 0 & -y_p^{-1}a \\ y_p a^{-1} & 2y_p^{-2}x_{k+p+1}z \end{bmatrix} \begin{bmatrix} 0 & 1 \\ 1 & y_{p-1}a^{-1} \end{bmatrix} \\
&= \begin{bmatrix} y_p^{-1} & y_p^{-1}y_{p-1}a^{-1} \\ a(-y_{p+1}y_p^{-1} + 2y_p^{-2}x_{k+p+1}z) & y_p - y_{p+1}y_p^{-1}y_{p-1} + 2y_p^{-2}y_{p-1}x_{k+p+1}z \end{bmatrix}.
\end{aligned}$$

This confirms (5.4).

Let now $x_{k+p} \neq 0$. Then the matrix G_1 satisfies the conditions of Theorem 4.1 with λ, ν, α, a, $b_1^{(1)}$, $b_2^{(1)}$, $b_1^{(2)}$ changed to $\widetilde{\lambda} = \nu$, $\widetilde{\nu}$, $\widetilde{\alpha}$, $\widetilde{a} = x_{k+p}a^{-1}$, $\widetilde{b}_1^{(1)} = -x^{-1}_{k+p}y_p$, $\widetilde{b}_2^{(1)} = -x^{-1}_{k+p}z$, $\widetilde{b}_1^{(2)} = -x^{-1}_{k+p}x_{k+p+1}$, respectively, $\widetilde{p} = 1$ and all other coefficients $b_i^{(j)}$ set at zero. The rest is computations. □

6. Trinomial f

We now turn to the consideration of the matrices (1.4) with f given by (2.2).

As in Case 3 we suppose that $\alpha, \nu > 0$ and $\beta = \frac{\nu}{\alpha}$ is irrational; otherwise we would be in the setting of Case 2 or Case 1, respectively. The difference from Case 3 is that we now drop the condition $\lambda = \alpha + \nu$. It was shown in [6] that the matrix under consideration is always AP-factorable if $\lambda < \alpha + \nu$. The case $\lambda > \alpha + \nu$ in its full generality is still open. However, Theorems 4.1, 4.2 and 5.2 allow us to consider the cases $\lambda = \alpha + p\nu$, $\frac{\alpha}{\nu} > p - 2$ ($p \in \mathbb{N}$) and $\lambda = m\nu < 2\alpha + \nu$ ($m \in \mathbb{N}$). We start with a statement resulting from Theorem 4.1.

Theorem 6.1. *Let G be given by (1.4) with f of the form (2.2) such that $0 < \alpha, \nu < \lambda = \alpha + p\nu$, $p \in \mathbb{N}$, $\beta = \dfrac{\nu}{\alpha}$ is irrational, and $p < 1 + \dfrac{\alpha}{\nu}$. Then:*

(1) *G is AP-factorable if and only if $c_0 = 0$ or*

$$|(pc_1)^\beta c_{-1}| \neq |c_0|^{1+\beta}. \tag{6.1}$$

(2) *The partial AP-indices of G equal*

$$\begin{aligned} &\pm\lambda \quad \textit{if} \quad c_{-1} = c_0 = c_1 = 0, \\ &\pm\alpha \quad \textit{if} \quad c_{-1} = c_0 = 0, \ c_1 \neq 0, \\ &\pm\nu \quad \textit{if} \quad c_{-1} \neq 0, \ c_0 = 0, \ \textit{and either } c_1 = 0 \textit{ or } p \neq 1, \\ &\ \ 0 \quad \textit{if} \quad (6.1) \textit{ holds and } c_0 \neq 0 \textit{ or } p = 1. \end{aligned}$$

(3) *In the case of zero partial AP-indices, $\mathbf{d}(G)$ is the matrix*

$$\begin{bmatrix} -p^{-1}c_1^{-1}c_0^{1-p}c_{-1}^p & (1-p^{-1})c_0^{-1} \\ -(1+p^{-1})c_0 & -p^{-1}c_1c_0^{p-1}c_{-1}^{-p} \end{bmatrix} \quad \textit{if} \quad |(pc_1)^\beta c_{-1}| > |c_0|^{1+\beta},$$

and the matrix

$$\begin{bmatrix} 0 & c_0^{-1} \\ -c_0 & 0 \end{bmatrix} \quad \textit{if} \quad |(pc_1)^\beta c_{-1}| < |c_0|^{1+\beta}.$$

Proof. Suppose first that $c_{-1} \neq 0$. Setting $a = c_{-1}$,

$$b_1^{(1)} = c_0 c_{-1}^{-1}, \qquad b_1^{(2)} = -c_1 c_{-1}^{-1}, \tag{6.2}$$

and $b_i^{(j)} = 0$ for all $i > 1$, $j = 1, 2$, we represent f as a special case of (4.2). Formulas (4.5) and (4.6) for this special case take the form

$$x_s = {b_1^{(1)}}^s, \qquad y_s = s b_1^{(2)} {b_1^{(1)}}^{s-1}. \tag{6.3}$$

Condition (4.7) can be rewritten as

$$\left| {b_1^{(1)}}^{(k+p+1)\xi + k + p} \right| \neq \left| p b_1^{(2)} {b_1^{(1)}}^{p-1} \right|^{1+\xi}$$

or, equivalently,

$$\left|b_1^{(1)}{}^{(k+2)\xi+k+1}\right| \neq \left|pb_1^{(2)}\right|^{1+\xi}.$$

Using (6.2), we rewrite the latter condition again:

$$\left|c_0^{(k+2)\xi+k+1}c_{-1}^{-((k+2)\xi+k+1)}\right| \neq \left|(pc_1)^{1+\xi}c_{-1}^{-(1+\xi)}\right|$$

or

$$\left|c_0^{(k+2)\xi+k+1}\right| \neq \left|(pc_1)^{1+\xi}c_{-1}^{(k+1)\xi+k}\right|.$$

Since

$$\frac{\xi+1}{(k+1)\xi+k} = \frac{\frac{\nu}{(k+1)\nu-\alpha}}{\frac{(k+1)\alpha-k(k+1)\nu}{(k+1)\nu-\alpha}+k} = \frac{\nu}{\alpha} = \beta,$$

taking the $((k+1)\xi+k)$-th root leads to (6.1). It remains to apply Theorem 4.1 and to observe that $x_{k+p} = x_{k+p+1} = y_p = 0$ if $c_0 = 0$ and either $c_1 = 0$ or $p \neq 1$; $x_{k+p} = x_{k+p+1} = 0$, $y_p \neq 0$ if $c_0 = 0$, $c_1 \neq 0$ and $p = 1$; and $x_{k+p}x_{k+p+1} \neq 0$ if $c_0 \neq 0$.
In the case $c_{-1} = 0$ the function f is a binomial and one can apply the corresponding results from [3]. To keep the exposition self-contained, we remark that instead of invoking [3] we could rewrite f in the form

$$f(x) = c_{-1}e^{-i\nu' x} - c_0 + c_1 e^{i\alpha x}$$

and use Case 3 with $\nu' = p\nu$ in place of ν and $\beta' = \frac{\nu'}{\alpha} = p\beta$ in place of β. □

Theorem 6.1 (along with a more general subcase of Theorem 4.1 in which only $b_2^{(1)}, \dots, b_{k+p+1}^{(1)}$ are set at zero) was stated earlier in [10].

The next two results can be derived from Theorems 4.2 and 5.2 in the same spirit as Theorem 6.1 was derived from Theorem 4.1; we omit the tiresome proofs.

Theorem 6.2. *Let G be as in Theorem 6.1 with $0 < \alpha, \nu < \lambda = \alpha + p\nu$, $p = \left\lfloor \frac{\alpha}{\nu} \right\rfloor + 2$ and $\beta = \frac{\nu}{\alpha} \notin \mathbb{Q}$. Then:*
(1) G is AP-factorable if and only if either $c_0 = 0$ or

$$|R_{\widetilde{k}+2}(p)|^{\beta-(\widetilde{k}-1)(1-k\beta)}\,|R_{\widetilde{k}+1}(p)|^{\widetilde{k}(1-k\beta)-\beta}\,|c_1^{\beta}c_{-1}| \neq |c_0|^{1+\beta}, \tag{6.4}$$

where

$$R_s(p) = \sum_{n_1+2n_2=s} \frac{(n_1+n_2)!}{n_1!n_2!}\,p^{n_1}(-1)^{n_2}$$

and $\widetilde{k}$ is the same as in Theorem 5.2.

(2) *The partial AP-indices of G equal*

$$\begin{aligned} &\pm\lambda \quad \text{if } c_{-1} = c_0 = c_1 = 0, \\ &\pm\alpha \quad \text{if } c_{-1} = c_0 = 0, c_1 \neq 0, \\ &\pm\nu \quad \text{if } c_{-1} \neq 0,\ c_0 = c_1 = 0, \\ \pm\left(\lambda - \left\lfloor \frac{\lambda+\nu}{\alpha+\nu} \right\rfloor (\alpha+\nu)\right) &\quad \text{if } c_0 = 0, c_{-1}c_1 \neq 0, \\ 0 &\quad \text{if (6.4) holds with } c_0 \neq 0. \end{aligned}$$

(3) *In the case of zero partial AP-indices,* $\mathbf{d}(G)$ *equals*

$$\begin{bmatrix} 0 & c_0^{-1} \\ -c_0 & \omega p c_1 c_0^{p-1} c_{-1}^{-p} \end{bmatrix}$$

if

$$|R_{\widetilde{k}+2}(p)|^{\beta-(\widetilde{k}-1)(1-k\beta)}\, |R_{\widetilde{k}+1}(p)|^{\widetilde{k}(1-k\beta)-\beta}\, |c_1^{\beta} c_{-1}| < |c_0|^{1+\beta}$$

and is equal to

$$\begin{bmatrix} -\Theta(p) c_1^{-1} c_0^{1-p} c_{-1}^{p} & (p-1)\Theta(p) c_0^{-1} \\ c_0(\Theta(p)+2) & \left(-(p-1)(\Theta(p)+2) - \Theta^{-1}(p)\right) c_1 c_0^{p-1} c_{-1}^{-p} \end{bmatrix}$$

if

$$|R_{\widetilde{k}+2}(p)|^{\beta-(\widetilde{k}-1)(1-k\beta)}\, |R_{\widetilde{k}+1}(p)|^{\widetilde{k}(1-k\beta)-\beta}\, |c_1^{\beta} c_{-1}| > |c_0|^{1+\beta},$$

Here $\Theta(p) = R_{\widetilde{k}+2}^{-1}(p) R_{\widetilde{k}+1}(p)$, $\omega = 1$ *if* $c_{-1} \neq 0$, *and* $\omega = 0$ *if* $c_{-1} = 0$.

Theorem 6.3. *Let G be as in Theorem 6.1 with* $0 < \alpha,\ \nu < \lambda = (p+k+1)\nu < 2\alpha+\nu$ *for* $k = \left\lfloor \frac{\alpha}{\nu} \right\rfloor$, $p = \left\lfloor \frac{\lambda-\alpha}{\nu} \right\rfloor$. *Let also* $\beta = \frac{\nu}{\alpha}$ *be irrational. Then:*

(1) *G is AP-factorable if and only if either* $c_0 = 0$ *or*

$$(p+1)^{(k+1)\beta-1} p^{1-k\beta} |c_1^{\beta} c_{-1}| \neq |c_0|^{1+\beta}. \tag{6.5}$$

(2) *The partial AP-indices of G equal*

$$\begin{aligned} &\pm\lambda \quad \text{if } c_{-1} = c_0 = c_1 = 0, \\ &\pm\alpha \quad \text{if } c_{-1} = c_0 = 0, c_1 \neq 0, \\ &\pm\nu \quad \text{if } c_{-1} \neq 0, c_0 = c_1 = 0, \\ \pm\left(\lambda - \left\lfloor \frac{\lambda+\nu}{\alpha+\nu} \right\rfloor (\alpha+\nu)\right) &\quad \text{if } c_0 = 0,\ c_{-1}c_1 \neq 0, \\ 0 &\quad \text{if (6.5) holds with } c_0 \neq 0. \end{aligned}$$

(3) *In the case of zero partial AP-indices,*

$$\mathbf{d}(G) = \begin{bmatrix} (c_{-1}c_0^{-1})^{k+p+1} & c_0^{-1} \\ -c_0 & 0 \end{bmatrix}$$

if $(p+1)^{(k+1)\beta-1}p^{1-k\beta}|c_1^\beta c_{-1}| < |c_0|^{1+\beta}$ *and*

$$\mathbf{d}(G) = \begin{bmatrix} 0 & c_0^{-1}(p+1)^{-1}p \\ -c_0 p^{-1}(p+1) & -(c_{-1}c_0^{-1})^{-(k+p+1)}p^{-1}(p+1)^{-1} \end{bmatrix}$$

if $(p+1)^{(k+1)\beta-1}p^{1-k\beta}|c_1^\beta c_{-1}| > |c_0|^{1+\beta}$.

7. Block matrix generalizations

Consider a block matrix analogue of (1.4):

$$G(x) = \begin{bmatrix} e^{i\lambda x}I_N & 0 \\ f(x) & e^{-i\lambda x}I_N \end{bmatrix}, \tag{7.1}$$

where f is an $N \times N$ matrix AP polynomial,

$$f(x) = \sum_{j=1}^{r} c_j e^{i\mu_j x} \qquad (c_j \in \mathbb{C}^{\mathrm{N}\times\mathrm{N}},\ \mu_j \in \mathbb{R}). \tag{7.2}$$

We will suppose that

(7.3) all the matrices c_j commute with each other.

If, in addition, the coefficient $a = \widehat{f}(-\nu)$ corresponding to $-\nu = \min\{\mu_j : \mu_j \in (-\lambda, 0)\}$ is invertible, the reduction procedure introduced in Section 3 transfers to this situation almost literally (with 1's in formulas changed to I_N). For $N > 1$, of course, the invertibility of a does not follow automatically from the fact that $a \neq 0$. If, however,

(7.4) $-\nu$ is the only point in $\Omega(f) \cap (-\lambda, 0)$,

consideration of canonical AP-factorization can always be reduced to the case of a being invertible.

To formulate (and prove) the corresponding statement explicitly, recall that the commutativity of the family $\{c_j\}$ implies the existence of a non-singular matrix $T_0 \in \mathbb{C}^{\mathrm{N}\times\mathrm{N}}$ such that for all j

$$T_0^{-1}c_jT_0 = \operatorname{diag}[c_{j1}, \dots, c_{jm}] \tag{7.5}$$

(see [9, Section 4.4]). Here each block c_{jk} is upper triangular, its size N_k does not depend on j, all its diagonal entries equal the same λ_{jk}, and the family of matrices

$\{c_{jk}\}_{j=1}^r$ is commutative for each $k = 1, \dots, m$. This means, in particular, that $\sigma(c_j) = \{\lambda_{jk}\}_{k=1}^m$, and for each fixed k the matrices c_j share a common eigenvector x_k such that $c_j x_k = \lambda_{jk} x_k$ $(j = 1, \dots, r)$. For this reason, we call the eigenvalues $\lambda_{1k}, \dots, \lambda_{rk}$ *bonded.*

Theorem 7.1. *Let G be given by* (7.1), (7.2) *and let conditions* (7.3), (7.4) *be satisfied. Then G admits a canonical AP-factorization if and only if*

(1) *zero is not a bonded eigenvalue of $c_{-1} = \widehat{f}(-\nu)$ and $c_0 = \widehat{f}(0)$, and*

(2) *a canonical AP-factorization exists for the matrix*

$$\widetilde{G}(x) = \begin{bmatrix} e^{i\lambda x} I_{\widetilde{N}} & 0 \\ \widetilde{f}(x) & e^{-i\lambda x} I_{\widetilde{N}} \end{bmatrix},$$

where $\widetilde{f}(x) = \sum_j \widetilde{c}_j e^{i\mu_j x}$ and the matrices $\widetilde{c}_j$ are obtained from $\operatorname{diag}[c_{j1}, \dots, c_{jm}]$ *by deleting all the blocks c_{jk} for which $\lambda_{-1,k} = 0$.*

Proof. From (7.5) it follows that

$$G(x) = \begin{bmatrix} T_0 & 0 \\ 0 & T_0 \end{bmatrix} \begin{bmatrix} e^{i\lambda x} I & 0 \\ \operatorname{diag}[\sum_j c_{jk} e^{i\mu_j x}]_{k=1}^m & e^{-i\lambda x} I \end{bmatrix} \begin{bmatrix} T_0 & 0 \\ 0 & T_0 \end{bmatrix}^{-1}.$$

Applying, in addition to the similarity $T_0 \oplus T_0$, an appropriate permutation Π we see that

$$G(x) = T \operatorname{diag}[G_1, \dots, G_m] T^{-1},$$

where

$$T = \begin{bmatrix} T_0 & 0 \\ 0 & T_0 \end{bmatrix} \Pi, \qquad G_k(x) = \begin{bmatrix} e^{i\lambda x} I_{N_k} & 0 \\ \sum_j c_{jk} e^{i\mu_j x} & e^{-i\lambda x} I_{N_k} \end{bmatrix}.$$

Hence, the matrix G admits a canonical AP-factorization if and only if all the blocks G_k have this property, $k = 1, \dots, m$. Therefore, it suffices to consider the particular cases

(i) $\sigma(c_0) = \sigma(c_{-1}) = \{0\}$,

(ii) $\sigma(c_{-1}) = \{0\}$, c_0 is invertible,

and to prove that a canonical AP-factorization exists in the latter case and does not exist in the former. Note that condition (1) of Theorem 7.1 fails in the first case and holds in the second.

In the setting (i), we use the similarity T_0 to put all the coefficients c_j in upper triangular form. Since the diagonal elements of c_0 and c_{-1} are then zeros, the first column of G is

$$[e^{i\lambda x}, \underbrace{0, \dots, 0}_{N-1 \text{ times}}, g, \underbrace{0, \dots, 0}_{N-1 \text{ times}}]^T,$$

where $\Omega(g) \subset (0, \lambda)$. Let $\alpha = \min\{\lambda, \Omega(g)\}$. Then for every column

$$\phi = [\eta, \underbrace{0, \dots, 0}_{2N-1 \text{ times}}]^T$$

with $\Omega(\eta) \subset (-\alpha, 0)$: $\phi \in AP^-$, $G\phi \in AP^+$. The vector functions $h^- = -\frac{1}{z-i}\phi$ and $h^+ = \frac{1}{z+i}G\phi$ lie in the Hardy spaces $H^2_- = P_-L^2(\mathbb{R})$ and $H^2_+ = P_+L^2(\mathbb{R})$, respectively. If we let $F(x) = \frac{x-i}{x+i}G(x)$, $x \in \mathbb{R}$, then, of course, $h^+ + Fh^- = 0$. Hence, $\operatorname{Ker} R_F$ is infinite-dimensional, and the operator R_F cannot be Fredholm. Since G differs from F by a nonzero multiple continuous on $\mathbb{R} \cup \{\infty\}$, the operator R_G is not Fredholm either. Due to Theorem 1.1, a canonical AP-factorization of G does not exist.

In the case (ii) we may again suppose that all the coefficients c_j are upper triangular. Along with G, let us consider the matrices

$$G_\epsilon = T_\epsilon G T_\epsilon^{-1}, \tag{7.6}$$

where

$$T_\epsilon = \operatorname{diag}[\epsilon^{N-1}, \epsilon^{N-2}, \dots, 1, \epsilon^{N-1}, \epsilon^{N-2}, \dots, 1], \qquad \epsilon \neq 0, \tag{7.7}$$

and

$$G_0 = \lim_{\epsilon \to 0} G_\epsilon = \begin{bmatrix} e^{i\lambda x} I & 0 \\ f_0 & e^{-i\lambda x} I \end{bmatrix}. \tag{7.8}$$

Obviously, the block f_0 of G_0 is obtained from f by substituting all the off-diagonal elements of c_j by zeros. Under this transformation, the coefficient c_{-1} disappears completely, so that $\Omega(f_0) \subset [0, \lambda)$. On top of that, $\widehat{f_0}(0)$ is invertible. According to [6, Theorem 2.1], G_0 admits a canonical AP-factorization. Since the latter property is stable under small perturbations, the same is true for the matrices G_ϵ when $|\epsilon|$ is sufficiently small. Finally, (7.6) implies that G is AP-factorable with zero partial AP-indices simultaneously with G_ϵ. □

Theorem 7.1 along with the matrix reformulation of Theorem 3.1 (which will not be formulated here but will simply be referred to as the *reduction procedure* in what follows) allow one to establish necessary and sufficient conditions for the existence of a canonical factorization and to compute $\mathbf{d}(G)$ for block analogues of matrices G considered in Sections 4–6. We restrict ourselves to the discussion of a (relatively simple) matrix version of Theorem 6.1; it gives sufficient understanding of additional difficulties caused by the matrix setting and ways to overcome them.

For commuting matrices $c_j \in \mathbb{C}^{N \times N}$ we keep the notation (7.5). We also introduce the *generalized inverse* $c_j^{[-1]}$ according to the formula

$$c_j^{[-1]} = T_0 \operatorname{diag}[z_{j1}, \dots, z_{jm}] T_0^{-1},$$

where

$$z_{jk} = \begin{cases} c_{jk}^{-1} & \text{if } \lambda_{jk} \neq 0, \\ 0 & \text{otherwise,} \end{cases}$$

and we let $c_j^{[-l]} = \left(c_j^{[-1]}\right)^l$, $l \in \mathbb{N}$.

Theorem 7.2. *Let G be given by (7.1) with f of the form*

$$f(x) = c_{-1}e^{-i\nu x} - c_0 + c_1 e^{i\alpha x},$$

where $c_j \in \mathbb{C}^{\mathbb{N}\times\mathbb{N}}$ satisfy (7.3), $\alpha, \nu > 0$, $\lambda = \alpha + p\nu$, $p \in \mathbb{N}$, $\beta = \frac{\nu}{\alpha} \notin \mathbb{Q}$ and $p < 1 + \frac{\alpha}{\nu}$. Then G admits a canonical AP-factorization if and only if for all bonded eigenvalue triples $\{\lambda_{jk}\}_{j=1}^{3}$ of c_j the following two conditions hold:

$$|(p\lambda_{1,k})^{\beta}\lambda_{-1,k}| \neq |\lambda_{0,k}|^{1+\beta}, \qquad k = 1, \dots, m, \tag{7.9}$$

and

$$c_0 \text{ is invertible if } p \neq 1. \tag{7.10}$$

If these conditions are satisfied, then

$$\mathbf{d}(G) = \begin{bmatrix} -p^{-1}c_{-1}^{p}c_0^{1-p}c_1^{[-1]}\chi & c_0^{[-1]}(I_N - p^{-1}\chi) \\ -c_0(I_N + p^{-1}\chi) & -p^{-1}c_1 c_0^{p-1}c_{-1}^{[-p]}\chi \end{bmatrix}, \tag{7.11}$$

where χ is the Riesz projection:

$$\begin{aligned} \chi &= T_0 \operatorname{diag}[P_1, \dots, P_m]T_0^{-1}, \\ P_k &= \begin{cases} I_{N_k} & \text{if } |(p\lambda_{1,k})^{\beta}\lambda_{-1,k}| > |\lambda_{0,k}|^{1+\beta}, \\ 0 & \text{otherwise.} \end{cases} \end{aligned}$$

Proof. The statement under consideration is invariant under similarities and taking direct sums. Therefore, it suffices to consider the situation when $\sigma(c_j) = \{\lambda_j\}$ are singletons and to check each of the following cases separately.

1) $\lambda_{-1} = 0$. Condition (2) of Theorem 7.1 is in this case satisfied automatically (since there is no matrix $\widetilde{G}$). Hence, G admits a canonical AP-factorization if and only if condition (1) of this theorem holds, that is, $\lambda_0 \neq 0$. This confirms (7.9) and (7.10).

2) $\lambda_{-1} \neq 0$. In this case c_{-1} is invertible, and the reduction procedure allows us to substitute f by the matrix function (4.4) with the coefficients x_s ($s = k+p, k+p+1$), y_p computed according to (6.3). Obviously, these matrices commute with each other and their spectra are singletons $\left(\frac{\lambda_0}{\lambda_{-1}}\right)^{s}$ for x_s and $-p\frac{\lambda_1\lambda_0^{p-1}}{\lambda_{-1}^{p}}$ for y_p, respectively. Further consideration of this case splits into several subcases.

2a) $\lambda_0 = 0$, $p \neq 1$, or $\lambda_0 = \lambda_1 = 0$. In these cases 0 is a bonded eigenvalue of x_{k+p} and y_p. Applying Theorem 7.1 to

$$G_1(x) = \begin{bmatrix} e^{i\nu x}I_N & 0 \\ f_1(x) & e^{-i\nu x}I_N \end{bmatrix}$$

we conclude that G_1 does not have a canonical AP-factorization. Due to the reduction procedure, the same is true for G. Observe that (7.10) fails for $\lambda_0 = 0$, $p \neq 1$ and (7.9) fails for $\lambda_0 = \lambda_1 = 0$.
2b) $\lambda_0 = 0$, $\lambda_1 \neq 0$, $p = 1$ or $\lambda_0 \neq \lambda_1 = 0$. Following ideas of the proof of Theorem 7.1, we introduce $G_\epsilon = T_\epsilon G_1 T_\epsilon^{-1}$, where T_ϵ is given by (7.7) and G_0 by the formula (7.8). The block f_0 equals $-p\frac{\lambda_1\lambda_0^{p-1}}{\lambda_{-1}^p}I_N$ in the first case, and

$$\left(\left(\frac{\lambda_0}{\lambda_{-1}}\right)^{k+p} e^{i(k\nu-\alpha)x} + \left(\frac{\lambda_0}{\lambda_{-1}}\right)^{k+p+1} e^{i((k+1)\nu-\alpha)x}\right) I_N$$

in the second. In both cases, G_0 is similar to a direct sum of N identical 2×2 matrices falling into Case 3. Since these matrices admit a canonical AP-factorization, the same is true for G_0 and therefore G. At the same time, conditions (7.9), (7.10) hold as well.
2c) $\lambda_0\lambda_1 \neq 0$. The coefficients of f_1 are then invertible. Applying [6, Theorem 5.1], we conclude that G_1 admits a canonical AP-factorization if and only if the spectrum of the matrix $x_{k+p+1}^{\beta} y_p^{-(1+\beta)} x_{k+p}$ (which is a singleton in our case) does not intersect the unit circle. The same reasoning as in the proof of Theorem 6.1 shows that the latter condition is equivalent to (7.9).
We now pass to the proof of (7.11). In the case of invertible c_j and $p = 1$ this formula was established in [6]. If $p \neq 1$ (but the c_j are still invertible) it can be used to compute $\mathbf{d}(G_1)$. From here and the reduction procedure we obtain (7.11) under the additional assumption that the c_j are invertible (of course, c_j^{-1} can be used in place of $c_j^{[-1]}$ in this particular setting). To relax the invertibility assumption, consider the set $\mathfrak{A}$ of all triples $\{c_j\}_{j=-1}^{1}$ of pairwise commuting $N \times N$ matrices such that (7.9), (7.10) hold. By what was already proved, all matrices G with $\{c_j\}_{j=-1}^{1} \in \mathfrak{A}$ admit a canonical AP-factorization. Hence, by [11], the mapping $\omega : \{c_j\}_{j=-1}^{1} \mapsto \mathbf{d}(G)$ is continuous on $\mathfrak{A}$ if only $\mathfrak{A}$ is considered as a topological subspace of $\mathbb{C}^{N\times N} \oplus \mathbb{C}^{N\times N} \oplus \mathbb{C}^{N\times N}$. The same, apparently, is true for $\widetilde{\omega} : \{c_j\}_{j=-1}^{1} \mapsto \tilde{\mathbf{d}}(G)$, where $\tilde{\mathbf{d}}(G)$ is simply the right hand side of (7.11). Invertibility of c_j singles out a dense subset of $\mathfrak{A}$. Coinciding on a dense subset of $\mathfrak{A}$ and being continuous on $\mathfrak{A}$, the functions ω and $\widetilde{\omega}$ have to coincide everywhere. □

8. Final remarks

1. As observed in [3], the matrix functions (7.1) and

$$\widetilde{G}(x) = \begin{bmatrix} e^{i\lambda x} I_N & 0 \\ f^*(x) & e^{-i\lambda x} I_N \end{bmatrix}$$

are AP-factorable only simultaneously, the sets of their partial AP-indices are symmetric with respect to the origin, and if they equal zero,

$$\mathbf{d}(\widetilde{G}) = \begin{bmatrix} 0 & I_N \\ I_N & 0 \end{bmatrix} \mathbf{d}(G)^* \begin{bmatrix} 0 & I_N \\ I_N & 0 \end{bmatrix}.$$

With the use of this fact, every statement of this paper implies its double having $\Omega(f)$ reflected about the origin. In particular, Theorems 6.1–6.3 lead to factorization results in the case when (2.2) holds with $\lambda = p\alpha + \nu$ or $\lambda = m\alpha$ $(m, p \in \mathbb{N})$.

2. It can be extracted from the proof of Theorem 7.2 that in the case of invertible c_0 and $p = 1$ the corresponding matrix G either admits a canonical AP-factorization or is not at all AP-factorable. This is obviously not true for general c_0 (or if $p \neq 1$). Moreover, an arbitrary family of $l \times l$ matrices c'_j can be embedded into the *commuting* family

$$c_j = \begin{bmatrix} 0 & c'_j \\ 0 & 0 \end{bmatrix}$$

of singular matrices of the doubled size $N = 2l$. Hence, AP-factorization of a $4l \times 4l$ matrix (7.1) with the block f having commuting coefficients is a problem of the same complexity as factorization of the $2l \times 2l$ matrix (7.1) with arbitrary (not necessarily commuting) coefficients of f.

Acknowledgements

The work on this paper was supported in part by JNICT Grant PBIC/C/CEN/1040/92 (Bastos), *NATO Collaborative Research Grant 95-0332* (Karlovich, Spitkovsky) *and NSF Grant DMS 94-01848* (Spitkovsky). *The authors would also like to thank the referee for a number of useful suggestions.*

References

[1] Böttcher, A., Karlovich, Yu.I., Spitkovsky, I.M.: Toeplitz operators with semi-almost periodic symbols on spaces with Muckenhoupt weight; Integral Equations Operator Theory 18 (1994), 261–276.

[2] Corduneanu, C.: Almost periodic functions; J. Wiley & Sons, 1968.

[3] Karlovich, Yu.I., Spitkovsky, I.M.: Factorization of almost periodic matrix valued functions and the Noether theory for certain classes of equations of convolution type; Math. USSR-Izv. 34 (1990), 281–316.

[4] ———, (Semi)-Fredholmness of convolution operators on the spaces of Bessel potentials; Operator Theory: Adv. Appl. 71 (1994), 122–152.

[5] ———, Almost periodic factorization: An analogue of Chebotarev's algorithm; Contemp. Math. 189 (1995), 327–352.

[6] ———, Factorization of almost periodic matrix functions; J. Math. Anal. Appl. 193 (1995), 209–232.

[7] Levitan, B.M.: Almost periodic functions; GITTL, Moscow 1953 (in Russian).

[8] LEVITAN, B.M., ZHIKOV, V.V.: Almost periodic functions and differential equations; Cambridge University Press, Cambridge 1982.

[9] LITVINCHUK, G.S., SPITKOVSKY, I.M.: Factorization of measurable matrix functions; Birkhäuser Verlag, Basel Boston 1987.

[10] SPITKOVSKY, I.M., TISHIN, P.M.: Factorization of new classes of almost-periodic matrix functions; Reports of the extended sessions of a seminar of the I. N. Vekua Institute for Applied Mathematics 3, No. 1 (1989), 170–173 (in Russian).

[11] SPITKOVSKY, I.M.: On the factorization of almost periodic matrix functions; Math. Notes 45:5–6 (1989), 482–488.

Departamento de Matemática
Instituto Superior Técnico
Av. Rovisco Pais
1096 Lisboa
Portugal

Fakultät Mathematik
TU Chemnitz-Zwickau
09107 Chemnitz
Germany

Department of Mathematics
The College of William & Mary
Williamsburg, VA 23187-8795
USA
ilya@math.wm.edu

Hydroacoustic Department
Marine Hydrophysical Institute
Ukrainian Academy of Sciences
270100 Odessa
Ukraine

1991 Mathematics Subject Classification: Primary 47A68; Secondary 42A75

Submitted: May 20,1996

Operator Theory:
Advances and Applications, Vol. 103

On the normal solvability of cohomological equations on compact topological spaces

G. BELITSKII and YU.I. LYUBICH

The functional equation

$$\varphi(Fx) - \varphi(x) = \gamma(x)$$

in continuous functions on a compact topological space X is considered, $F : X \to X$ is a continuous mapping. It is proved that the equation is normally solvable in $C(X)$ if and only if F is preperiodic, i.e. $F^{p+l} = F^l$ for some $p \geq 1$, $l \geq 0$. The solvability problem in measurable functions is also investigated.

1. Introduction

Let X be a topological space and let $F : X \to X$ be a continuous mapping. The pair (X, F) is called *a topological dynamical* system on X. Its *trajectories* (or *orbits*) are generated by iterations

$$F^n = F \circ \cdots \circ F \qquad (n \text{ times}),$$

namely, the *orbit* $O(x)$ *with initial point* $x \in X$ is the sequence $(F^n x)_0^\infty$.

If F is a homeomorphism, the dynamical system (X, F) is called *invertible*.

Every topological dynamical system generates the following *cohomological equation* (*c.e.*) on X:

$$\varphi(Fx) - \varphi(x) = \gamma(x), \tag{1.1}$$

where γ is a given complex valued function and φ is an unknown function. This equation is connected with various properties of the dynamical system (see [2], [3], [8], [12], [13], [20]). A basic formula for these connections is

$$\varphi(F^n x) - \varphi(x) = \sum_{k=0}^{n-1} \gamma(F^k x), \qquad x \in X. \tag{1.2}$$

For example, this shows that (1.1) *has no bounded solutions if* $\inf \gamma > 0$ *or if* $\sup \gamma < 0$. In particular, this is true for *the Abel equation*

$$\varphi(Fx) - \varphi(x) = 1. \tag{1.3}$$

(This is a generalization of an equation introduced by N.H. ABEL [1]).

Another useful conclusion from (1.2) is:

Lemma 1.1. *If the c.e.* (1.1) *has a solution* φ, *then*

$$\sum_{k=0}^{p-1} \gamma(F^k z) = 0 \tag{1.4}$$

for every periodic point z *of a period* $p \geq 1$.

The periodicity means that $F^p z = z$. In particular, $p = 1$ is just the case of a fixed point z and then the necessary condition (1.4) takes the simplest form $\gamma(z) = 0$.

A more general form of condition (1.4) is

$$\sum_{k=0}^{p-1} \gamma(F^{k+l} z) = 0 \tag{1.5}$$

for any *preperiodic* point z with a *period* p and a *preperiod* $l \geq 0$: $F^{p+l} z = F^l z$. Indeed, the latter means that $F^l z$ is a periodic point of the period p.

We see that (1.1) *with strictly positive (negative)* γ *has no solutions in presence of at least one preperiodic point.* In particular, this regards the Abel equation.

A mapping F is called *preperiodic* of a *period* $p \geq 1$ and a *preperiod* $l \geq 0$ if $F^{p+l} = F^l$. In this case (1.5) must be fulfilled for all z in order that (1.1) is solvable. It turns out that this condition is also sufficient in the preperiodic case.

Lemma 1.2. *Let* $F^{p+l} = F^l$ *for some* $l \geq 0$ *and* $p \geq 1$, *and let*

$$\sum_{k=0}^{p-1} \gamma(F^{k+l} x) = 0, \qquad x \in X. \tag{1.6}$$

Then the function

$$\varphi_\gamma(x) = -\sum_{k=0}^{l-1} \gamma(F^k x) + \frac{1}{p} \sum_{k=1}^{p-1} k\gamma(F^{k+l} x) \tag{1.7}$$

is a solution of the c.e. (1.1).

Proof. The proof of the statement is a trivial verification. □

Remark 1.3. In any case a solution φ of (1.1) is not unique since $\varphi + c$, $c = \text{const}$, is also a solution.

In this paper we investigate the c.e. on a compact space X, so that our dynamical system (X, F) is *compact*. Any given function γ is supposed to be continuous, $\gamma \in C(X)$. A desired solution φ is from $C(X)$ or a wider functional space. The continuity of γ may result in some conditions of solvability, besides the above mentioned ones.

Recall that a regular Borel measure μ on X is called *invariant* for the dynamical system (X, F) if

$$\int_X \varphi(Fx)\, d\mu = \int_X \varphi(x)\, d\mu \tag{1.8}$$

for all $\varphi \in C(X)$. Actually, (1.8) is valid for all $\varphi \in L_1(X, \mu)$, in particular, for indicator functions of measurable sets, so that $\mu(F^{-1}Y) = \mu(Y)$ for any such set Y and its preimage $F^{-1}Y$. Conversely, the last property implies the invariance.

Lemma 1.4. *Let μ be an invariant measure. If the c.e.* (1.1) *has a measurable solution φ then*

$$\int_X \gamma\, d\mu = 0. \tag{1.9}$$

Proof. The assertion is obvious for $\varphi \in L_1(X, \mu)$. The general case was proved in [2] by the BIRKHOFF-KHINTCHIN ergodic theorem (see [9] or [17]). □

Condition (1.4) is the simplest particular case of (1.9) corresponding to the measure μ_z uniformly distributed on the cycle $C_z = \{z, Fz, \ldots, F^{p-1}z\}$, $\mu\{F^k z\} = \frac{1}{p}$ $(0 \le k \le p-1)$. If even F has no periodic points, an invariant measure does exist by a well-known BOGOLYUBOV-KRYLOV theorem (see [17] or [24]).

Some further dynamical lemmas we need are collected in Section 2 of the present paper.

In Section 3 we consider the c.e. in $C(X)$, i.e. we require that $\varphi \in C(X)$. In this case (1.1) can be rewritten in the operator form

$$S\varphi = \gamma \tag{1.10}$$

where $S = T - I$, $T\varphi = \varphi \circ F$ and I is the identity operator. All these operators in the Banach space $C(X)$ are linear and bounded. The conjugate operators act in the space $M(X) = C(X)^*$ consisting of the regular Borel (complex) measures. The invariant ones form a closed subspace $M(X, F)$. In operator terms, (1.8) means that $M(X, F) = \operatorname{Ker} S^*$. Being restricted to $\varphi \in C(X)$, Lemma 1.4 means that $\operatorname{Im} S \subset (\operatorname{Ker} S^*)_\perp$. Actually, the closure $\overline{\operatorname{Im} S}$ coincides with $(\operatorname{Ker} S^*)_\perp$ irrespective to any concrete form of S. Following a standard operator terminology (which goes back to [4] and [6]) we call the c.e. (1.1) *normally solvable* if $\operatorname{Im} S$ is closed or, equivalently, $\operatorname{Im} S = (\operatorname{Ker} S^*)_\perp$. This exactly means that (1.1) has a continuous solution φ for any continuous γ which are orthogonal to all invariant measures, so that (1.9) is not only necessary but also sufficient for the solvability in $C(X)$.

We prove the following:

Main Theorem *The c.e.* (1.1) *is normally solvable in $C(X)$ if and only if F is preperiodic.*

Thus, the normal solvability of the c.e. in $C(X)$ is a rare phenomenon. Let us distinguish an important particular case.

Corollary 1.5. *If F is surjective and the corresponding c.e. is normally solvable in $C(X)$, then F is periodic.*

A fortiori, F turns out to be periodic if the dynamical system (X, F) is invertible and the corresponding c.e. is normally solvable in $C(X)$.

It was shown in [16] for any aperiodic homeomorphism F with an invariant measure μ that there exists $\gamma \in C(X)$ such that (1.1) has a measurable solution $\varphi \notin L_1(X, \mu)$. If F is ergodic then all measurable solutions are $\varphi + c$, $c = \text{const.}$, (see [17]). So, there are no solutions in $L_1(X, \mu)$ and, a fortiori, in $C(X)$ while (1.9) is fulfilled by Lemma 1.4. In such a way the Main Theorem follows for ergodic homeomorphisms.

Another particular case of the Main Theorem is contained in a theorem on absence of measurable solutions (see Appendix).

Our proof of the Main Theorem is based on some dynamical statements collected in Section 2 and on some results from the spectral and ergodic theory [11], [18]. The ergodic approach to equations like (1.10) in Banach spaces (in particular, to the c.e.) was developed by many authors (see [19] for patterns and references). Note that KOOPMAN [15] was the first who connected the linear operator T with a dynamical system (X, F) to get its ergodic properties. On this base VON NEUMANN [25] established the first abstract ergodic theorem.

2. Dynamical lemmas

For a topological dynamical system (X, F) on a compact space X we consider the following constructions:

$$\Omega = \bigcap_{n=0} F^n X \tag{2.1}$$

going back to BIRKHOFF [5] and

$$\Omega_0 = \bigcup_{\mu} \{\operatorname{supp} \mu\} \tag{2.2}$$

where μ runs over all invariant measures. Recall that the *support* of a measure μ is the smallest compact set K such that $\mu(X \setminus K) = 0$. Any point $x \in \operatorname{supp} \mu$ is just such that $\mu(U) > 0$ for every neighborhood U of x. If $\mu \neq 0$ then supp $\mu \neq \emptyset$.

Lemma 2.1. *For any invariant measure μ the compact set $K = \operatorname{supp} \mu$ is invariant and the restriction $F|K$ is surjective.*

Proof. Let $x \in K$, $y = Fx$. For any neighborhood V of y the preimage $F^{-1}V$ is a neighborhood of x and $\mu(V) = \mu(F^{-1}V) > 0$ and hence $y \in K$. Thus, K is invariant.

Suppose that $FK \neq K$. Then there exists a function $\varphi \in C(X)$ such that $\varphi | FK = 0$, $\varphi \neq 0$ and $\varphi(x) \geq 0$ everywhere. Then

$$\int_X \varphi(Fx)\, d\mu = \int_K \varphi(Fx)\, d\mu = 0$$

while

$$\int_X \varphi(x)\, dx > 0.$$

This contradicts (1.8). □

Corollary 2.2. *The set Ω_0 is invariant and the restriction $F \mid \Omega_0$ is surjective.*

On the other hand, we have:

Lemma 2.3. *Every invariant set Y such that $F|Y$ is surjective is contained in Ω.*

Proof. $Y = F^nY \subset F^nX$ for all n, so $Y \subset \Omega$ by definition (2.1). □

As an immediate conclusion of the previous statements we get:

Lemma 2.4. $\Omega_0 \subset \Omega$.

Note that in general, $\Omega_0 \neq \Omega$.

Example 2.5. Let X be the closed interval $[0,1] \subset \mathbb{R}$ and let $Fx = x^2$. Then $\Omega = X$ but Ω_0 is the two-point set $\{0,1\}$.

With respect to the c.e. (1.1) the set Ω_0 plays an important role because of the following theorem.

Theorem 2.6. *Let Y be a nonempty compact invariant set and let $\Omega_0 \subset Y$. If the c.e. is normally solvable in $C(X)$, then it is normally solvable in $C(Y)$.*

Proof. Up to a trivial extension of measures we have $M(Y,F) = M(X,F)$ since $\Omega_0 \subset Y$. Let $\gamma \in C(Y)$ and let

$$\int_Y \gamma\, d\omega = 0$$

for all $\omega \in M(Y,F)$. Choose an extension $\widetilde{\gamma} \in C(X)$, $\widetilde{\gamma} \mid Y = \gamma$. Then

$$\int_X \widetilde{\gamma}\, d\omega = 0$$

for all $\omega \in M(X,F)$. By assumption, there exists a continuous solution of the c.e. $\varphi(Fx) - \varphi(x) = \widetilde{\gamma}(x)$, $x \in X$. Then $\psi = \varphi \mid Y$ is the continuous solution of the c.e. $\psi(Fx) - \psi(x) = \gamma(x)$, $x \in Y$. □

Turning to the set Ω we note, first of all, that Ω is a nonempty compact invariant set since this is true for all F^nX and $F^nX \supset F^{n+1}X$, $n = 0, 1, 2, \ldots$.

Lemma 2.7. *The mapping $F \mid \Omega$ is surjective.*

Proof. Suppose that $F\Omega \neq \Omega$. Let $x \in \Omega \setminus F\Omega$. Then $F^{-1}\{x\} \subset X \setminus \Omega$ because Ω is invariant. Thus,

$$F^{-1}\{x\} \subset \bigcup_{n \geq 0} G_n, \qquad G_n = X \setminus F^nX.$$

Since the G_n are open and $G_n \subset G_{n+1}$, the compact set $F^{-1}\{x\}$ is contained in some G_m, so that $F^{-1}\{x\} \bigcap F^mX = \emptyset$ and hence $x \notin F^{m+1}X$ in contradiction to $x \in \Omega$. □

Taking Lemma 2.3 into account we get:

Corollary 2.8. *Ω is the largest invariant set such that F is surjective on it.*

In the following sense Ω is a *trap* for the dynamical system (X, F).

Lemma 2.9. *For any neighborhood $G \supset \Omega$ there exists an n such that $F^nX \subset G$.*

Proof. It follows from $G \supset \Omega$ that

$$\bigcap_{n \geq 0} (F^nX \setminus G) = \Omega \setminus G = \emptyset.$$

Since all $F^nX \setminus G$ are compact, there exists an n such that $F^nX \setminus G = \emptyset$, i.e. $F^nX \subset G$. □

3. Proof of the Main Theorem

This theorem was formulated in the introduction. Its "if" part follows directly from Lemma 1.2 which yields the explicit expression (1.7) for a solution φ_γ for any $\gamma \in C(X)$ satisfying the necessary conditions (1.6). The set of such $\gamma's$ is closed and, obviously, $\varphi_\gamma \in C(X)$.

For the "only if" part we have to suppose (1.1) to be normally solvable in $C(X)$. We will write (1.1) in the operator form (1.10) keeping in mind that $T\varphi = \varphi \circ F$, $S = T - I$. Let us consider two cases.

1) *F is surjective.* Then T is an isometry in $C(X)$. Its spectrum $\sigma(T)$ contains the point 1 since $T\varphi = \varphi$ for any constant φ. By assumption, Im S is closed. Therefore the point 1 is isolated in $\sigma(T)$ (see [26], p.330). We conclude that T is invertible since the spectrum of any noninvertible isometry is the whole unit disk $|\lambda| \leq 1$. On the other hand, T is an endomorphism of the Banach algebra $C(X)$.

Thus, T is an automorphism of the algebra $C(X)$. This algebra is semisimple. By the KAMOWITZ-SHEINBERG theorem [11] T is periodic because its spectrum is not the whole unit circle $|\lambda| = 1$. As a result, F is periodic.

2) *F is not surjective.* Then $F \mid \Omega$ is surjective by Lemma 2.7. Besides, the c.e. is normally solvable in $C(\Omega)$ because of Lemma 2.4 and Theorem 2.6. Hence, $F \mid \Omega$ is periodic. It remains to prove that $\Omega = F^l X$ for some $l \geq 1$.

Now the operator T is a contraction in $C(X)$ and Im S is closed. Then there exists a bounded linear operator P in $C(X)$ such that

$$\lim_{n\to\infty} \left\| \frac{1}{n+1} \sum_{k=0}^{n} T^k - P \right\| = 0$$

by a version of the uniform ergodic theorem [18]. This means that for every $\varepsilon > 0$ there exists an $n(\varepsilon)$ such that

$$\left| \frac{1}{n+1} \sum_{k=0}^{n} \gamma(F^k x) - (P\gamma)(x) \right| < \varepsilon \|\gamma\|, \qquad x \in X, \tag{3.1}$$

for $n \geq n(\varepsilon)$ and for all $\gamma \in C(X)$.

Let $\gamma \mid \Omega = 0$. Then $\gamma(F^k x) = 0$ for $x \in \Omega$ and for all k. Therefore $|(P\gamma)(x)| < \varepsilon \|\gamma\|$, $x \in \Omega$, which, in turn, yields $P\gamma \mid \Omega = 0$ as $\varepsilon \to 0$. Since $P = PT^j$ for all j, we have $(P\gamma)(x) = (P\gamma)(F^j x)$ for all $x \in X$. Hence,

$$\|P\gamma\| \leq \max\{|(P\gamma)(z)| : z \in F^j X\}.$$

Letting $j \to \infty$ we get $P\gamma = 0$ because of Lemma 2.9. Now (3.1) takes the form

$$\left| \frac{1}{n+1} \sum_{k=0}^{n} \gamma(F^k x) \right| < \varepsilon \|\gamma\|, \qquad x \in X,$$

for $n \geq n(\varepsilon)$ and $\gamma \in C(X)$ with $\gamma \mid \Omega = 0$. Hence,

$$\left| \frac{1}{l+1} \sum_{k=0}^{l} \gamma(F^k x) \right| < \|\gamma\|, \qquad x \in X, \tag{3.2}$$

for $l = n(1)$. Let us show that $F^l X = \Omega$.

Suppose there exists $z \in F^l X \setminus \Omega$, $z = F^l x$. Then $F^k x \notin \Omega$ for $0 \leq k \leq l$. This contradicts (3.2) with γ such that $\gamma(F^k x) = 1 \ (0 \leq k \leq l)$, $\gamma \mid \Omega = 0$. The Main Theorem is completely proved.

Now we can extend this result to the c.e. on a noncompact space X. In this case we take the Banach space $CB(X)$ of continuous bounded functions as a receptacle for γ and φ.

Corollary 3.1. *Let X be a completely regular Hausdorff topological space. Then the c.e.* (1.1) *is normally solvable in $BR(X)$ if and only if F is preperiodic.*

Proof. The "if" part is considered as before. For the "only if" part we use the Czech-Stone compactification $\overline{X}$ (see [14]). This allows us to consider X as a dense subset in $\overline{X}$ and, in this way, F is the restriction to X of a continuous mapping $\overline{F} : \overline{X} \to \overline{X}$, $CB(X) = CB(\overline{X}) \mid X$. For this reason the Banach spaces $CB(X)$ and $C(\overline{X})$ are isometric. The normal solvability of (1.1) in $CB(X)$ implies the same property of the c.e.

$$\psi(\overline{F}z) - \psi(z) = \theta(z)$$

in $C(\overline{X})$. By the Main Theorem $\overline{F}$ is preperiodic, a fortiori, so is F. □

Corollary 3.2. *Let X be a nonempty set. The c.e.* (1.1) *is normally solvable in the space $B(X)$ of bounded functions if and only if F is preperiodic.*

Proof. In fact, the set X is a discrete topological space, so that $B(X) = CB(X)$. □

In conclusion we consider the c.e. on a topological semigroup X generated by a shift $x \mapsto sx$,

$$\varphi(sx) - \varphi(x) = \gamma(x). \tag{3.3}$$

Corollary 3.3. *Let X be a completely regular Hausdorff topological semigroup. Then the c.e.* (3.3) *is normally solvable in $CB(X)$ if and only if the subsemigroup generated by s is finite.*

Every topological group (with the separability axiom T_0) is Hausdorff and completely regular by the Birkhoff-Kakutani theorem (see [10] or [22]). Therefore we have:

Corollary 3.4. *Let X be a T_0-topological group. Then the c.e.*

$$\varphi(gx) - \varphi(x) = \gamma(x)$$

is normally solvable in $CB(X)$ if and only if the element g is of a finite order.

4. Appendix

Here we prove a theorem on the absence of measurable solutions of the c.e. (1.1) with a continuous γ orthogonal to the invariant measures. This is an extension of results in [7] and [21] concerning the case $X = \mathbb{T}$, the unit circle. Our proof is based on the closed graph method introduced in [21] and on a theory of dissipative compact dynamical systems developed in [23]. A similar theorem on the absence of bounded solutions was obtained in [18] on the base of the uniform ergodic theory.

Another theorem on the absence of measurable solutions of the c.e. (1.1) on a smooth manifold X with $\gamma \in C^\infty(X)$ was established in [12] under assumptions on F which are quite different from ours.

From now on we suppose that X *is a metric compact space.* Let $d(x_1, x_2)$ be the metric in X. A dynamical system (X, F) is called *uniformly stable* if for every $\varepsilon > 0$ there exists $\delta > 0$ such that

$$d(x_1, x_2) < \delta \quad \Longrightarrow \quad d(F^k x_1, F^k x_2) < \varepsilon \tag{4.1}$$

for $k = 1, 2, \dots$. A system (X, F) is called *dissipative* if F is nonexpansive, i.e.

$$d(Fx_1, Fx_2) \le d(x_1, x_2). \tag{4.2}$$

Obviously, every dissipative dynamical system is uniformly stable.

Lemma 4.1. *Every uniformly stable dynamical system (X, F) is dissipative with respect to a metric $\widetilde{d}$ which is topologically equivalent to the metric d.*

Proof. We define $\widetilde{d}$ as

$$\widetilde{d}(x_1, x_2) = \sup_{k \ge 0} \frac{d(F^k x_1, F^k x_2)}{1 + d(F^k x_1, F^k x_2)}.$$

Obviously, $\widetilde{d}(Fx_1, Fx_2) \le \widetilde{d}(x_1, x_2)$. Since

$$\widetilde{d}(x_1, x_2) \ge \frac{d(x_1, x_2)}{1 + d(x_1, x_2)},$$

the $\widetilde{d}$-topology on X is stronger than the d-topology. On the other hand, by (4.2)

$$d(x_1, x_2) < \delta \quad \Longrightarrow \quad \widetilde{d}(x_1, x_2) < \varepsilon,$$

so the d-topology is stronger than the $\widetilde{d}$-topology. □

A dynamical system (X, F) is called *conservative* if F is *isometric*, i.e.

$$d(Fx_1, Fx_2) = d(x_1, x_2). \tag{4.3}$$

Theorem 4.2. [23] *If a dynamical system (X, F) is dissipative, then its restriction to Ω is conservative and invertible.*

Recall that a dynamical system (X, F) is called *topologically transitive* if there exists an orbit $O(x)$ which is dense in X. It is called *topologically Ω-transitive* if its restriction to Ω is topologically transitive.

Theorem 4.3. [23] *Any topologically transitive conservative dynamical system is topologically equivalent to a topologically transitive shift on a compact Abelian group.*

The proof of this theorem in [23] is based on a theory of almost periodic operators [22] which works because of almost periodicity of the corresponding operator T.

Corollary 4.4. *Any topologically Ω-transitive dissipative dynamical system (X, F) is uniquely ergodic.*

The last property means the uniqueness of a normalized invariant measure μ. In this case F is automatically ergodic with respect to μ [17].

Proof. By Lemma 2.4 supp $\mu \subset \Omega$. By Theorem 4.2 the restriction $(\Omega, F \mid \Omega)$ satisfies the conditions of Theorem 4.3. Hence, up to topological equivalence, Ω is a compact Abelian group and F is a topologically transitive shift. In this case μ is the only Haar measure. □

Now we can prove the theorem mentioned in the beginning of this section.

Theorem 4.5. *Let a dynamical system (X, F) be uniformly stable and topologically Ω-transitive. If the set Ω is infinite, then there exists $\gamma \in C(X)$ such that* (1.9) *is fulfilled but the c.e.* (1.1) *has no measurable solutions.*

Speaking about a measurable solution φ we assume that φ satisfies (1.1) almost everywhere (a.e.).

Proof. By the above preliminaries we can only consider a topologically transitive shift $x \mapsto gx$ on a compact Abelian group X. As aforesaid, the only normalized Haar measure μ appears in this context. Thus, we consider the c.e.

$$\varphi(gx) - \varphi(x) = \gamma(x), \qquad \gamma \in C(X), \tag{4.4}$$

under the condition

$$\int_X \gamma \, d\mu = 0. \tag{4.5}$$

Suppose that (4.4) has a measurable solution for every γ satisfying (4.5). All these $\gamma' s$ form a closed subspace $L \subset C(X)$. On the other hand, let us denote by M the F-space of (mod 0)–classes of measurable functions provided with the metric

$$\rho(\varphi_1, \varphi_2) = \int_X \frac{\mid \varphi_1 - \varphi_2 \mid}{1+ \mid \varphi_1 - \varphi_2 \mid} \, d\mu.$$

For simplicity we will speak about functions (mod 0) instead of their classes. Clearly, this metric determines the convergence in measure:

$$\lim_n \rho(\varphi_n, \varphi) = 0 \quad \Longleftrightarrow \quad \forall\, \varepsilon > 0 : \ \lim_n \mu\{x : |\varphi_n(x) - \varphi(x)| \geq \varepsilon\} = 0.$$

Finally, we denote by $\widehat{M}$ the factor space of M with respect to the 1-dimensional subspace of constants. $\widehat{M}$ is also an F-space.

For any given γ a mesurable solution φ is unique (mod 0) up to an additive constant. Therefore its coset $\widehat{\varphi} \in \widehat{M}$ is uniquely determined. For this reason, there exists a linear operator $R : L \to \widehat{M}$ such that $R\gamma = \widehat{\varphi}$. Let us check that R is closed.

Let a sequence $(\gamma_n) \subset L$ tend to $\gamma \in L$ uniformly, and let $R\gamma_n = \widehat{\varphi}_n$ tend to $\widehat{\varphi}$ in the space $\widehat{M}$. The functions φ_n can be changed by addition of suitable constants in order to get $\lim_n \varphi_n = \varphi$ in M. Passing to a subsequence we can assume that $\lim_n \varphi_n(x) = \varphi(x)$ a.e. Let $E \subset X$ be the subset where the convergence is violated. Then $\mu(E) = 0$ and hence $\mu(g^{-1}E) = 0$. Consequently, the limit pair (φ, γ) satisfies (4.4) outside $E_1 = E \cap g^{-1}E$. Since $\mu(E_1) = 0$, we get $R\gamma = \widehat{\varphi}$.

By the closed graph theorem the operator R is continuous. To disprove that let us consider the dual group X^* and its homomorphism into $\mathbb{T} : \chi \mapsto \chi(g)$. (In other words, we consider g as a charachter of G^* by duality). Its image is a subgroup of $\mathbb{T}$ and it is infinite because the element g is of infinite order. Therefore this image is dense in $\mathbb{T}$. In particular, one can choose a sequence $(\chi_n) \subset X^*$ such that $\lim_n \chi_n(g) = 1$. If we set

$$\gamma_n(x) = (\chi_n(g) - 1)\chi_n(x),$$

then χ_n turns out to be a solution of (4.4) for $\gamma = \gamma_n$. Obviously, $\lim_n \gamma_n = 0$ in L and hence $\widehat{\chi}_n = R\gamma_n$ tends to zero in $\widehat{M}$. This means that $\chi_n - c_n$ with some constants c_n tends to zero in measure. The sequence (c_n) is bounded and

$$\int_X (\chi_n - c_n)\, d\mu = 0.$$

However,

$$\int_X \chi_n \, d\mu = 0$$

by the orthogonality of characters. We get $\lim_n c_n = 0$ and then $\lim_n \chi_n = 0$ in measure. This contradicts the equality

$$\int_X |\chi_n| \, d\mu = \int_X d\mu = 1.$$

The proof is complete. □

References

[1] ABEL, N.H.: Détermination d'une fonction au moyen d'une équation qui ne contient qu'une seule variable; Oeuvres complètes, II, Christiania (1881), 36–39.

[2] ANOSOV, D.V.: On an additive functional homological equation connected with an ergodic rotation on the circle; Math. USSR-Izv. 7 (1973), 1257–1271.

[3] ARNOLD, V.I.: Chapitres Supplémentaires de la Théorie des Equations Différentielles Ordinaires; Mir, Moscow 1980.

[4] ATKINSON, F.W.: The normal solvability of linear equations in normal spaces; Math. USSR-Sb. 28:1 (1951), 3–13.

[5] BIRKHOFF, G.D.: Dynamical Systems; Amer. Math. Soc. Colloq. Publ. 9, Providence, R.I. 1927.

[6] GOHBERG, I.C.: On linear equations in normed spaces, Dokl. Akad. Nauk SSSR 76 (1951), 477–480.

[7] GORDON, A.YA.: Sufficient condition for unsolvability of the additive functional homological equation connected with ergodic rotation of the circle; Funktsional Anal. i Prilozhen 9:4 (1975), 71–72 (in Russian).

[8] GOTTSCHALK, W.H., HEDLUND G.A.: Topological Dynamics; Amer. Math. Soc. Colloq. Publ. 36, Providence, R.I. 1955.

[9] HALMOS, P.: Lectures on Ergodic Theory; Math. Soc. of Japan, Tokyo 1953.

[10] HEWITT, E., ROSS, K.A.: Abstract Harmonic Analysis I; Springer Verlag, Berlin Heidelberg New York 1963.

[11] KAMOWITZ, H., SHEINBERG, S.: The spectrum of automorphisms of Banach algebras; J. Funct. Anal. 4 (1969), 268–276.

[12] KATOK, A.: Constructions in Ergodic Theory; preprint, 1983.

[13] KATOK, A., HASENBLATT, B.: Introduction to the Modern Theory of Dynamical Systems; Cambridge University Press, Cambridge 1995.

[14] KELLEY, J.L.: General Topology; Van Nostrand, 1957.

[15] KOOPMAN, B.O.: Hamiltonian systems and transformations in Hilbert space; Proc. Nat. Acad. Sci. U.S.A. 17 (1931), 315–318.

[16] KORNFELD, I.: On the additive homological equation; Funct. Anal. Appl. 10:2 (1976), 73–74.

[17] KORNFELD, I., SINAI, YA.G., FOMIN, S.V.: Ergodic Theory; Nauka, Moscow 1980 (in Russian).

[18] LIN, M.: On the uniform ergodic theorem; Proc. Amer. Math. Soc. 43 (1974), 337–340.

[19] LIN, M., SINE, R.: Ergodic theory and the functional equation $(I - T)x = y$; J. Operator Theory 10 (1983), 153–166.

[20] LYUBICH, M.YU.: The dynamics of rational transforms: the topological picture; Russian Math. Surveys 41:4 (1986), 43–117.

[21] LYUBICH, YU.I.: Method of closed graph for the additive homological equation on the circle; Theory of Functions of Several Real Variables, Yaroslavl State University 1980, 123–125.

[22] LYUBICH, YU.I.: Introduction to the Theory of Banach Representations of Groups; Birkhäuser, Basel 1988.

[23] LYUBICH, YU.I.: Dissipative actions and almost periodic representations of Abelian semigroups; Ukrainian Math. J. 40:4 (1988), 58–62.

[24] LYUBICH, YU.I.: Linear Functional Analysis; Encyclopedia Math. Sci. 19, Springer Verlag, Berlin Heidelberg New York 1992.

[25] VON NEUMANN, J.: Zur Operatoren Methode in der klassischen Mechanik; Ann. of Math. 33 (1932), 587–642.

[26] TAYLOR, A.E., LAY, D.C.: Introduction to Functional Analysis; John Wiley & Sons, New York 1980.

Department of Mathematics and Computer Science
Ben-Gurion University of the Negev
P.O.Box 653
Beer-Sheva 84105
Israel
genrich@cs.bgu.ac.il

Department of Mathematics
Technion
Israel Institute of Technology
Haifa 32000
Israel
lyubich@techunix.technion.ac.il

1991 Mathematics Subject Classification: 39B22

Submitted: May 20, 1996

Operator Theory:
Advances and Applications, Vol. 103

On nonnegative realizations of rational matrix functions and nonnegative input–output systems

K.-H. FÖRSTER and B. NAGY

We consider realizations (A,B,C) of rational $\ell \times k$ matrix functions W, i.e. $W(\lambda) = C(\lambda - A)^{-1}B$ if $\lambda \notin \sigma(A)$. Generalizing the scalar case we prove that W has a realization with nonnegative matrices iff for each of its minimal realizations (A_0, B_0, C_0) there exists an A_0-invariant polyhedral proper cone $\mathcal{P}_0$ with $B_0\mathbb{R}^k_+ \subset \mathcal{P}_0$ and $C_0\,\mathcal{P}_0 \subset \mathbb{R}^\ell_+$. For such functions W the McMillan degree is less than or equal to the size of nonnegative-minimal realizations, and the latter is less than or equal to the minimum of the number of extreme rays of polyhedral cones $\mathcal{P}_0$ with the properties mentioned above. We give an example for which both these inequalities are strict.
Consider the input system $\dot{x} = Ax + Bu$ on $[0,\infty[$, $x(0) = x_0$. Let $\mathcal{P}$ be an A-invariant proper polyhedral cone and let $\mathcal{K}$ be a polyhedral cone with $B\mathcal{K} \subset \mathcal{P}$. Generalizing the case of the nonnegative cones we prove that $B\mathcal{K} = \mathcal{P}$ iff the system is approximately $\mathcal{K}$-controllable in positive time t, i.e., $\left\{ \int_0^t e^{sA}Bu(s)ds \;:\; u \in L^\infty([0,t],\mathbb{R}^k), u(s) \in \mathcal{K} \text{ a.e. on } [0,t] \right\}$ is dense in $\mathcal{P}$. For output systems a dual result is true.

1. Introduction

Let W be a rational $\ell \times k$ matrix function which vanishes at infinity. It is well known that there exist realizations of W, i.e. we have

$$W(\lambda) = C(\lambda I - A)^{-1}B, \qquad \lambda \notin \sigma(A),$$

for some complex matrices A, B and C where these matrices are of sizes $n \times n$, $n \times k$ and $\ell \times n$, respectively ([GLR], Theorem 7.1.2). Of special interest are the minimal realizations, i.e. realizations where the size of A is minimal; such minimal realizations of W exist and the size of A in these cases is called the McMillan degree of W ([GLR], §7.1).

In the second section of this paper we deal with rational matrix functions which have at least one realization where the matrices A, B and C are (entrywise) nonnegative. Our main result is that a rational matrix function has such a realization if and only if for each of its minimal realizations (A_0, B_0, C_0) (see [GLR], p. 218) there exists an A_0-invariant polyhedral proper cone $\mathcal{P}_0$ with $B_0\mathbb{R}^k_+ \subset \mathcal{P}_0$ and $C_0\mathcal{P}_0 \subset \mathbb{R}^\ell_+$. This is a generalization of a corresponding result for scalar-valued rational functions by Y. Ohta, H. Maeda and S. Kodama ([OMK], Theorem 5). We discuss the connections between the McMillan degree n_0 (the sizes of all minimal realizations), the size $n_\oplus$ of nonnegative-minimal realizations and the number

r of extreme rays of the polyhedral cones $\mathcal{P}_0$. In general $n_0 \leq n_\oplus \leq r$ is true, and we shall give an example for which all these inequalities are strict.

In the appendix we characterize realizations which are intertwined with each minimal realization. For such a nonnegative realization we can always find a cone $\mathcal{P}_0$ with at most as many extreme rays as the size of the nonnegative realization.

In the third section we deal with input-output systems which have system-invariant polyhedral cones; more precisely, we assume for a triple (A, B, C) of matrices that there exists a triple of polyhedral cones $(\mathcal{P}, \mathcal{K}, \mathcal{L})$ such that $A\mathcal{P} \subset \mathcal{P}$, $B\mathcal{K} \subset \mathcal{P}$ and $C\mathcal{P} \subset \mathcal{L}$. For example, generalizing results of T. Schanbacher ([S], §4, 4.1), we show that $B\mathcal{K} = \mathcal{P}$ if and only if the system is approximately $\mathcal{K}$-controllable in positive time t, i.e., $\left\{ \int_0^t e^{sA} Bu(s)\, ds : u \in L^\infty([0,t], \mathbb{R}^k),\ u(s) \in \mathcal{K} \text{ a.e. on } [0,t] \right\}$ is dense in $\mathcal{P}$.

In the following we use the notations and concepts of the monograph of I.C. Gohberg, P. Lancaster and L. Rodman ([GLR], §7.a, 8.2) on realizations of rational matrix functions and linear input-output systems. For the concepts and results on cones we refer to the monographs of R.T. Rockafellar ([R]) and of J. Stoer and C. Witzgall ([SW]).

At this place a short review of related problems and results (mainly for scalar-valued rational functions) might be appropriate. From the engineering and system theoretic point of view it was H. Maeda, S. Kodama and F. Kajiya [MKK] who first (or among the first) studied the realization of so called compartmental linear systems, where the realizing matrices are (essentially) nonnegative. They gave examples where the minimal dimension of an (essentially) nonnegative realization was strictly greater than the McMillan degree, and showed that for the special case of tree-compartmental systems they must be equal. H. Maeda and S. Kodama [MK] emphasized the need for establishing a system theory with nonnegative constraints and the importance of a realization theory of such transfer functions. L. Farina and L. Benvenuti [FB] gave a discrete-time version of the main result of Ohta, Maeda and Kodama [OMK] (for scalar-valued rational functions), and also characterized the existence of positive reachable and positive almost reachable realizations of transfer functions. They also characterized positive realizability of transfer functions of McMillan degree not greater than 3 in the discrete and continuous-time cases. For further results in this direction see the results by L. Farina [F1], [F2], [F3] and also by B. Anderson, M. Deistler, L. Farina and L. Benvenuti [ADFB], where different spectral conditions for nonnegative realizability of rational (scalar-valued) transfer functions are presented, as well as the contribution by S. Rinaldi and L. Farina [RF]. Another type of the positive realization problem is studied and solved with the help of the notion of the k-shift invariance of a cone in the thesis of J. van den Hof (see also [vdH]). For connections with the theory of stochastic processes see, e.g., the paper by M. Fliess [Fli]. Two early and very valuable contributions to the theory of convex polyhedral cones are D. Gale [Ga] and M. Gerstenhaber [Ge].

2. Nonnegative realizations of rational matrix functions

In the present section we characterize the minimal realizations of such rational matrix functions that have at least one nonnegative realization (Theorem 2.4). Throughout this paper we consider rational matrix functions which vanish at infinity and which have realizations with real matrices, and we consider such real realizations only.

Let W be such a rational $\ell \times k$ matrix function and let (A, B, C) be a real realization of W, where A, B and C are real matrices of sizes $n \times n$, $n \times k$ and $\ell \times n$, respectively. If $\mathcal{P}$, $\mathcal{K}$ and $\mathcal{L}$ are closed convex cones in $\mathbb{R}^n$, $\mathbb{R}^k$ and $\mathbb{R}^\ell$, respectively, the triple (A, B, C) is called $(\mathcal{P}, \mathcal{K}, \mathcal{L})$-positive if $\mathcal{P}$ is invariant under A, B maps $\mathcal{K}$ into $\mathcal{P}$ and C maps $\mathcal{P}$ into $\mathcal{L}$. An $(\mathbb{R}^n_+, \mathbb{R}^k_+, \mathbb{R}^\ell_+)$-positive triple is called nonnegative; this is exactly the case if all three matrices have nonnegative entries.

It is easy to see that for a $(\mathcal{P}, \mathcal{K}, \mathcal{L})$-positive triple (A, B, C) we have:

(I) if $\mathcal{K}$ is generating, then $\operatorname{span}\mathcal{P} \supset \mathcal{J}(A, B)$, where $\mathcal{J}(A, B)$ is the minimal A-invariant subspace of $\mathbb{R}^n$ containing the image of B ([GLR], p.215), and

(II) if $\mathcal{L}$ is pointed, then $\mathcal{P} \cap (-\mathcal{P}) \subset \mathcal{K}(C, A)$, where $\mathcal{K}(C, A)$ is the maximal A-invariant subspace of $\mathbb{R}^n$ contained in the kernel of C ([GLR], p.215).

A realization (A, B, C) of a rational matrix function is minimal if and only if $\mathcal{J}(A, B) = \mathbb{R}^n$ and $\mathcal{K}(C, A) = \{0\}$ ([GLR], Theorem 7.1.5); therefore the next proposition follows immediately.

Proposition 2.1. *Let (A, B, C) be a real minimal realization of a rational matrix function (vanishing at infinity). If (A, B, C) is $(\mathcal{P}, \mathcal{K}, \mathcal{L})$-positive, where $\mathcal{K}$ is generating and $\mathcal{L}$ is pointed, then $\mathcal{P}$ is generating and pointed.*

Proposition 2.2. *Let (A, B, C) be a $(\mathcal{P}, \mathbb{R}^k_+, \mathbb{R}^\ell_+)$-positive realization of an $\ell \times k$ rational matrix function W, where $\mathcal{P}$ is a polyhedral cone in $\mathbb{R}^n$, then there exists a nonnegative realization (A_+, B_+, C_+) of W, where the size of A_+ is the number of the extreme rays of $\mathcal{P}$.*

Proof. Let n_+ be the number of the extreme rays of $\mathcal{P}$, then a real $n \times n_+$ matrix S exists such that $\mathcal{P} = S\mathbb{R}^{n+}_+$. Now $A\mathcal{P} \subset \mathcal{P}$ is equivalent to the existence of a nonnegative $n_+ \times n_+$ matrix A_+ with $AS = SA_+$, $B\mathbb{R}^k_+ \subset \mathcal{P}$ is equivalent to the existence of a nonnegative $n_+ \times k$ matrix B_+ with $B = SB_+$, and $C\mathcal{P} \subset \mathbb{R}^\ell_+$ implies that $C_+ = CS$ is nonnegative. Therefore (A_+, B_+, C_+) is a nonnegative triple which is a realization of W, since $(\lambda - A)^{-1}S = S(\lambda - A_+)^{-1}$ for all $\lambda \notin \sigma(A) \cup \sigma(A_+)$. □

Proposition 2.3. *Let (A, B, C) be a $(\mathcal{P}, \mathcal{K}, \mathcal{L})$-positive realization of a rational matrix function W where $\mathcal{P}$ is a polyhedral cone, and let (A_0, B_0, C_0) be a reduction of (A, B, C) in the following sense: A, B and C have block matrix representations* *of the form*

$$\begin{bmatrix} A_{11} & A_{12} & A_{13} \\ 0 & A_0 & A_{23} \\ 0 & 0 & A_{33} \end{bmatrix}, \begin{bmatrix} B_1 \\ B_0 \\ 0 \end{bmatrix} \quad \text{and} \quad \begin{bmatrix} 0 & C_0 & C_3 \end{bmatrix}, \text{ respectively.}$$

Then there exists a polyhedral cone $\mathcal{P}_0$ such that (A_0, B_0, C_0) is a $(\mathcal{P}_0, \mathcal{K}, \mathcal{L})$-positive realization of W.

Proof. Let n_0, n_1 and n be the sizes of A_0, A_{11} and A, respectively. We identify $\mathbb{R}^{n_1+n_0}$ with the subspace of $\mathbb{R}^n$ of those n-tuples which have zeros as the last $n-(n_1+n_0)$ coordinates. Then $\mathcal{P}_1 = \mathbb{R}^{n_1+n_0} \cap \mathcal{P}$ is a polyhedral cone in $\mathbb{R}^{n_1+n_0}$ and it follows easily that

$$\left(\begin{bmatrix} A_{11} & A_{12} \\ 0 & A_0 \end{bmatrix}, \begin{bmatrix} B_1 \\ B_0 \end{bmatrix}, [0 \quad C_0] \right)$$

is a $(\mathcal{P}_1, \mathcal{K}, \mathcal{L})$-positive realization of W. Therefore $\mathcal{P}_1 = S\,\mathbb{R}_+^{n_+}$ for some $n_+ \in \mathbb{N}$ and some real $(n_1+n_0) \times n_+$ matrix S. Let $S = \begin{bmatrix} S_1 \\ S_0 \end{bmatrix}$, where S_1 and S_0 are $n_1 \times n_+$ and $n_0 \times n_+$ real matrices, respectively, and define $\mathcal{P}_0 = S_0\,\mathbb{R}_+^{n_+}$. Then $\mathcal{P}_0$ is a polyhedral cone in $\mathbb{R}^{n_0}$ ([R], Theorem 19.3), and it follows easily that (A_0, B_0, C_0) is a $(\mathcal{P}_0, \mathcal{K}, \mathcal{L})$-positive realization of W. □

For scalar rational functions the next results were proved in [OMK], Theorem 5; see also [BNS], Theorem 8.4.6. Our proof is even in that special case more direct than the proofs in these papers.

Theorem 2.4. *Let W be a rational $\ell \times k$ matrix function (vanishing at infinity). The following two assertions are equivalent:*

(I) *W has a nonnegative realization.*

(II) *For each minimal real realization of W there exists a polyhedral cone $\mathcal{P}$ such that this realization is $(\mathcal{P}, \mathbb{R}_+^k, \mathbb{R}_+^\ell)$-positive.*

Proof. Let (A_+, B_+, C_+) be a nonnegative realization of W, and let (A_0, B_0, C_0) be a minimal realization of W. From [GLR], Theorems 7.1.3 and 7.1.4 it follows that there exists an invertible, real $n_+ \times n_+$ matrix U such that $A = UA_+U^{-1}$, $B = UB_+$ and $C = C_+U^{-1}$ have block matrix representations as in the formulation of Proposition 2.3. Now $\mathcal{P} = U\mathbb{R}_+^{n_+}$ is a polyhedral cone in $\mathbb{R}^{n_+}$ and (A, B, C) is a $(\mathcal{P}, \mathbb{R}_+^k, \mathbb{R}_+^\ell)$-positive realization of W. Therefore (I) implies (II). For the proof of the reverse implication we apply Proposition 2.2. □

We remark that the cone $\mathcal{P}$ in Theorem 2.4 (II) is a proper one, since polyhedral cones are closed and, by Proposition 2.1, $\mathcal{P}$ is generating and pointed. So $\mathcal{P}$ is a closed, pointed and solid cone.

Let W be a rational matrix function which has at least one nonnegative realization. Then the minimum $n_\oplus$ of the size of the nonnegative matrices A_+ such that for some (B_+, C_+) the triple (A_+, B_+, C_+) is a nonnegative realization of W is well-defined. For all minimal realizations (A_0, B_0, C_0) of W the size n_0 of A_0 is constant, and n_0 is called the McMillan degree of W. Now let $r_\oplus$ be the minimum of the number of the extreme rays of any polyhedral cone $\mathcal{P}_0$ in $\mathbb{R}^{n_0}$ such that (A_0, B_0, C_0) is $(\mathcal{P}_0, \mathbb{R}^k_+, \mathbb{R}^\ell_+)$-positive. This number $r_\oplus$ does not depend on the special minimal realization of W, since two minimal realizations of a rational matrix function are similar in the sense of [GLR], Theorem 7.1.4. The inequalities $n_0 \le n_\oplus \le r_\oplus$ always hold, the last one follows from Proposition 2.2. Now $n_0 = n_\oplus$ implies $n_0 = r_\oplus$ (see Theorem 2.8 below). To show that the cases $n_0 < n_\oplus = r_\oplus$ and $n_0 < n_\oplus < r_\oplus$ are possible, we will give examples. The first of the following examples is a modification of that in [OMK], p. 178. The second example shows that $n_\oplus - n_0$ can be arbitrarily large.

Example 2.5. Let $W(\lambda) = (2\lambda^2 - \lambda + 2)(\lambda - 1)^{-1}(\lambda^2 + 1)^{-1}$ for $\lambda \neq \{1, \pm i\}$. W has the 3-dimensional realization

$$(A_0, B_0, C_0) = \left(\begin{bmatrix} 0 & 0 & 1 \\ 1 & 0 & -1 \\ 0 & 1 & 1 \end{bmatrix}, \begin{bmatrix} 1 \\ 0 \\ 0 \end{bmatrix}, \begin{bmatrix} 1 & 1 & 1 \end{bmatrix} \right).$$

This realization is minimal since, for each realization (A, B, C) of W, the poles of W are eigenvalues of A, thus $n_0 = 3$. W has the 4-dimensional nonnegative realization

$$\left(\begin{bmatrix} 0 & 1 & 0 & 0 \\ 0 & 0 & 1 & 0 \\ 0 & 0 & 0 & 1 \\ 1 & 0 & 0 & 0 \end{bmatrix}, \begin{bmatrix} 1 \\ 0 \\ 0 \\ 1 \end{bmatrix}, \begin{bmatrix} 1 & 1 & 0 & 1 \end{bmatrix} \right).$$

W does not have 3-dimensional nonnegative realizations. Indeed, the trace of any positive integer power p of a nonnegative matrix, and therefore the sum of its eigenvalues to any positive integer power p, has to be nonnegative, but this is not the case for $p = 2$ in our example.
Using the usual notation of the canonical basis in $\mathbb{R}^3$, we define $\mathcal{P}_0$ to be the cone generated by $\{e_1, e_2, e_3, e_1 - e_2 + e_3\}$. Then the minimal realization (A_0, B_0, C_0) from above is $(\mathcal{P}_0, \mathbb{R}^k_+, \mathbb{R}^\ell_+)$-positive. We have $n_0 = 3 < 4 = n_\oplus = r_\oplus$ for this example.

Example 2.6. Let $n \in \mathbb{N}$, $n \ge 3$. There is a rational matrix function W, vanishing at infinity, of McMillan degree 3 such that W has nonnegative realizations, and the minimal degree of such realizations is not less than n.

Let $\lambda = \lambda(n) = e^{i\,2\pi/n}$, let α and β denote the real and imaginary parts of λ, and let

$$A = A(n) := \begin{pmatrix} 1 & 0 & 0 \\ 0 & \alpha & \beta \\ 0 & -\beta & \alpha \end{pmatrix}.$$

Then A is a real matrix with spectrum $\{1, \lambda, \overline{\lambda}\}$, hence it satisfies the Perron-Schaefer condition, and every eigenvalue of A is an n-th root of unity. By ([TS], Theorem 7.9), there is a proper polyhedral cone $\mathcal{P}$ in $\mathbb{R}^3$ such that $A\mathcal{P} \subset \mathcal{P}$. In fact, the cone $\mathcal{P}$ may be

$$\mathcal{P} := \operatorname{conv}\left\{ \begin{pmatrix} 1 \\ \cos(2\pi k/n) \\ \sin(2\pi k/n) \end{pmatrix} \in \mathbb{R}^3;\ k = 0, 1, \ldots, n-1 \right\}.$$

Let $B = B(n) := \begin{pmatrix} 1 \\ 1 \\ 0 \end{pmatrix} \in M^{3\times 1}(\mathbb{R})$. Then the matrix

$$(\,B \quad AB \quad A^2B\,) = \begin{pmatrix} 1 & 1 & 1 \\ 1 & \alpha & \alpha^2 - \beta^2 \\ 0 & -\beta & -2\alpha\beta \end{pmatrix}$$

is clearly nonsingular, hence $\mathcal{J}(A, b) = \mathbb{R}^3$. Further we have $B\mathbb{R}_+ \subset \mathcal{P}$. Let

$$C = C(n) := (\,1 \quad 1 \quad 0\,) \in M^{1\times 3}(\mathbb{R}).$$

Then $C\mathcal{P} \subset \mathbb{R}_+$. Hence, by Proposition 2.2, there exists a nonnegative realization of the function

$$W(\lambda) := C(\lambda - A)^{-1}B.$$

Further, the matrix

$$\begin{pmatrix} C \\ CA \\ CA^2 \end{pmatrix} = \begin{pmatrix} 1 & 1 & 0 \\ 1 & \alpha & \beta \\ 1 & \alpha^2 - \beta^2 & 2\alpha\beta \end{pmatrix}$$

is clearly nonsingular. Hence the realization (C, A, B) of W is minimal. Therefore the spectrum of W coincides with that of A, i.e. is the set $\{1, \lambda, \overline{\lambda}\}$. If (C_+, A_+, B_+) is any nonnegative realization of W, then the spectrum of A_+ must contain $\lambda = e^{(i2\pi/n)}$, and if the realization above is minimal among the nonnegative ones, then the spectral radius of A_+ is 1, see [FN]. By, e.g., ([HJ], Corollary 8.4.10), then the size of A_+ is not less than n.

Example 2.7. Let $n(\lambda) = \lambda^2 + \lambda + 1$ and $d(\lambda) = \lambda^3 - 1$. We consider the following rational matrix function

$$W(\lambda) = \frac{1}{d(\lambda)} \begin{bmatrix} \lambda^2 & 1 & \lambda & \lambda^2 & 1 & \lambda \\ \lambda & \lambda^2 & 1 & \lambda & \lambda^2 & 1 \\ 1 & \lambda & \lambda^2 & 1 & \lambda & \lambda^2 \\ 0 & 0 & 0 & n(\lambda) & n(\lambda) & n(\lambda) \\ 0 & 0 & 0 & 0 & 0 & 0 \\ n(\lambda) & n(\lambda) & n(\lambda) & 0 & 0 & 0 \end{bmatrix}.$$

We show that $n_0 = 4$, $n_\oplus = 5$ and $r_\oplus = 6$ for this function W. W has the following minimal realization: $W(\lambda) = C_0(\lambda - A_0)^{-1}B_0$, where $(A_0, B_0, C_0) =$

$$\left(\begin{bmatrix} 0 & 0 & 1 & 0 \\ 1 & 0 & 0 & 0 \\ 0 & 1 & 0 & 0 \\ 0 & 0 & 0 & 1 \end{bmatrix}, \begin{bmatrix} 1 & 0 & 0 & 1 & 0 & 0 \\ 0 & 1 & 0 & 0 & 1 & 0 \\ 0 & 0 & 1 & 0 & 0 & 1 \\ 0 & 0 & 0 & 1 & 1 & 1 \end{bmatrix}, \begin{bmatrix} 1 & 0 & 0 & 0 \\ 0 & 1 & 0 & 0 \\ 0 & 0 & 1 & 0 \\ 0 & 0 & 0 & 1 \\ 0 & 0 & 0 & 0 \\ 1 & 1 & 1 & -1 \end{bmatrix} \right).$$

The McMillan degree of W is 4. Indeed, $x \in \mathcal{K}(C_0, A_0)$ implies $C_0 x = 0$. Since the kernel of C_0 is trivial, we obtain $\mathcal{K}(C_0, A_0) = \{0\}$. Further, the subspace $\mathcal{J}(A_0, B_0)$ contains the image of B_0, which is $\mathbb{R}^4$. Hence the (not nonnegative) realization (C_0, A_0, B_0) is observable and controllable, i.e. is minimal as stated. It is also immediately seen that the cone $\mathcal{P}_0 \subset \mathbb{R}^4_+$ defined, by using the usual notation of the canonical unit vectors, as follows:

$$\mathcal{P}_0 := \text{cone}[e_1, e_2, e_3, e_1 + e_4, e_2 + e_4, e_3 + e_4]$$

is a pointed solid cone in $\mathbb{R}^4$. Further, $\mathcal{P}_0$ has exactly the 6 edges occurring as generators in its definition, for none of them is a nonnegative linear combination of the remaining 5.

It is clear that B_0 maps the cone $\mathbb{R}^6_+$ onto the cone $\mathcal{P}_0$, the main operator A_0 maps $\mathcal{P}_0$ onto $\mathcal{P}_0$, and the operator C_0 maps $\mathcal{P}_0$ into the cone $\mathbb{R}^6_+$. Moreover, any other cone $\mathcal{P}$ satisfying the similar statement in place of $\mathcal{P}_0$, after mildening "onto" everywhere into "into", must clearly satisfy

$$\mathcal{P}_0 \subset \mathcal{P} \subset C_0^{-1}[\mathbb{R}^6_+],$$

where the last symbol denotes full inverse image.

It is well-known (see [R], Theorem 19.3) that the right-hand side cone is also polyhedral and, denoting the canonical unit vectors of $\mathbb{R}^6$ by $w_1, \ldots, w_6$, we see that x is an element of this cone if and only if

$$(x, C_0' w_j) = (C_0 x, w_j) \geq 0, \qquad j = 1, \ldots, 6.$$

Here $(\cdot,\cdot)$ denotes the scalar product in $\mathbb{R}^6$, and M' denotes the transpose of the matrix M. Denoting the components of any $x \in \mathbb{R}^4$ by $x_1, \ldots, x_4$, the condition above is equivalent to

$$x_1 + x_2 + x_3 - x_4 \geq 0 \quad \text{and} \quad x_j \geq 0, \qquad j = 1, \ldots, 4.$$

The edges of the polyhedral cone above can be determined by the technique described in [GT], p. 23, and we find that they coincide with the 6 edges of $\mathcal{P}_0$. Hence

$$\mathcal{P}_0 = \mathcal{P} = C_0^{-1}[\mathbb{R}_+^6],$$

and the minimal realization (C_0, A_0, B_0) has the single "system-invariant" cone $\mathcal{P}_0$, which has exactly 6 extreme rays. Since minimal realizations of the same transfer function are similar in the well-known sense, each of them has a unique "system-invariant" cone, which, being a nonsingular transformation of $\mathcal{P}_0$, must also have exactly 6 extreme rays; see also Theorem 3.2 in the next section.

The degree of a minimal nonnegative realization of W cannot be 4, i.e. the McMillan degree of W. For, assuming the contrary, every realization of degree 4 would have a "system-invariant" cone with 4 edges, which is impossible by the preceding paragraph.

On the other hand, we show that there is a nonnegative realization of W of degree 5. Indeed, define the matrices

$$A := A_0 \oplus I_1 = \begin{bmatrix} 0 & 0 & 1 & 0 & 0 \\ 1 & 0 & 0 & 0 & 0 \\ 0 & 1 & 0 & 0 & 0 \\ 0 & 0 & 0 & 1 & 0 \\ 0 & 0 & 0 & 0 & 1 \end{bmatrix}, \qquad T := \begin{bmatrix} 1 & 0 & 0 & 0 & 0 \\ 0 & 1 & 0 & 0 & 0 \\ 0 & 0 & 1 & 0 & 0 \\ 0 & 0 & 0 & 1 & 0 \\ 1 & 1 & 1 & -1 & 1 \end{bmatrix}.$$

Then $AT = TA$, and T is nonsingular. Define the matrices

$$D := [0 \;\; 0 \;\; 0 \;\; 0 \;\; 0 \;\; 1]^t, \quad C := [C_0 \;\; D], \quad B := \begin{bmatrix} B_0 \\ 0 \end{bmatrix}, \qquad B_+ := TB.$$

Then B^+ is a nonnegative matrix, and for every $k \in \mathbb{N}_0$ we have

$$CA^kB = [C_0 \;\; D] \begin{bmatrix} A_0^k & 0 \\ 0 & I_1 \end{bmatrix} \begin{bmatrix} B_0 \\ 0 \end{bmatrix} = C_0 A_0^k B_0.$$

Hence (C, A, B) is a (not nonnegative) realization of degree 5 of W. Moreover, the matrix $C_+ := CT^{-1}$ is nonnegative:

$$C^+ = \begin{bmatrix} 1 & 0 & 0 & 0 & 0 \\ 0 & 1 & 0 & 0 & 0 \\ 0 & 0 & 1 & 0 & 0 \\ 0 & 0 & 0 & 1 & 0 \\ 0 & 0 & 0 & 0 & 0 \\ 0 & 0 & 0 & 0 & 1 \end{bmatrix}.$$

For every k$\in \mathbb{N}_0$ we clearly have

$$C_+A^kB_+ = CT^{-1}TA^kT^{-1}TB = CA^kB.$$

Hence (C_+, A, B_+) is a nonnegative realization of degree 5 of W.
Summarizing, the rational matrix function W has McMillan degree 4, the smallest degree $n_\oplus$ of its nonnegative realizations is 5, and every minimal realization of W has a single "system-invariant" cone, which has exactly 6 extreme rays, therefore $r_\oplus = 6$.

Theorem 2.8. *Let W be a rational $\ell \times k$ matrix function (vanishing at infinity), then the following two assertions are equivalent:*

(I) *W has a nonnegative realization (A_+, B_+, C_+) with $\mathcal{K}(C_+, A_+)+ \mathcal{J}(A_+, B_+) = \mathbb{R}^{n_+}$, where n_+ is the size of A_+.*

(II) *For any minimal realization there exists a polyhedral cone $\mathcal{P}$ with at most n_+ extremal rays such that this realization is $(\mathcal{P}, \mathbb{R}^k_+, \mathbb{R}^\ell_+)$-positive.*

Proof. Let (A, B, C) be a minimal realization of W. The condition $\mathcal{K}(C_+, A_+) + \mathcal{J}(A_+, B_+) = \mathbb{R}^{n_+}$ implies that there exists an invertible real $n_+ \times n_+$ matrix U such that UA_+U^{-1}, UB_+ and CU^{-1} have block matrix representations of the form

$$\begin{bmatrix} A_{11} & A_{12} \\ 0 & A_0 \end{bmatrix}, \quad \begin{bmatrix} B_1 \\ B_0 \end{bmatrix} \quad \text{and} \quad [0 \quad C_0],$$

respectively (see [GLR], Theorems 7.1.3 and 7.1.4). Let $U = \begin{bmatrix} U_1 \\ U_0 \end{bmatrix}$ where U_0 is a real $n_0 \times n_+$ matrix, where n_0 is the size of A. Now $\mathcal{P} = U_0 \mathbb{R}^{n_+}$ is a polyhedral cone with at most n_+ extremal rays and (A, B, C) is $(\mathcal{P}, \mathbb{R}^k_+, \mathbb{R}^\ell_+)$-positive. Thus (I) implies (II). For the reverse implication let S, A_+, B_+ and C_+ be as in the proof of Proposition 2.2 and apply the Proposition in the Appendix. □

The nonnegative realization of the rational function W of Example 2.5 has not the property (I) of Theorem 2.8, but

$$\left(\begin{bmatrix} 0 & 0 & 0 & 1 \\ 1 & 0 & 0 & 0 \\ 0 & 1 & 0 & 0 \\ 0 & 0 & 1 & 0 \end{bmatrix}, \begin{bmatrix} 1 \\ 0 \\ 0 \\ 0 \end{bmatrix}, [2 \quad 1 \quad 1 \quad 2] \right)$$

is a nonnegative realization of W which has this property.

3. Nonnegative input–output systems

For the rational matrix function in Example 2.7 each of its minimal realizations has a unique "system-invariant" polyhedral cone. In this section we will discuss this problem in the context of time-invariant linear input-output systems.

In the following (A, B, C) is a triple of matrices of sizes $n \times n$, $n \times k$ and $\ell \times n$, respectively. For this triple we consider the input-output system

$$\begin{aligned} \frac{dx}{dt}(t) &= Ax(t) + Bu(t), \\ y(t) &= Cu(t), \qquad x = x_0,\ t \geq 0. \end{aligned}$$

If u is a k-dimensional vector-valued function on $\mathbb{R}_+$ which is Lebesgue measurable and locally essentially bounded, then the solution of the system is unique and given by the formula

$$x(t) = x(t, x_0, u) = e^{tA}x_0 + \int_0^t e^{(t-s)A}Bu(s)\,ds, \qquad t \geq 0.$$

We now assume that the matrices A, B and C are real. For a convex cone $\mathcal{K}$ in $\mathbb{R}^k$ and $t \geq 0$ the set $\mathcal{A}_t(A, B, \mathcal{K})$ of attainable (reachable) states from the origin in time t by means of $\mathcal{K}$-valued inputs is defined as (see [BNS], p. 36; [S], p. 464)

$$\begin{aligned} \mathcal{A}_t(A, B, \mathcal{K}) &= \Big\{ \int_0^t e^{(t-s)A}Bu(s)\,ds : u \in L^\infty([0,t], \mathbb{R}^k) \\ &\qquad u(s) \in \mathcal{K} \text{ for a.a. } s \in [0,t] \Big\}. \end{aligned}$$

If $0 \leq t_1 \leq t_2$ then $\mathcal{A}_{t_1}(A, B, \mathcal{K}) \subset \mathcal{A}_{t_2}(A, B, \mathcal{K})$, since

$$\int_0^{t_1} e^{(t_1-s)A}Bu_1(s)\,ds = \int_0^{t_2} e^{(t_2-s)A}Bu_2(s)\,ds,$$

where

$$u_2(s) = \begin{cases} 0 & \text{if } 0 \leq s \leq t_2 - t_1, \\ u_1(s - t_2 + t_1) & \text{if } t_2 - t_1 < s \leq t_2. \end{cases}$$

The set $\mathcal{A}(A, B, \mathcal{K})$ of attainable states from the origin in arbitrary time by means of $\mathcal{K}$-valued inputs is defined as $\mathcal{A}(A, B, \mathcal{K}) = \bigcup_{t\geq 0} \mathcal{A}_t(A, B, \mathcal{K})$.

$\mathcal{A}_t(A, B, \mathcal{K})$ for $t > 0$ and $\mathcal{A}(A, B, \mathcal{K})$ are convex cones in $\mathbb{R}^n$, but they are not necessarily closed even if $\mathcal{K}$ is closed (see [S], p. 464). For a convex cone $\mathcal{L}$ in $\mathbb{R}^\ell$ and $t \geq 0$ the set $\mathcal{S}_t(C, A, \mathcal{L})$ of initial states which cause $\mathcal{L}$-valued outputs on $[0, t]$ is defined as (see [BNS], p. 146)

$$\mathcal{S}_t(C, A, \mathcal{L}) = \{x \in \mathbb{R}^n : Ce^{sA}x \in \mathcal{L} \text{ for all } s \in [0, t]\}.$$

The set $\mathcal{S}(C, A, \mathcal{L})$ of initial states which cause $\mathcal{L}$-valued outputs for all positive times is defined as $\mathcal{S}(C, A, \mathcal{L}) = \bigcap_{t\geq 0} \mathcal{S}_t(C, A, \mathcal{L})$. The proof of the following proposition is straightforward.

Proposition 3.1. *Let (A, B, C) be a triple of real matrices of sizes $n \times n$, $n \times k$ and $\ell \times n$, respectively, and let $(\mathcal{P}, \mathcal{K}, \mathcal{L})$ be a triple of closed convex cones in $\mathbb{R}^n$, $\mathbb{R}^k$ and $\mathbb{R}^\ell$, respectively. Then the following three assertions are equivalent:*
(I) *(A, B, C) is $(\mathcal{P}, \mathcal{K}, \mathcal{L})$-positive,*
(II) *for all initial states $x_0 \in \mathcal{P}$ and all $\mathcal{K}$-valued inputs u we have $x(t, x_0, u) \in \mathcal{P}$, $Ax(t, x_0, u) \in \mathcal{P}$ and $y(t, x_0, u) \in \mathcal{L}$ for all $t \geq 0$.*
(III) *$\mathcal{P}$ is invariant under A and*

$$\mathcal{A}(A, B, \mathcal{K}) \subset \mathcal{P} \subset \mathcal{S}(C, A, \mathcal{L}).$$

In the following we consider only system matrices A which are nonnegative with respect to a polyhedral cone. Similar results are true for input-output systems for a matrix triple (A, B, C) where A is exponentially nonnegative with respect to a polyhedral cone $\mathcal{P}$ (i.e. $e^{tA}\mathcal{P} \subset \mathcal{P}$ for all $t \geq 0$), $B\mathcal{K} \subset \mathcal{P}$ and $C\mathcal{P} \subset \mathcal{L}$, where $\mathcal{K}$ and $\mathcal{L}$ are polyhedral cones, respectively. Note that in the case of a polyhedral cone $\mathcal{P}$ for a matrix A

(I) A is exponentially nonnegative with respect to $\mathcal{P}$

and

(II) $A + \alpha I$ is nonnegative with respect to $\mathcal{P}$ for some real α

are equivalent, see, e.g. ([BNS]; Theorem 3.4.1). Such versions of our results have applications in compartmental models ([BH]).

The main result of this section is:

Theorem 3.2. *Let (A, B, C) be a triple of real matrices of sizes $n \times n$, $n \times k$ and $\ell \times n$, respectively, let $(\mathcal{P}, \mathcal{K}, \mathcal{L})$ be a triple of polyhedral cones in $\mathbb{R}^n$, $\mathbb{R}^k$ and $\mathbb{R}^\ell$, respectively. We assume that $\mathcal{P}$ is pointed and (A, B, C) is $(\mathcal{P}, \mathcal{K}, \mathcal{L})$-positive. Then:*

(I) closure $\mathcal{A}_t(A, B, \mathcal{K}) = \mathcal{P}$ *for some $t > 0$ if and only if $\mathcal{P} = B\mathcal{K}$, and*
(II) $\mathcal{S}_t(C, A, \mathcal{L}) = \mathcal{P}$ *for some $t > 0$ if and only if $\mathcal{P} = C^{-1}\mathcal{L}$.*

Part (I) of the theorem was proved for $\mathcal{P} = \mathbb{R}^n_+$ and $\mathcal{K} = \mathbb{R}^k_+$ in [S], §4.1.; systems with closure $\mathcal{A}_t(A, B, \mathbb{R}^\ell_+) = \mathbb{R}^n_+$ are called approximately positive controllable in time t (see [S], Definition 4.4). The proof of the theorem given below is an extension of the ideas developed in [S] to the case of polyhedral cones. We will prove the theorem after formulating Corollary 3.3. Using Proposition 3.1 we obtain from the theorem immediately the following:

Corollary 3.3. *Let (A, B, C) be a triple of matrices as in the theorem, and let $\mathcal{K}$ and $\mathcal{L}$ be polyhedral cones in $\mathbb{R}^k$ and $\mathbb{R}^\ell$, respectively. Assume $B\mathcal{K} = C^{-1}\mathcal{L}$, then there exists at most one polyhedral cone $\mathcal{P}$ in $\mathbb{R}^n$ such that (A, B, C) is $(\mathcal{P}, \mathcal{K}, \mathcal{L})$-positive; and such a cone exists if and only if $B\mathcal{K} = C^{-1}\mathcal{L}$ is invariant under A.*

Note that each minimal realization of the rational matrix function in Example 2.7 has the properties of Corollary 3.3.

Proof (of Theorem 3.2). First we remark that (I) and (II) correspond by duality; indeed, we have for convex cones $\mathcal{K}$ in $\mathbb{R}^k$ and $\mathcal{L}$ in $\mathbb{R}^\ell$ and $t > 0$

$$\mathcal{A}_t(A,B,\mathcal{K})' = \mathcal{S}_t(B',A',\mathcal{K}'), \quad \mathcal{A}(A,B,\mathcal{K})' = \mathcal{S}(B',A',\mathcal{K}'),$$

and

$$\mathcal{S}_t(C,A,\mathcal{L})' = \text{closure } \mathcal{A}_t(A',C',\mathcal{L}'), \quad \mathcal{S}(C,A,\mathcal{L})' = \text{closure } \mathcal{A}(A',C',\mathcal{L}'),$$

where M' denotes the transpose of the matrix M and $\mathcal{K}'$ denotes the dual cone of $\mathcal{K}$, i.e. $\mathcal{K}' = \{y \in \mathbb{R}^k : (y,x) \geq 0 \text{ for all } x \in \mathcal{K}\}$; these relations were proved for $\mathcal{A}$ and $\mathcal{S}$ in [BNS], Theorem 8.3.5, but the same proof works for $\mathcal{A}_t$ and $\mathcal{S}_t$, $t > 0$. Further, it is well-known that for a polyhedral cone its dual (and its polar) cone is polyhedral (see [SW], p. 56), and that $\mathcal{P} = C^{-1}\mathcal{L}$ is equivalent to $\mathcal{P}' = C'\mathcal{L}'$ for closed convex cones $\mathcal{L}$ and $\mathcal{L}'$ ([R], Corollary 16.3.2).
Hence we will only prove part (I) of the last theorem. $\mathcal{P} = B\mathcal{K}$ is sufficient for closure $\mathcal{A}_t(A,B,\mathcal{K}) = \mathcal{P}$ for all $t > 0$, since the closure of $\mathcal{A}_t(A,B,\mathcal{K})$ is the closed convex hull of

$$\{e^{sA}Bu : u \in \mathcal{K},\ 0 \leq s \leq t\};$$

this has been proved for $\mathcal{K} = \mathbb{R}^k_+$ in [S], Proposition 4.7(c), and for arbitrary convex cones $\mathcal{K}$ for $\mathcal{A}(A,B,\mathcal{K})$ in [BNS], Lemma 8.3.4. It is easy to see that the proofs given there work in our slightly more general case.
Since $\mathcal{P}$ is a pointed closed convex cone, it is the convex hull of 0 and its extreme directions ([R], Theorem 18.5). Therefore, to prove that closure $\mathcal{A}_t(A,B,\mathcal{K}) = \mathcal{P}$ implies $\mathcal{P} = B\mathcal{K}$, it is sufficient to show that each extreme direction of $\mathcal{P}$ is in $B\mathcal{K}$. Therefore the proof is complete after having applied the next theorem. □

Theorem 3.4. *Let the assumptions be as in Theorem (3.2), and let $\mathcal{K} = U\,\mathbb{R}^m_+$ for some m and some $k \times m$ matrix U. If for some $t > 0$ the closure of $\mathcal{A}_t(A,B,\mathcal{K})$ is equal to $\mathcal{P}$, then for each extreme direction h of $\mathcal{P}$ there exists a vector e_j in the canonical basis of $\mathbb{R}^m$ and a $\mu > 0$ such that $h = \mu BUe_j$.*

We shall prove this theorem after some preparatory remarks. First we shall need a generalization of Lemma A1 in [S]. The next lemma was proved in [S] for the case when $\mathcal{P}$ is the positivity cone in a Banach lattice, but the proof works without any change for closed convex cones in a normed space. Since we are only interested in the finite dimensional case, we formulate this extension of [S], Lemma A1, only for cones in $\mathbb{R}^n$.

Lemma 3.5. *Let $\mathcal{P}$ be a closed convex cone in $\mathbb{R}^n$, let $\mathcal{D}$ be a compact subset of $\mathcal{P}$, and let $y \in \mathcal{P}'\backslash\{0\}$. Then $\mathcal{D}$ and* Ker $y = \{x \in \mathbb{R}^n : (y,x) = 0\}$ *have a common vector if there exists a nonzero vector in* Ker y *which belongs to the closed convex cone generated by $\mathcal{D}$.*

Now let $\mathcal{P}$ be a polyhedral cone in $\mathbb{R}^n$, let $\mathcal{F}$ be a non empty face of $\mathcal{P}$ and let $\{x_1, \ldots, x_m\} \subset \mathcal{P}\backslash\mathcal{F}$ for some $m \in \mathbb{N}$. Then there exists a $y \in \mathcal{P}'$ such that

$$(y, x) = 0 \text{ for all } x \in \mathcal{F} \text{ and } (y, x_j) > 0, \qquad j = 1, \ldots, m.$$

To see this choose $q \in \mathbb{N}$ and $y_1, \ldots, y_q \in \mathbb{R}^n$ such that

$$\mathcal{P} = \{x \in \mathbb{R}^n : (y_i, x) \geq 0,\ i = 1, \ldots, q\}.$$

Then $\mathcal{F}$ is the intersection of $\mathcal{P}$ with some bounding hyperplanes (see [SW], p. 38), say $\mathcal{F} = \{x \in \mathbb{R}^n : (y_i, x) = 0, i = 1, \ldots, p\} \cap \mathcal{C}$, where $p \leq q$. Since $x_j \in \mathcal{P}\backslash\mathcal{F}$, there exists at least one $i_j \in \{1, \ldots, q\}$ with $(y_{i_j}, x_j) > 0$, and $y = y_{i_1} + \cdots + y_{i_m}$ has the property wanted.

Finally, a simple remark: $x \in \mathcal{P}$, $y \in \mathcal{P}'$, $s \geq 0$ and $(y, e^{sA}x) = 0$ imply $(y, x) = 0$, since $A\mathcal{P} \subset \mathcal{P}$ and therefore $e^{sA}x - x \in \mathcal{P}$.

Now we are ready for the:

P r o o f (of Theorem 3.4). Let h be an extreme direction of $\mathcal{P}$, then $\mathcal{F} = \{\lambda h : \lambda \geq 0\}$ is a face. Assume that Theorem 3.4 is not true, then $\{BUe_1, \ldots, BUe_m\} \subset \mathcal{P}\backslash\mathcal{F}$. Choose $y \in \mathcal{P}'$ as described above, i.e.

$$(y, BUe_j) > 0, \quad j = 1, \ldots, m, \quad \text{and} \quad (y, x) = 0 \quad \text{for all} \quad x \in \mathcal{F}.$$

Since $\mathcal{P}$ is the closure of $\mathcal{A}_t(A, B, \mathcal{K})$ by assumption, $\mathcal{P}$ is the closed convex cone generated by

$$\mathcal{D} = \{e^{sA}BUe_j : j = 1, \ldots, m,\ 0 \leq s \leq t\}$$

(see [S], Proposition 4.7.(b)). Therefore h is a nonzero vector in $\operatorname{Ker} y$ and the closed convex cone generated by $\mathcal{D}$. Now $\mathcal{D}$ is a compact subset of $\mathcal{P}$, therefore by Lemma 3.5 there exist a $j \in \{1, \ldots, m\}$ and a $s \in [0, t]$ such that $(y, e^{sA}BUe_j) = 0$. But then, by the preceding remark, we have $(y, BUe_j) = 0$, which is a contradiction to the choice of y. Theorem 3.4 is proved. □

4. Appendix

The proposition proved below is needed in the proof of Theorem 2.8. We separate this proposition from the rest of the paper, because it is not restricted to nonnegative or real realizations. In the following $\mathbb{K}$ denotes either the field $\mathbb{C}$ of the complex numbers or the field $\mathbb{R}$ of the real numbers.

Let W be a rational matrix function vanishing at infinity. Let (A, B, C) be an arbitrary realization of W and let (A_0, B_0, C_0) be a minimal realization of W. If n and n_0 denote the sizes of A and A_0, respectively, we have $n_0 \leq n$. Therefore $\operatorname{col}\big[C_0 A_0^j\big]_{j=0}^{n-1}$ is left invertible and $\operatorname{row}\big[A_0^j B_0\big]_{j=0}^{n-1}$ is right invertible (see [GLR],

p. 216). We take an arbitrary left inverse L_0 of $\operatorname{col}\left[C_0A_0^j\right]_{j=0}^{n-1}$ and an arbitrary right inverse R_0 of $\operatorname{row}\left[A_0^jB_0\right]_{j=0}^{n-1}$, and define

$$U = L_0 \operatorname{col}\left[CA^j\right]_{j=0}^{n-1} \quad \text{and} \quad V = \operatorname{row}\left[A^jB\right]_{j=0}^{n-1} R_0.$$

Since

$$\sum_{k=0}^{\infty} \lambda^{-k-1} CA^kB = C(\lambda - A)^{-1}B = W(\lambda) = \sum_{k=0}^{\infty} \lambda^{-k-1} C_0A_0^kB$$

for all $\lambda \in \mathbb{K}$ with sufficiently large modulus, we have $CA^kB = C_0A_0^kB_0$ for $k = 0, 1, \ldots$. Therefore $UA^kV = A_0^k$ for $k = 0, 1, \ldots$. For $k = 0$ we obtain $UV = I = n_0 \times n_0$ identity matrix, and then $Q = VU$ is a projection in $\mathbb{K}^n$ which projects onto $\operatorname{Im} V$ along $\operatorname{Ker} U$. We will show that

$$(*) \qquad \mathcal{K}(C, A) \dot{+} \operatorname{Im} Q = \mathcal{K}(C, A) + \mathcal{J}(A, B).$$

The sum of the left-hand side is direct, since $\mathcal{K}(C, A) = \operatorname{Ker} \operatorname{col}\left[CA^j\right]_{j=0}^{n-1} \subset \operatorname{Ker} U = \operatorname{Ker} Q$, and it is contained in right-hand side, since $\operatorname{Im} Q = \operatorname{Im} V \subset \operatorname{Im} \operatorname{row}\left[A^jB\right]_{j=0}^{n-1} = \mathcal{J}(A, B)$. To prove the equality in $(*)$ it is sufficient to show that $\mathcal{J}(A, B) \subset \mathcal{K}(C, A) \dot{+} \operatorname{Im} Q$. Using

$$\operatorname{col}\left[CA^j\right]_{j=0}^{n-1} \cdot \operatorname{row}\left[A^jB\right]_{j=0}^{n-1} = \operatorname{col}\left[C_0A_0^j\right]_{j=0}^{n-1} \cdot \operatorname{row}\left[A_0^jB_0\right]_{j=0}^{n-1},$$

by direct computation we obtain

$$\operatorname{col}\left[CA^j\right]_{j=0}^{n-1} Q \operatorname{row}\left[A^jB\right]_{j=0}^{n-1} = \operatorname{col}\left[CA^j\right]_{j=0}^{n-1} \cdot \operatorname{row}\left[A^jB\right]_{j=0}^{n-1}.$$

Let $y = \operatorname{row}\left[A^jB\right]_{j=0}^{n-1} x$, then $y = Qy + (I - Q) \operatorname{row}\left[A^jB\right]_{j=0}^{n-1} x \in \operatorname{Im} Q + \mathcal{K}(C, A)$ since $\mathcal{K}(C, A) = \operatorname{Ker} \operatorname{col}\left[CA^j\right]_{j=0}^{n-1}$. This proves $(*)$.

Proposition 4.1. *Let (A, B, C) be a realization of a rational matrix function W vanishing at infinity. Then the following assertions are equivalent:*

(I) $\mathcal{J}(A, B) + \mathcal{K}(C, A) = \mathbb{K}^n$, *where n is the size of A.*

(II) *For each minimal realization (A_0, B_0, C_0) of W exists an $n_0 \times n$ matrix S such that $SA = A_0S$, $SB = B_0$ and $C = C_0S$.*

Proof. With the notation from above we obtain from the relations above that $QAQ = VA_0U$, $CQ = CVU = C_0U$ and $UB = B_0$. Let (I) be true: Then $(*)$ and $\mathcal{K}(C, A) \subset \operatorname{Ker} Q$ show that $\mathcal{K}(C, A) = \operatorname{Ker} Q$. Therefore $\operatorname{Ker} Q$ is A-invariant, which is equivalent to $QAQ = QA$. Therefore $QA = VUAVU = VA_0U$. Multiplying this equality by U from the left, we get $UA = A_0U$. Since $\operatorname{Ker} Q =$

$\mathcal{K}(A,C) \subset \operatorname{Ker} C$, we have $C = CQ = C_0U$. Take $S = U$, then (II) follows. Let (II) be true. Then we obtain

$$S \operatorname{row}\left[A^j B\right]_{j=0}^{n-1} = \operatorname{row}\left[A_0^j B_0\right]_{j=0}^{n-1}, \quad \operatorname{col}\left[CA^j\right]_{j=0}^{n-1} = \operatorname{col}\left[C_0 A_0^j\right]_{j=0}^{n-1}.$$

Since (A_0, B_0, C_0) is a minimal realization, we have

$$\operatorname{Im} \operatorname{row}\left[A_0^j B_0\right]_{j=0}^{n-1} = \mathcal{J}(A_0, B_0) = \mathbb{K}^{n_0}, \quad \operatorname{Ker} \operatorname{col}\left[C_0 A_0^j\right]_{j=0}^{n-1} = \mathcal{K}(C_0, A_0) = \{0\}.$$

Thus $\mathcal{K}(C,A) = \operatorname{Ker} S$ and $S\mathcal{J}(A,B) = \mathbb{K}^{n_0}$, and (I) follows immediately. □

Acknowledgements

The work of the second author was supported by the Hungarian OTKA grant No. T 016925.

The authors thank Prof. Dr. I.C. Gohberg for bringing the topic of nonnegative realizations of rational matrix functions to their attention. They are also grateful to a referee for calling their attention to some of the related work cited in the introduction.

References

[ADFB] Anderson, B.D.O., Deistler, M., Farina, L., Benvenuti, L.: Nonnegative realizations of a linear system with nonnegative impulse response; IEEE Trans. Circuits and Systems 43 (1996), 134–142.

[BH] Bernstein, D.S., Hyland, D.C.: Compartmental modeling and second-moment analysis of state space systems; SIAM J. Matrix Anal. 14 (1993), 880–901.

[BNS] Berman, A., Neumann, M., Stern, R.J.: Nonnegative Matrices in Dynamical Systems; John Wiley & Sons, New York 1989.

[F1] Farina, L.: A note on discrete-time positive realizations; Systems Control Lett. 22 (1994), 467–469.

[F2] Farina, L.: Necessary conditions for positive realizability of continuous-time linear systems; Systems Control Lett. 25 (1995), 121–124.

[F3] Farina, L.: On the existence of a positive realization; Systems Control Lett. 28 (1996), 219–226.

[FB] Farina, L., Benvenuti, L.: Positive realizations of linear systems; Systems Control Lett. 26 (1995), 1–9.

[Fli] Fliess, M.: Séries rationelles positives et processus stochastiques; Ann. Inst. H. Poincaré 11 (1975), 1–21.

[FN] Förster, K.-H., Nagy, B.: Spectral properties of nonnegative realizations of rational matrix functions; in preparation.

[Ga] Gale, D.: Convex polyhedral cones and linear inequalities; in: Koopmans, Tjalling C. (ed.): Activity Analysis of Production and Allocation; pp. 287–297, John Wiley & Sons, New York 1951.

[Ge] GERSTENHABER, M.: Theory of convex polyhedral cones; in: KOOPMANS, TJALLING C. (ed.): Activity Analysis of Production and Allocation; pp. 298–316, John Wiley & Sons, New York 1951.

[GLR] GOHBERG, I.C., LANCASTER, P., RODMAN, L.: Invariant Subspaces of Matrices with Applications; John Wiley & Sons, New York 1986.

[GT] GOLDMAN, A.J., TUCKER, A.W.: Polyhedral cones; in: KUHN, H.W., TUCKER, A.W. (eds.): Linear inequalities and related systems; Ann. of Math. Stud. 38, Princeton University Press, Princeton, N.J. 1956.

[HJ] HORN, R.A., JOHNSON, C.R.: Matrix Analysis; Cambridge University Press, Cambridge London New York 1986.

[MK] MAEDA, H., KODAMA, S.: Positive realization of difference equations; IEEE Trans. Circuits Systems 28 (1981), 39–47.

[MKK] MAEDA, H., KODAMA, S., KAJIYA F.: Compartmental system analysis: Realization of a class of linear systems with physical constraints; IEEE Trans. Circuits Systems 24 (1977), 8–14.

[OMK] OHTA, Y., MAEDA, H., KODAMA, S.: Reachability, observability and realizability of continuous-time positive systems; SIAM J. Control Optim. 22 (1984), 171–180.

[R] ROCKAFELLAR, T.R.: Convex Analysis; Princeton University Press, Princeton, N.J. 1970.

[RF] RINALDI, S., FARINA, L.: Positive Linear Systems: Theory and Applications; Città Studi Edizione, Milano 1995 (in Italian).

[S] SCHANBACHER, T.: Aspects of Positivity in Control Theory; SIAM J. Control Optim. 27 (1989), 457–475.

[SW] STOER, J., WITZGALL, C.: Convexity and Optimization in Finite Dimension; Springer Verlag, Berlin 1970.

[TS] TAM, B.S., SCHNEIDER, H.: On the core of a cone-preserving map; Trans. Amer. Math. Soc. 343 (1994), 479–524.

[vdH] VAN DEN HOF, J.M.: Realizations of positive linear systems; Report BS-R9532, CWI, Amsterdam 1955; Linear Algebra Appl., to appear.

Technische Universität Berlin
Fachbereich Mathematik, MA 6-4
Strasse d. 17. Juni 135
D-10623 Berlin
Germany
foerster@math.tu-berlin.de

University of Technology
Institute of Mathematics
Department of Analysis
H-1521 Budapest
Hungary

1991 Mathematics Subject Classification: Primary 93B20, 93B05; Secondary 15A48

Submitted: March 29, 1996

Operator Theory:
Advances and Applications, Vol. 103

On the geometric structure of regular dilations

D. GAŞPAR and N. SUCIU

An important role in the spectacular progress made in the understanding of the spectral theory of nonselfadjoint operators is played by the results of B. Sz.-Nagy and C. Foiaş regarding the isometric dilation of a Hilbert space contraction, its geometric structure and the functional model with the aid of the characteristic function. Similar results in the study of n-tuples of commutative operators were in some respect more difficult to obtain. The results from the single operator case were easily extended for n-tuples consisting of doubly commuting operators (B. Sz.-Nagy, C. Foiaş [14], M. Słocinski [10]). An intermediate situation is that where the n-tuple of contractions has a regular (or $*$-regular) isometric dilation. It is our aim to obtain structure results for Hilbert space multicontractions having this property. For simplicity we shall work in the case $n = 2$.

1. Notations and preliminaries

Let $\mathcal{H}$ be a complex Hilbert space and $\mathcal{B}(\mathcal{H})$ the C^*-algebra of all bounded linear operators on $\mathcal{H}$, where A^* denotes the adjoint of the element $A \in \mathcal{B}(\mathcal{H})$. An element of the (closed) unit ball of $\mathcal{B}(\mathcal{H})$ will be called a contraction on $\mathcal{H}$.

A bicontraction on $\mathcal{H}$ is a commuting pair $T = (T_0, T_1)$, where T_j are contractions on $\mathcal{H}$. It is clear that $T^* = (T_0^*, T_1^*)$ is a bicontraction too. We shall also use the multiindex notation $T^n = T_0^{n_0} T_1^{n_1}$, where $n = (n_0, n_1) \in \mathbb{Z}_+^2$, $\mathbb{Z}(\mathbb{Z}_+)$ being the set of all (nonnegative) integers. A bicontraction T on $\mathcal{H}$ will be briefly denoted by $[\mathcal{H}, T]$. A bicontraction consisting of isometric (unitary) operators will be called an *isometric* (*unitary*) *commuting pair.*

We shall say that the commuting isometric pair $[\mathcal{K}, U]$ is a *minimal isometric dilation* of the bicontraction $[\mathcal{H}, T]$ if $\mathcal{K}$ contains $\mathcal{H}$ as a closed subspace and the bicontraction $U = (U_0, U_1)$ consists of isometric operators such that $T^n = P_{\mathcal{H}} U^n \mid \mathcal{H}$, $n \in \mathbb{Z}_+^2$, where $P_{\mathcal{H}} = P_{\mathcal{K},\mathcal{H}}$ is the orthogonal projection of $\mathcal{K}$ on $\mathcal{H}$, and $\mathcal{K}$ satisfies the minimality property $\mathcal{K} = \bigvee_{n \in \mathbb{Z}_+^2} U^n \mathcal{H}$. When the pair U consists of unitary operators and $\mathcal{K}$ satisfies another minimality property, namely $\mathcal{K} = \bigvee_{n \in \mathbb{Z}^2} U^n \mathcal{H}$, then $[\mathcal{K}, U]$ will be called *a minimal unitary dilation* of $[\mathcal{H}, T]$. Note that the existence of the isometric and unitary dilation for each bicontraction is proved in [1], but the minimality condition does not assure the uniqueness of the dilation (up to unitary equivalence). In the following we shall suppose that the minimality properties are fulfilled, without further mentioning.

An isometric (unitary) dilation $[\mathcal{K}, U]$ of $[\mathcal{H}, T]$ is said to be *regular* if

$$T^{*m_-} T^{m_+} = P_{\mathcal{H}} U^{*m_-} U^{m_+} \mid \mathcal{H}, \qquad m \in \mathbb{Z}^2, \tag{1.1}$$

and $*$-*regular* if

$$T^{m_+}T^{*m_-} = P_{\mathcal{H}}U^{*m_-}U^{m_+} \mid \mathcal{H}, \qquad m \in \mathbb{Z}^2, \tag{1.2}$$

where for $m = (m_0, m_1)$,

$$m_+ = (\max\{m_0, 0\}, \max\{m_1, 0\}), \qquad m_- = (\max\{-m_0, 0\}, \max\{-m_1, 0\}).$$

If such dilations do exist, then the minimality property implies their uniqueness (up to unitary equivalence; see [14]). It is easy to see from (1.1) and (1.2) that $[\mathcal{K}, U]$ is a regular unitary dilation of $[\mathcal{H}, T]$ iff $[\mathcal{K}, U^*]$ is a $*$-regular unitary dilation of $[\mathcal{H}, T^*]$. Moreover, if $[\mathcal{K}, U]$ is a regular unitary dilation of $[\mathcal{H}, T]$ and we put

$$\begin{aligned} \mathcal{K}^+ &:= \bigvee_{n\in\mathbb{Z}_+^2} U^n\mathcal{H}, & \mathcal{K}_*^+ &:= \bigvee_{n\in\mathbb{Z}_+^2} U^{*n}\mathcal{H}, \\ V_i &:= U_i \mid \mathcal{K}^+, & V_{*i} &:= U_i^* \mid \mathcal{K}_*^+, \qquad i = 0, 1, \end{aligned}$$

then $[\mathcal{K}^+, V]$ with $V := (V_0, V_1)$ is a regular isometric dilation of $[\mathcal{H}, T]$, whereas $[\mathcal{K}_*^+, V_*]$ with $V_* := (V_{*0}, V_{*1})$ is a $*$-regular isometric dilation of $[\mathcal{H}, T]$.

It is well known (see [14], [15]) that $[\mathcal{H}, T]$ has a regular isometric (unitary) dilation iff T fulfills the condition

$$\Delta_T := I - T_0^*T_0 - T_1^*T_1 + T_0^*T_1^*T_0T_1 \geq 0. \tag{1.3}$$

We shall say (following [3]) that T is a *bidisc isometry* when $\Delta_T = 0$. If T^* is a bidisc isometry we shall say that T is a *bidisc coisometry.* It is easily seen that if T_0 or T_1 is an isometry, then T is a bidisc isometry. Now $I - T_0^*T_0 - T_1^*T_1 \geq 0$ implies $\Delta_T \geq 0$. If, in particular, $I - T_0^*T_0 - T_1^*T_1 = 0$, we shall say that T is a *spherical isometry* (see [15]). Correspondingly, we define a *spherical coisometry.* Finally, we remark that if T is a doubly commuting bicontraction (i.e $T_0^*T_1 = T_1T_0^*$), then $\Delta_T = (I - T_0^*T_0)(I - T_1^*T_1) \geq 0$ and consequently T has an isometric regular dilation, which is also $*$-regular. This is the natural isometric dilation of $[\mathcal{H}, T]$ (see [10]). Conversely, if $[\mathcal{H}, T]$ has a regular and $*$-regular isometric dilation, then T is doubly commuting. It can be shown that an isometric dilation $[\mathcal{K}, U]$ of $[\mathcal{H}, T]$ is doubly commuting iff $[\mathcal{K}, U]$ is a $*$-regular isometric dilation of $[\mathcal{H}, T]$ (see [7]).

We recall the Wold–Słocinski geometric structure of a doubly commuting isometric pair $[\mathcal{H}, V]$, where $V = (V_0, V_1)$. If $\mathcal{L}_i = \mathcal{H} \ominus V_i\mathcal{H}$ is the defect space of V_i, $i = 0, 1$, then the subspace $\mathcal{R} = \mathcal{L}_0 \cap \mathcal{L}_1$ is wandering for V (i.e. $V^n\mathcal{R} \perp V^m\mathcal{R}$, $n, m \in \mathbb{Z}_+^2$, $n \neq m$) and $\mathcal{R}_i = \bigcap_{p\in\mathbb{Z}_+} V_{1-i}^p\mathcal{L}_i$ is wandering for V_i, $i = 0, 1$. Moreover, if we denote

$$\mathcal{H}^+(\mathcal{R}) = \bigoplus_{n\in\mathbb{Z}_+^2} V^n\mathcal{R}, \qquad \mathcal{H}_i^+(\mathcal{R}_i) = \bigoplus_{n_i\in\mathbb{Z}_+} V_i^{n_i}\mathcal{R}, \qquad i = 0, 1,$$

then the following orthogonal decomposition of $\mathcal{H}$ into four reducing summands (for V) holds:

(1.4) $$\mathcal{H} = \mathcal{H}_u \oplus \mathcal{H}^+(\mathcal{R}) \oplus \mathcal{H}_0^+(\mathcal{R}_0) \oplus \mathcal{H}_1^+(\mathcal{R}_1),$$

where V_i maps $\mathcal{H}_u \oplus \mathcal{H}_{1-i}^+(\mathcal{R}_{1-i})$ unitarily on itself and acts as a shift operator on $\mathcal{H}^+(\mathcal{R}) \oplus \mathcal{H}_i^+(\mathcal{R}_i)$, $i = 0, 1$, and

$$\mathcal{H}_u = \{h \in \mathcal{H} : ||h|| = ||V^{*n}h||, n \in \mathbb{Z}_+^2\} = \bigcap_{n \in \mathbb{Z}_+^2} V^n\mathcal{H}.$$

If $\mathcal{H} = \mathcal{H}^+(\mathcal{R})$, then the isometric pair $[\mathcal{H}, V]$ is called a *shift pair* (see [4], [5]), or a *bishift*. An isometric dilation $[\mathcal{K}, U]$ of a bicontraction $[\mathcal{H}, T]$ is said of *bishift type* if U is a shift pair on $\mathcal{K}$. We mention that a bishift type dilation consists of doubly commuting isometries and, consequently, by the above mentioned remark, it is a $*$-regular isometric dilation.

2. The structure of regular and $*$-regular isometric dilations

At the beginning we refer to the structure of regular dilations.

Theorem 2.1. *Let $T = (T_0, T_1)$ be a bicontraction on $\mathcal{H}$ which has the regular isometric dilation $[\mathcal{K}, (U_0, U_1)]$. Then the subspaces*

$$\mathcal{L} = \overline{(U_0U_1 - U_0T_1 - U_1T_0 + T_0T_1)\mathcal{H}}, \qquad \mathcal{L}_i = \overline{(U - T_i)\mathcal{H}}, \qquad i = 0, 1,$$

are wandering for $U = (U_0, U_1)$ and for U_i, $i = 0, 1$, respectively, and we have

(2.1) $$\mathcal{K} = \mathcal{H} \oplus \mathcal{K}^+(\mathcal{L}) \oplus \mathcal{K}_0^+(\mathcal{L}_0) \oplus \mathcal{K}_1^+(\mathcal{L}_1).$$

Moreover, $\dim \mathcal{L} = \dim \overline{\Delta_T^{1/2}\mathcal{H}}$.

Proof. We first verify that the subspace $\mathcal{L}$ is wandering for U and in this respect it is enough to prove that $U^m\mathcal{L} \perp U^n\mathcal{L}$ for $m = (p,q) \in \mathbb{Z}_+^2$, $p \geq 1$ and $n = (0,0)$ and for $m = (p, 0)$ and $n = (0, q)$ with $p, q \geq 1$. Let $p, q \in \mathbb{Z}_+$, $p \neq 0$. Then for $h, h' \in \mathcal{H}$ we obtain (using (1.1)):

$$\begin{aligned}
&(U_0^pU_1^q(U_0U_1 - U_0T_1 - U_1T_0 + T_0T_1)h, (U_0U_1 - U_0T_1 - U_1T_0 + T_0T_1)h') \\
&= (U_0^pU_1^qh, h') - (U_0^pU_1^{q+1}h, T_1h') - (U_0^{p+1}U_1^qh, T_0h') + (U_0^{p+1}U_1^{q+1}h, T_0T_1h') \\
&\quad -(U_0^pU_1^qT_1h, U_1h') + (U_0^pU_1^qT_1h, T_1h') + (U_0^{p+1}U_1^qT_1h, U_1T_0h') \\
&\quad -(U_0^{p+1}U_1^qT_1h, T_0T_1h') - (U_0^{p-1}U_1^qT_0h, h') + (U_0^{p-1}U_1^{q+1}T_0h, T_1h') \\
&\quad -(U_0^pU_1^qT_0h, T_0h') - (U_0^pU_1^{q+1}T_0h, T_0T_1h') + (U_0^{p-1}U_1^qT_0T_1h, U_1h') \\
&\quad -(U_0^{p-1}U_1^qT_1h, T_1h') - (U_0^pU_1^qT_0T_1h, U_1T_0h') + (U_0^pU_1^qT_0T_1h, T_0T_1h') \\
&= -(U_0^pU_1^qT_1h, U_1h') + (U_0^{p+1}U_1^qT_1h, U_1T_0h') + (U_0^{p-1}U_1^qT_0T_1h, U_1h') \\
&\quad -(U_0^pU_1^qT_0T_1h, U_1T_0h') \\
&= -(T_0^pT_1^{q+1}h, T_1h') + (T_0^{p+1}T_1^{q+1}h, T_0T_1h') + (T_0^pT_1^{q+1}h, T_1h') \\
&\quad -(T_0^{p+1}T_1^{q+1}h, T_0T_1h') = 0.
\end{aligned}$$

For $p, q \in \mathbb{Z}_+ \backslash \{0\}$ and $h, h' \in \mathcal{H}$ we also have

$$\begin{aligned}
&(U_0^p(U_0U_1 - U_0T_1 - U_1T_0 + T_0T_1)h, U_1^q(U_0U_1 - U_0T_1 - U_1T_0 + T_0T_1)h') \\
&= (U_0^p h, U_1^q h') - (U_0^p h, U_1^{q-1}T_1 h') - (U_0^{p+1}h, U_1^q T_0 h') \\
&\quad +(U_0^{p+1}h, U_1^{q-1}T_0T_1h') - (U_0^pT_1h, U_1^{q+1}h') + (U_0^pT_1h, U_1^qT_1h') \\
&\quad +(U_0^{p+1}T_1h, U_1^{q+1}T_0h') - (U_0^{p+1}T_1h, U_1^qT_0T_1h') - (U_0^{p-1}T_0h, U_1^qh') \\
&\quad +(U_0^{p-1}T_0h, U_1^{q-1}T_1h') + (U_0^pT_0h, U_1^qT_0h') - (U_0^pT_0h, U_1^{q-1}T_0T_1h') \\
&\quad +(U_0^{p-1}T_0T_1h, U_1^{q+1}h') - (U_0^{p-1}T_0T_1h, U_1^qT_1h') - (U_0^pT_0T_1h, U_1^{q+1}T_0h') \\
&\quad +(U_0^pT_0T_1h, U_1^qT_0T_1h') \\
&= (T_0^ph, T_1^qh') - (T_0^ph, T_1^qh') - (T_0^{p+1}h, T_0T_1^qh') + (T_0^{p+1}h, T_0T_1^qh') \\
&\quad -(T_0^pT_1h, T_1^{q+1}h') + (T_0^pT_1h, T_1^{q+1}h') + (T_0^{p+1}T_1h, T_0T_1^{q+1}h') \\
&\quad -(T_0^{p+1}T_1h, T_0T_1^{q+1}h') - (T_0^ph, T_1^qh') + (T_0^ph, T_1^qh') \\
&\quad +(T_0^{p+1}h, T_0T_1^qh') - (T_0^{p+1}h, T_0T_1^qh') + (T_0^pT_1h, T_1^{q+1}h') \\
&\quad -(T_0^pT_1h, T_1^{q+1}h') - (T_0^{p+1}T_1h, T_0T_1^{q+1}h') + (T_0^{p+1}T_1h, T_0T_1^{q+1}h') = 0.
\end{aligned}$$

It follows that $\mathcal{L}$ is wandering for $U = (U_0, U_1)$.
Now observe that the spaces $\mathcal{L}$ and $\mathcal{L}_i$ can be written as

$$\mathcal{L} = \overline{P_{\widetilde{\mathcal{H}}}(U_0U_1 - U_0T_1 - U_1T_0)\mathcal{H}}, \qquad \mathcal{L}_i = \overline{P_{\widetilde{\mathcal{H}}}U_i\mathcal{H}}, \qquad i = 0, 1,$$

where $\widetilde{\mathcal{H}} = \mathcal{K} \ominus \mathcal{H}$ and $P_{\widetilde{\mathcal{H}}} = P_{\mathcal{K},\widetilde{\mathcal{H}}}$. Since (U_0, U_1) is a minimal isometric dilation of (T_0, T_1) we have $U_i^* \mid \mathcal{H} = T_i^*$, so that $U_i\widetilde{\mathcal{H}} \subset \widetilde{\mathcal{H}}$, $i = 0, 1$. Thus $\mathcal{K}^+(\mathcal{L}) = \bigoplus_{n \in \mathbb{Z}_2^+} U^n\mathcal{L} \subset \widetilde{\mathcal{H}}$. But $\mathcal{L}_i$ is wandering for U_i since for $q \in \mathbb{Z}_+$, $q \neq 0$, and $h, h' \in \mathcal{H}$ we have

$$(U_i^q P_{\widetilde{\mathcal{H}}}U_ih, P_{\widetilde{\mathcal{H}}}U_ih') = (U_i^{q-1}P_{\widetilde{\mathcal{H}}}U_ih, h') = 0.$$

It follows that $\mathcal{K}_i^+(\mathcal{L}) = \bigoplus_{q \geq 0} U_i^q\mathcal{L}_i \subset \widetilde{\mathcal{H}}$, $i = 0, 1$. Denoting $V_i = U_i \mid \widetilde{\mathcal{H}}$, $i = 0, 1$, we have that V_0 and V_1 are commuting isometries on $\widetilde{\mathcal{H}}$.
Now we show that the subspaces $\mathcal{K}^+(\mathcal{L})$, $\mathcal{K}_0^+(\mathcal{L}_0)$ and $\mathcal{K}^+(\mathcal{L}_1)$ are pairwise orthogonal in $\widetilde{\mathcal{H}}$. We shall verify first that $U_1^p\mathcal{L} \subset \widetilde{\mathcal{H}} \ominus \mathcal{K}_0^+(\mathcal{L}_0)$, $p \geq 0$. Using (1.1), we have for all $p, q \geq 0$ and $h, h' \in \mathcal{H}$

$$\begin{aligned}
&((U_1^q(U_0U_1 - U_0T_1 - U_1T_0 + T_0T_1)h, U_0^p(U_0 - T_0)h') \\
&= (U_1^{q+1}h, U_0^ph') - (U_0U_1^{q+1}h, U_0^pT_0h') - (U_1^qT_1h, U_0^ph') + (U_0U_1^qT_1h, U_0^pT_0h') \\
&\quad -(U_1^{q+1}T_0h, U_0^{p+1}h') + (U_1^{q+1}T_0h, U_0^pT_0h') + (U_1^qT_0T_1h, U_0^{p+1}h') \\
&\quad -(U_1^qT_0T_1h, U_0^pT_0h') \\
&= (T_1^{q+1}h, T_0^ph') - (T_0T_1^{q+1}h, T_0^{p+1}h') - (T_1^{q+1}h, T_0^ph') + (T_0T_1^{q+1}h, T_0^{p+1}h') \\
&\quad -(T_0T_1^{q+1}h, T_0^{p+1}h') + (T_0T_1^{q+1}h, T_0^{p+1}h') + (T_0T_1^{q+1}h, T_0^{p+1}h') \\
&\quad -(T_0T_1^{q+1}h, T_0^{p+1}h') = 0.
\end{aligned}$$

$\mathcal{K}_0^+(\mathcal{L}_0)$ is invariant for V_0^* because $V_0^*\mathcal{L}_0 = \{0\}$. It follows that $\widetilde{\mathcal{H}} \ominus \mathcal{K}_0^+(\mathcal{L}_0)$ is invariant for V_0 and U_0, thus $U^n\mathcal{L} \subset \widetilde{\mathcal{H}} \ominus \mathcal{K}_0^+(\mathcal{L}_0)$ for $n \in \mathbb{Z}_+^2$ and the subspaces $\mathcal{K}^+(\mathcal{L})$ and $\mathcal{K}_0^+(\mathcal{L}_0)$ are orthogonal in $\widetilde{\mathcal{H}}$. By symmetry the subspaces $\mathcal{K}^+(\mathcal{L})$ and $\mathcal{K}_1^+(\mathcal{L}_1)$ are also orthogonal. It remains to verify only that $\mathcal{K}_1^+(\mathcal{L}_1)$ and $\mathcal{K}_0^+(\mathcal{L}_0)$ are orthogonal. But this is true, since for $p, q \geq 0$ and $h, h' \in \mathcal{H}$ we have:

$$\begin{aligned}
&(U_0^p((U_0 - T_0)h, U_1^q(U_1 - T_1)h') \\
&= (U_0^{p+1}h, U_1^{q+1}h') - (U_0^{p+1}h, U_1^qT_1h') - (U_0^pT_0h, U_1^{q+1}h') + (U_0^pT_0h, U_1^qT_1h') \\
&= (T_0^{p+1}h, T_1^{q+1}h') - (T_0^{p+1}h, T_1^{q+1}h') - (T_0^{p+1}h, T_1^{q+1}h') + (T_0^{p+1}h, T_1^{q+1}h') \\
&= 0.
\end{aligned}$$

Next we show that $\widetilde{\mathcal{H}} = \mathcal{K}^+(\mathcal{L}) \oplus \mathcal{K}_0^+(\mathcal{L}_0) \oplus \mathcal{K}_1^+(\mathcal{L}_1)$, which implies (2.1). Let $k \in \widetilde{\mathcal{H}}$ be orthogonal on $U^n\mathcal{L}, U_0^p\mathcal{L}_0$ and $U_1^q\mathcal{L}_1$ for all $n \in \mathbb{Z}_+^2$ and $p, q \in \mathbb{Z}_+$. First note that $k \perp U_0^p\mathcal{H}, k \perp U_1^q\mathcal{H}$, $p, q \in \mathbb{Z}_+$, and $k \perp U_0U_1\mathcal{H}$. By induction, from $k \perp U_0^p\mathcal{L}$ it follows $k \perp U_0^pU_1\mathcal{H}$, $p > 1$, and by a symmetry argument $k \perp U_0U_1^q\mathcal{H}$, $q > 1$. Next $k \perp U_0U_1\mathcal{L}$ implies $k \perp U_0^2U_1^2\mathcal{L}$ and for the same reasons as before we have $k \perp U_0^pU_1^2\mathcal{H}$ and $k \perp U_0^2U_1^q\mathcal{H}$ if $p, q > 1$. Repeating this argument we finally obtain that $k \perp U^n\mathcal{H}$ for all $n \in \mathbb{Z}_+^2$, so $k = 0$ because of the minimality of $\mathcal{K}$.
In order to obtain the dimension of $\mathcal{L}$, observe that for $h \in \mathcal{H}$,

$$\begin{aligned}
&||(U_0U_1 - U_0T_1 - U_1T_0 + T_0T_1h||^2 \\
&= ||h||^2 - (U_1h, T_1h) - (U_0h, T_0h) + (U_0U_1h, T_0T_1h) - (T_1h, U_1h) + ||T_1h||^2 \\
&\quad +(U_0T_1h, U_1T_0h) - (U_0T_1h, T_0T_1h) - (T_0h, U_0h) + (U_1T_0h, U_0T_1h) + ||T_0h||^2 \\
&\quad -(U_1T_0h, T_0T_1h) + (T_0T_1h, U_0U_1h) - (T_0T_1h, U_0T_1h) - (T_0T_1h, U_1T_0h) \\
&\quad +||T_0T_1h||^2 \\
&= ||h||^2 - ||T_0h||^2 - ||T_1h||^2 + ||T_0T_1h||^2 \\
&= (\Delta_Th, h) = ||\Delta_T^{1/2}h||^2.
\end{aligned}$$

Thus, we can define the unitary operator ω from $\overline{\Delta_T^{1/2}\mathcal{H}}$ onto $\mathcal{L}$ by

$$\omega(\Delta_T^{1/2}h) = (U_0U_1 - U_0T_1 - U_1T_0 + T_0T_1)h, \qquad h \in \mathcal{H},$$

so $\dim \mathcal{L} = \dim(\overline{\Delta_T^{1/2}\mathcal{H}})$. □

Preserving the above notations, we observe that, for every $i = 0, 1$, $V_i = U_i \mid \mathcal{K} \ominus \mathcal{H}$, reduces the subspace $\mathcal{K}_i^+(\mathcal{L}_i)$ because $V_i^*\mathcal{L}_i = \{0\}$. Thus V_i reduces also the subspace $\mathcal{K}^+(\mathcal{L}) \oplus \mathcal{K}_{i-1}^+(\mathcal{L}_{1-i})$, but it does not necessarily reduce each of the subspaces $\mathcal{K}^+(\mathcal{L})$ and $\mathcal{K}_{1-i}^+(\mathcal{L}_{1-i})$. In fact, $\mathcal{K}^+(\mathcal{L})$ is invariant for V_i, so $\mathcal{K}_{1-i}^+(\mathcal{L}_{1-i})$ is also an invariant subspace for V_i. Since for all $h \in \mathcal{H}$ we have

$$U_i(U_{1-i} - T_{1-i})h = (U_0U_1 - U_0T_1 - U_1T_0 + T_0T_1)h + (U_{1-i} - T_{1-i})T_ih,$$

we obtain:

Corollary 2.2. *Let $T=(T_0,T_1)$ be a bicontraction on $\mathcal{H}$ having a regular isometric dilation $[\mathcal{K},(U_0,U_1)]$. Then the subspace $\mathcal{K}_{1-i}^+(\mathcal{L}_{1-i})$ is invariant for $V_i=U_i \mid \mathcal{K}\ominus\mathcal{H}$, $i=0,1$, iff T is a bidisc isometry on $\mathcal{H}$ or, equivalently, iff $\mathcal{K}=\mathcal{H}\oplus\mathcal{K}_0^+(\mathcal{L}_0)\oplus\mathcal{K}_1^+(\mathcal{L}_1)$.*

From the previous proof we also see that the subspace $\mathcal{L}$ is in fact wandering for the minimal unitary extension $[\widetilde{\mathcal{K}},(\widetilde{U}_0,\widetilde{U}_1)]$ of $[\mathcal{K},(U_0,U_1)]$, which means that $\widetilde{U}^n\mathcal{L}\perp\widetilde{U}^m\mathcal{L}$ for all $m,n\in\mathbb{Z}^2$, $m\neq n$. Thus the subspace $\widetilde{\mathcal{K}}(\mathcal{L}):=\bigoplus_{n\in\mathbb{Z}^2}\widetilde{U}^n\mathcal{L}\subset\widetilde{\mathcal{K}}$ reduces $\widetilde{U}_0$ and $\widetilde{U}_1$. We obtain:

Corollary 2.3. *Let $[\mathcal{H},T]$ be a bicontraction which is not a bidisc isometry on $\mathcal{H}$ and has a regular unitary dilation. Then this regular unitary dilation contains a (bilateral) shift pair of multiplicity* $\dim\overline{\Delta_T^{1/2}\mathcal{H}}$.

Regarding the Wold decomposition (1.4) of the $*$-regular isometric dilations we have:

Proposition 2.4. *Let $T=(T_0,T_1)$ be a bicontraction on $\mathcal{H}$ having a $*$-regular isometric dilation $[\mathcal{K},(V_0,V_1)]$. Then in the Wold decomposition (1.4) of $\mathcal{K}$ with respect to $V=(V_0,V_1)$ we have*

$$\mathcal{K}_u=\bigvee_{n\in\mathbb{Z}_+^2}V^nP_{\mathcal{K}_u}\mathcal{H}$$

and

$$\mathcal{R}=\overline{(I-V_0T_0^*-V_1T_1^*+V_0V_1T_0^*T_1^*)\mathcal{H}},\quad \mathcal{R}_i=\bigcap_{p\geq 0}\bigvee_{q\geq p}V_{1-i}^q\overline{(I-V_iT_i^*)\mathcal{H}},\quad i=0,1.$$

Moreover, $\dim\mathcal{R}=\dim\overline{\Delta_{T^*}^{1/2}\mathcal{H}}$.

Proof. Consider the decomposition (1.4) of $\mathcal{K}$. We have

$$\mathcal{K}_u=P_{\mathcal{K}_u}\bigvee_{n\in\mathbb{Z}_+^2}V^n\mathcal{H}\subset\bigvee_{n\in\mathbb{Z}_+^2}P_{\mathcal{K}_u}V^n\mathcal{H}=\bigvee_{n\in\mathbb{Z}_+^2}V^nP_{\mathcal{K}_u}\mathcal{H}\subset\mathcal{K}_u,$$

hence the desired equality for $\mathcal{K}_u$. To obtain the relation of $\mathcal{R}$, in fact, we have to show that $\mathcal{R}=\overline{\Delta_{V^*}\mathcal{H}}$. Let $[\widetilde{\mathcal{K}},(U_0,U_1)]$ be the minimal unitary extension of V. Since $[\widetilde{\mathcal{K}},(U_0^*,U_1^*)]$ is the minimal regular unitary dilation of T^*, it follows that the subspace $U_0^*U_1^*\overline{\Delta_{V^*}\mathcal{H}}$ is wandering for (U_0,U_1), so $\overline{\Delta_{V^*}\mathcal{H}}$ is wandering for (V_0,V_1). Then the subspace $\mathcal{M}=\mathcal{K}^+(\overline{\Delta_{V^*}\mathcal{H}})$ reduces both the isometries V_0 and V_1 because they are doubly commuting. It follows that $(V_0\mid\mathcal{M},V_1\mid\mathcal{M})$ is a shift pair and we obtain (see [4], Proposition 1):

$$\overline{\Delta_{V^*}\mathcal{H}}\subset\mathcal{R}=\mathcal{K}\ominus P_{\widetilde{\mathcal{K}},\mathcal{K}}\bigvee_{n\notin-\mathbb{Z}_+^2}U^n\mathcal{K}.$$

Let $k \in \mathcal{R}$, $k \perp \Delta_{V^*}\mathcal{H}$. For $h \in \mathcal{H}$ we have

$$0 = (k, (I - U_0T_0^* - U_1T_1^* + U_0U_1T_0^*T_1^*)h) = (k, h),$$

so $k \perp \mathcal{H}$. By the minimality of $\mathcal{K}$ and by the above structure of $\mathcal{R}$, we obtain $k \perp \mathcal{K}$ and hence $k = 0$. Then $\mathcal{R} = \overline{\Delta_{V^*}\mathcal{H}}$. But the U_i are unitary, so by Theorem 2.1 we have

$$\dim \mathcal{R} = \dim(U_0^*U_1^*\overline{\Delta_{V^*}\mathcal{H}}) = \dim \overline{\Delta_{T^*}^{1/2}\mathcal{H}}.$$

Denote $\mathcal{N}_i = \mathcal{K} \ominus V_i\mathcal{K}$, $i = 0, 1$. Since

$$\mathcal{K} = \bigvee_{q \geq 0} V_1^q\mathcal{H} \vee \bigvee_{p\geq 1, q\geq 0} V_0^pV_1^q\mathcal{H},$$

we have $\mathcal{N}_0 = P_{\mathcal{N}_0}\mathcal{K} = \overline{P_{\mathcal{N}_0}\mathcal{K}_1} = \overline{(I - V_0V_0^*)\mathcal{K}_1}$, where $\mathcal{K}_1 = \bigvee_{q\geq 0} V_1^q\mathcal{H}$. But V_0 and V_1 are doubly commuting and $V_0^* \mid \mathcal{H} = T_0^*$, so V_0^* leaves the subspace $\mathcal{K}_1$ invariant and thus $\mathcal{N}_0 = \bigvee_{q\geq 0} V_1^q(I - V_0T_0^*)\mathcal{H}$. A similar formula holds for $\mathcal{N}_1$ and using the structure of $\mathcal{R}_i$ from the decomposition (1.4), we obtain the formulas for $\mathcal{R}_i$, $i = 0, 1$, appearing in the conclusion of our statement. □

Remark 2.5. The geometrical structure (2.1) and the Wold decomposition for the natural isometric dilation of a pair $T = (T_0, T_1)$ of double commuting contractions on $\mathcal{H}$ were obtained by M. Słocinski in [10]. In this case, he deduces even a geometrical structure for the minimal unitary extension of the natural dilation, using the defect spaces from the decomposition (2.1) corresponding to the bicontractions T, T^*, (T_0, T_1^*) and (T_0^*, T_1). Using the Wold decomposition we point out some particular classes of bicontractions having $*$-regular isometric dilations.

Proposition 2.6. *For a bicontraction $T = (T_0, T_1)$ on $\mathcal{H}$ the following assertions are equivalent:*

(i) *T has an isometric dilation $[\mathcal{K}, V]$ which is a bidisc coisometry on $\mathcal{K}$;*

(ii) *T has a $*$-regular isometric dilation $[\mathcal{K}, V]$ having the Wold decomposition of the form $\mathcal{K} = \mathcal{K}_u \oplus \mathcal{K}_0^+(\mathcal{R}_0) \oplus \mathcal{K}_1^+(\mathcal{R}_1)$;*

(iii) *$[\mathcal{H}, T]$ is a bidisc coisometry.*

Proof. If $[\mathcal{K}, V]$ is an isometric dilation of T which is a bidisc coisometry, then $\Delta_{T^*} = P_{\mathcal{H}}\Delta_{V^*} \mid \mathcal{H} = 0$, so T^* is a bidisc isometry on $\mathcal{H}$. In this case, $\mathcal{R} = \{0\}$ in the Wold decomposition of the $*$-regular isometric dilation $[\mathcal{K}, U]$ of T. But then $\Delta_{U^*} \mid \mathcal{H} = 0$ and since the U_i are doubly commuting, the minimality of $\mathcal{K}$ implies that $\Delta_{U^*} = 0$, which means that U^* is a bidisc isometry on $\mathcal{K}$. We have thus verified that (i) $\Rightarrow$ (iii) $\Rightarrow$ (ii) $\Rightarrow$ (i). □

Corollary 2.7. *The $*$-regular isometric dilation of a pair of commuting coisometries consists of unitary operators.*

Proof. Suppose that $T = (T_0, T_1)$ is a pair of commuting isometries on $\mathcal{H}$, and let $[\mathcal{M}, (W_0, W_1)]$ be the $*$-regular isometric dilation of T^*. Since

$$||(I - W_i T_i)h||^2 = ||h||^2 - ||T_i h||^2 = 0, \qquad h \in \mathcal{H},$$

from Proposition 2.4 it follows that $\mathcal{R}_i = \{0\}, i = 0, 1$, in the Wold decomposition of $\mathcal{M}$. So $\mathcal{M} = \mathcal{M}_u$, which means that W_0 and W_1 are unitary on $\mathcal{M}$. □

Recall from [14] that a contraction A on $\mathcal{H}$ is of the class $C_{0\cdot}$ if $A^p h \to 0$ as $p \to \infty$ for all $h \in \mathcal{H}$. If $A^* \in C_{0\cdot}$, we say that $A \in C_{\cdot 0}$, and A is of class C_{00} if it is simultaneously of class $C_{0\cdot}$ and $C_{\cdot 0}$.

Proposition 2.8. *Let $T = (T_0, T_1)$ be a bicontraction on $\mathcal{H}$ having a $*$-regular isometric dilation $[\mathcal{K}, (V_0, V_1)]$. Then the decomposition (1.4) of $\mathcal{K}$ relatively to $V = (V_0, V_1)$ has the form*

$$\mathcal{K} = \mathcal{K}^+(\mathcal{R}) \oplus \mathcal{K}_0^+(\mathcal{R}_0)$$

iff $T_0 \in C_{\cdot 0}$. In particular, $\mathcal{K} = \mathcal{K}_0^+(\mathcal{R}_0)$ iff $T_0 \in C_{\cdot 0}$ and T_1 is a coisometry on $\mathcal{H}$.

Proof. Suppose $T_0 \in C_{\cdot 0}$, and let

$$\mathcal{K} = \mathcal{K}_u \oplus \mathcal{K}^+(\mathcal{R}) \oplus \mathcal{K}_0^+(\mathcal{R}_0) \oplus \mathcal{K}_1^+(\mathcal{R}_1)$$

be the Wold decomposition of the space $[\mathcal{K}, V]$. Then V_0 is unitary on $\mathcal{K}_1 = \mathcal{K}_u \oplus \mathcal{K}_1^+(\mathcal{R}_1)$ and V_0 is a unilateral shift on $\mathcal{K}_0 = \mathcal{K}^+(\mathcal{R}) \oplus \mathcal{K}_0^+(\mathcal{R}_0)$, so $V_0^{*p} k \to 0$ as $p \to \infty$ for any $k \in \mathcal{K}$. But for $h = k_1 + k_0 \in \mathcal{H}$, with $k_1 \in \mathcal{K}_1, k_0 \in \mathcal{K}_0$ and $p \geq 0$, we have

$$||T_0^{*p} h||^2 = ||V_0^{*p} h||^2 = ||V_0^{*p} k_1||^2 + ||V_0^{*p} k_0||^2 = ||k_1||^2 + ||V_0^{*p} k_0||^2,$$

and by a limit argument $k_1 = 0$. Thus $\mathcal{H} \subset \mathcal{K}_0$ and by the invariance of $\mathcal{K}_0$ to V and the minimality of $\mathcal{K}$, we obtain $\mathcal{K} = \mathcal{K}_0$.
Suppose now that $\mathcal{K} = \mathcal{K}_0^+(\mathcal{R}_0)$. Then V_1 is unitary on $\mathcal{K}$, so $T_1^* = V_1^* \mid \mathcal{H}$ is an isometry and therefore T_1 is a coisometry. Clearly $T_0 \in C_{\cdot 0}$. Conversely, if $T_0 \in C_{\cdot 0}$ and T_1 is a coisometry, we have $\mathcal{K} = \mathcal{K}_0$ and T is a bidisc coisometry, so by Proposition 2.6, $\mathcal{K} = \mathcal{K}_0^*(\mathcal{R}_0)$. □

Proposition 2.9. *For a bicontraction $T = (T_0, T_1)$ the following statements are equivalent:*

(i) *T has a bishift-type isometric dilation;*
(ii) *T has a $*$-regular isometric dilation and $T_i \in C_{\cdot 0}$, $i = 0, 1$;*
(iii) *T has a minimal unitary dilation $[\mathcal{K}, (U_0, U_1)]$ of the form*

$$\mathcal{K} = \bigoplus_{n \in \mathbb{Z}^2} U^n \mathcal{R} \quad \text{with} \quad \mathcal{H} \subset \bigoplus_{n \in \mathbb{Z}_+^2} U^n \mathcal{R},$$

where $\mathcal{R} = \overline{(I - U_0 T_0^* - U_1 T_1^* + U_0 U_1 T_0^* T_1^*)\mathcal{H}}$.

Proof. Condition (i) means that T has an isometric dilation $[\mathcal{K}, V]$ with $V = (V_0, V_1)$ a shift pair on $\mathcal{K}$, so the V_i are doubly commuting unilateral shifts on $\mathcal{K}$. In this case $T_i^* = V_i^* \mid \mathcal{H}$ so $T_i^* \in C_{0\cdot}$, $i = 0, 1$. Thus (i) $\Rightarrow$ (ii). Next if T and $[\mathcal{K}, V]$ are as in (ii), by Proposition 2.8, $\mathcal{K} = \mathcal{K}^+(\mathcal{R})$ and the minimal unitary extension of V has the form indicated in (iii). Finally, if $[\mathcal{K}, U]$ is a unitary dilation of T as in (iii), we get $\mathcal{R} \subset \mathcal{K}^+ = \bigvee_{n \in \mathbb{Z}_+^2} U^n \mathcal{H}$, so $\mathcal{K}^+ = \bigoplus_{n \in \mathbb{Z}_+^2} U^n \mathcal{R}$. It follows that $[\mathcal{K}^+, (U_0 \mid \mathcal{K}^+, U_1 \mid \mathcal{K}^+)]$ is a bishift-type isometric dilation of T. □

Corollary 2.10. *A bicontraction $[\mathcal{H}, T]$ has a bishift-type regular isometric dilation iff T_0 and T_1 are doubly commuting on $\mathcal{H}$ and $T_i \in C_{\cdot 0}$, $i = 0, 1$.*

Corollary 2.11. *Let $[\mathcal{H}, T]$ be a bicontraction. Then T and T^* have bishift-type isometric dilations iff T_0 and T_1 double commute on $\mathcal{H}$ and $T_i \in C_{00}$, $i = 0, 1$.*

3. Functional model and maximal function for a bicontraction having a $*$-regular dilation

Let us first recall some function theory ingredients. For a separable complex Hilbert space $\mathcal{G}, L^2(\mathbb{T}^2, \mathcal{G})$ denotes the $\mathcal{G}$-valued measurable functions f on the bitorus $\mathbb{T}^2$ for which

$$||f||^2 = \int_{\mathbb{T}^2} ||f(t)||_{\mathcal{G}}^2 \, dm_2(t) < +\infty$$

where m_2 in the normalized Lebesgue measure on $\mathbb{T}^2$. The Hardy space $H^2(\mathbb{T}^2, \mathcal{G})$ is the subspace of all "analytic" functions f of $L^2(\mathbb{T}^2, \mathcal{G})$, i.e. those functions having Fourier coefficients $\hat{f}(m) = 0$, for $m \in \mathbb{Z}^2 \setminus \mathbb{Z}_+^2$. Note that $H^2(\mathbb{T}^2, \mathcal{G})$ is invariant with respect to the unitary pair $Z = (Z_0, Z_1)$, where $Z_j, j = 0, 1$, are the multiplication operators by the coordinate functions on $L^2(\mathbb{T}^2, \mathcal{G})$, and their restrictions to $H^2(\mathbb{T}^2, \mathcal{G})$ give us a shift pair with $\mathcal{G}$ as defect space.

For a commuting isometric (unitary) pair $[\mathcal{H}, V]$, F_V denotes its corresponding semispectral (spectral) measure on $\mathbb{T}^2$, i.e.

$$(V^n h, h) = \int_{\mathbb{T}^2} t^n \, d(F_V(t)h, h)$$

for all $h \in \mathcal{H}$ and $n \in \mathbb{Z}_+^2$ ($n \in \mathbb{Z}^2$). If $[\mathcal{K}, U]$ is an isometric (unitary) dilation of a bicontraction $[\mathcal{H}, T]$, then $F_T(\cdot) = P_{\mathcal{H}} F_U(\cdot) \mid \mathcal{H}$ is called the *semispectral measure of T with respect to the dilation U*. Particularly, the semispectral measure of T with respect to its regular ($*$-regular) isometric dilation will be called *the regular ($*$-regular) semispectral measure of T.*

We first refer to the functional model of bishift-type isometric dilation, introduced by R.E. Curto and F.H. Vasilescu in [3].

Suppose that $\mathcal{H}$ is separable, $[\mathcal{H}, T]$ is a bicontraction having a $*$-regular unitary dilation such that $T_i \in C_{\cdot 0}$, $i = 0, 1$. Let $\mathcal{D}_* = \overline{\Delta_{T^*}^{1/2}\mathcal{H}}$, and $H^2(\mathbb{T}^2, \mathcal{D}_*)$ be the associated Hardy space. If $S_i = Z_i \mid H^2(\mathbb{T}^2, \mathcal{D}_*)$, $i = 0, 1$, then (S_0, S_1) is a shift pair on $H^2(\mathbb{T}^2, \mathcal{D}_*)$. But for $h \in \mathcal{H}$ we have

$$\sum_{q,k=0}^{p} ||\Delta_{T^*}^{1/2} T_0^{*q} T_1^{*k} h||^2 = ||h||^2 - ||T_0^{*(p+1)} h||^2 - ||T_1^{*(p+1)} h||^2$$

$$+ ||T_0^{*(p+1)} T_1^{*(p+1)} h||^2 \to 0 \qquad \text{for } \ p \to \infty,$$

so we can define the operator A from $\mathcal{H}$ into $H^2(\mathbb{T}^2, \mathcal{D}_*)$ by

$$Ah := \sum_{n \in \mathbb{Z}_+^2} S^n \Delta_{T^*}^{1/2} T^{*n} h, \qquad h \in \mathcal{H}. \tag{3.1}$$

Then A is an isometry and for $t = (t_0, t_1) \in \mathbb{T}^2$ and $h \in \mathcal{H}$ we have

$$\begin{aligned}(S_0^* Ah)(t) &= \bar{t}_0 \Big(\sum_{n \in \mathbb{Z}_+^2} t^n \Delta_{T^*}^{1/2} T^{*n} h - \sum_{k=0}^{\infty} t_1^k \Delta_{T^*}^{1/2} T_1^{*k} h \Big) \\ &= \sum_{p=1}^{\infty} \sum_{q=0}^{\infty} t_0^{p-1} t_1^q \Delta_{T^*}^{1/2} T_0^{*p} T_1^{*q} h = \sum_{n \in \mathbb{Z}_+^2} t^n \Delta_{t^*}^{1/2} T^{*n} (T_0^* h) \\ &= (AT_0^* h)(t).\end{aligned}$$

Here we have denoted successively $n = (p, q) \in \mathbb{Z}_+^2$ and $(p-1, q) = n$. So $S_0^* A = AT_0^*$ and by symmetry $S_1^* A = AT_1^*$. It follows that $S^{*n} A = AT^{*n}$ and hence $T^n = A^* S^n A$, for all $n \in \mathbb{Z}_+^2$. Also $\mathcal{D}_* \subset A\mathcal{H}$ implies that $H^2(\mathbb{T}^2, \mathcal{D}_*) = \bigvee_{n \in \mathbb{Z}_+^2} S^n A\mathcal{H}$. Identifying $\mathcal{H}$ with $A\mathcal{H}$ in $H^2(\mathbb{T}^2, \mathcal{D}_*)$, we conclude that $[H^2(\mathbb{T}^2, \mathcal{D}_*), S]$ is a bishift-type isometric dilation of T whose minimal unitary extension is $[L^2(\mathbb{T}^2, \mathcal{D}_*), Z]$. Thus we have proved the existence of a functional model for T contained in the following theorem.

Theorem 3.1. *If $\mathcal{H}$ is separable complex Hilbert space and $[\mathcal{H}, T]$ is a bicontraction having a $*$-regular unitary dilation with $T_i \in C_{\cdot 0}$, then T^* is unitary equivalent to the restriction of the backward bishift on the Hardy space $H^2(\mathbb{T}^2, \mathcal{D}_*)$ where $\mathcal{D}_*$ is the $*$-defect space of T.*

We mention that the operator A defined by (3.1) plays an important role not only in the construction of the functional model for T, but also in the so-called maximal function Θ_T of T. Indeed, for $h \in \mathcal{H}$ the Poisson representation $P[Ah]$ of the function Ah obviously has the Taylor expansion

$$P[Ah](z) = \sum_{n \in \mathbb{Z}_+^2} z^n \Delta_{T^*}^{1/2} T^{*n} h$$

in $\mathbb{D}^2$ which easily leads to

$$P[Ah](z) = \Delta_{T^*}^{1/2}(I - z_0T_0^*)^{-1}(I - z_1T_1^*)^{-1}h, \qquad z = (z_0, z_1) \in \mathbb{D}^2. \tag{3.2}$$

But the function of the right hand side of (3.2) also makes sense without the hypothesis $T_i \in C_{\cdot 0}$, $i = 0, 1$. We shall denote it by $\Theta_T(z)$.

On the other hand, to the $*$-regular semispectral measure F_T we can attach a maximal operator valued outer function $(\mathcal{H}, \mathcal{R}; \Theta)$ as in [8] in the following way. Let $\mathcal{R}$ be the defect space of V given by Proposition 2.4 and $\Phi_{\mathcal{R}}$ the Fourier transform of $\mathcal{K}^+(\mathcal{R}) = \bigoplus_{m\in\mathbb{Z}_+^2} V^m\mathcal{R}$ on $H^2(\mathbb{T}^2, \mathcal{R})$. Define the operator $A_0 : \mathcal{H} \longrightarrow H^2(\mathbb{T}^2, \mathcal{R})$ by $A_0 = \Phi_{\cdot\mathcal{R}} P_{\mathcal{K}^+(\mathcal{R})} \mid \mathcal{H}$. We also define the operator valued analytic function $(\mathcal{H}, \mathcal{R}; \Theta)$ on $\mathbb{D}^2$ by

$$\Theta(z)h = P[A_0h](z), \qquad z \in \mathbb{D}^2,\ h \in \mathcal{H},$$

where $P[\,\cdot\,]$ is the Poisson representation on $\mathbb{D}^2$ (see [9]). Then Θ is an L^2- bounded analytic function on $\mathbb{D}^2$ ([8], [13]), and it is also an outer function in the sense that

$$H^2(\mathbb{T}^2, \mathcal{R}) = \bigvee_{n\in\mathbb{Z}_+^2} Z^n A_0\mathcal{H}.$$

Moreover, the operator A_0 is uniquely determined by Θ, so we can define the ($\mathcal{B}(\mathcal{H})$-valued) semispectral measure F_Θ on $\mathbb{T}^2$ by

$$F_\Theta(\sigma) = A_0^* E_Z(\sigma) A_0, \qquad \sigma \in \mathrm{Bor}\,(\mathbb{T}^2),$$

where E_Z is the spectral measure of Z. By Theorem 2 and Corollary 3 of [8], $F_\Theta \le F_T$ and the function Θ is maximal with this property among the L^2-bounded analytic (on $\mathbb{D}^2$) functions of the form $(\mathcal{H}, \mathcal{R}'; \Theta')$.

We are now able to give the analytic expression of $(\mathcal{H}, \mathcal{R}; \Theta)$, up to a unitary equivalence.

Theorem 3.2. *Let $[\mathcal{H}, T]$ be a bicontraction having a $*$-regular isometric dilation. Then the maximal analytic function $(\mathcal{H}, \mathcal{R}; \Theta)$ coincides (in the sense of B. Sz.-Nagy and C. Foiaş [14]) with the analytic function $(\mathcal{H}, \mathcal{D}_*; \Theta_T)$ on $\mathbb{D}^2$ given by the right hand side of* (3.2).

Proof. Let $[\widetilde{\mathcal{K}}, U]$ be the minimal unitary extension of the $*$-regular isometric dilation $[\mathcal{K}, V]$ of T. Since $\mathcal{R} = \overline{\Delta_{V^*}\mathcal{H}}$, $\mathcal{L}_* = U_0^*U_1^*\mathcal{R}$ is the wandering subspace corresponding to T^* by Theorem 2.1. So we can define the unitary operator ω_* from $\mathcal{L}_*$ to $\mathcal{D}_*$ by

$$\omega_*(U_0^*U_1^* - U_0^*T_1^* - U_1^*T_0^* + T_0^*T_1^*)h = \Delta_{T^*}^{1/2}h, \qquad h \in \mathcal{H}.$$

Then $\omega_* U_0^*U_1^*$ is a unitary operator from $\mathcal{R}$ onto $\mathcal{D}_*$. Next we prove that $\Theta_T(z) = \omega_* U_0^*U_1^*\Theta(z)$, $z \in \mathbb{D}^2$, so that Θ coincides with Θ_T in the sense of [14].

Let $\{\Theta_n\}_{n\in\mathbb{Z}_+^2} \subset \mathcal{B}(\mathcal{H},\mathcal{R})$ be the sequence of the Taylor coefficients of the maximal analytic operator function Θ. By the definition of Θ, for $h \in \mathcal{H}$ the Taylor coefficients of $\Theta(\cdot)h \in H^2(\mathbb{D}^2,\mathcal{R})$ are the Fourier coefficients of the function $A_0h \in H^2(\mathbb{T}^2,\mathcal{R})$. So for $r \in \mathcal{R}$ and $n \in \mathbb{Z}_+^2$ we have

$$\begin{aligned}(\Theta_n h, r) &= \int_{\mathbb{T}^2} \bar{t}^n((A_0h)(t), r)\, dm_2(t) = \int_{\mathbb{T}^2}((A_0h)(t), t^n r)\, dm_2(t)\\ &= (A_0h, \Phi_{\mathcal{R}}V^n r)_{H^2(\mathbb{T}^2,\mathcal{R})} = (\Phi_{\mathcal{R}}P_{\mathcal{K}+(\mathcal{R})}h, \Phi_{\mathcal{R}}V^n r)\\ &= (h, V^n r) = (T^{*n}h, r).\end{aligned}$$

Then for $h, h' \in \mathcal{H}$ and $n = (p,q) \in \mathbb{Z}_+^2$ we have

$$\begin{aligned}&(\Theta_n h, (I - V_0T_0^* - V_1T_1^* + V_0V_1T_0^*T_1^*)h')\\ &= (T^{*n}h, h') - (T^{*n}h, T_0T_0^*h') - (T^{*n}h, T_1T_1^*h') + (T^{*n}h, T_0T_1T_0^*T_1^*h')\\ &= (h, T^n h') - (h, T_0^{p+1}T_1^qT_0^*h') - (h, T_0^pT_1^{q+1}T_1^*h') + (h, T_0^{p+1}T_1^{q+1}T_0^*T_1^*h').\end{aligned}$$

We have to show that

$$\Theta_n = U_0U_1\omega_*^{-1}\Delta_{T^*}^{1/2}T^{*n} = (I - V_0T_0^* - V_1T_1^* + V_0V_1T_0^*T_1^*)T^{*n}.$$

This follows from the above calculation and also from the following. For $h, h' \in \mathcal{H}$ and $n = (p,q) \in \mathbb{Z}_+^2$ we have (using (1.2))

$$\begin{aligned}&((I - V_0T_0^* - V_1T_1^* + V_0V_1T_0^*T_1^*)T^{*n}h, (I - V_0T_0^* - V_1T_1^* + V_0V_1T_0^*T_1^*)h')\\ &= (T^{*n}h, h') - (T^{*n}h, T_0T_0^*h') - (T^{*n}h, T_1T_1^*h') + (T^{*n}h, T_0T_1T_0^*T_1^*h')\\ &\quad -(T_0T_0^{*(p+1)}T_1^{*q}h, h') + (T_0^{*(p+1)}T_1^{*q}h, T_0^*h') + (T_1^*T_0^{*(p+1)}T_1^{*q}h, T_0^*T_1^*h')\\ &\quad -(T_0^{*(p+1)}T^{*q}h, T_1T_0^*T_1^*h') - (T_1T_0^{*p}T_1^{*(q+1)}h, h') + (T_0^*T_1^*T_0^{*p}T_1^{*q}h, T_1^*T_0^*h')\\ &\quad +(T_0^{*p}T_1^{*(q+1)}h, T_1^*h') - (T_0^{*p}T_1^{*(q+1)}h, T_0T_0^*T_1^*h') + (T_0T_1T_0^{*(p+1)}T_1^{*(q+1)}h, h')\\ &\quad -(T_1T_0^{*(p+1)}T_1^{*(q+1)}h, T_0^*h') - (T_0T_0^{*(p+1)}T_1^{*(q+1)}h, T_1^*h')\\ &\quad +(T^{*(p+1)}T_1^{*(q+1)}h, T_0^*T_1^*h')\\ &= (h, T^n h') - (h, T_0^{p+1}T_1^qT_0^*h') - (h, T_0^pT_1^{q+1}T_1^*h') + (h, T_0^{p+1}T_1^{q+1}T_0^*T_1^*h').\end{aligned}$$

This completes the proof. □

Regarding the maximality property of $(\mathcal{H},\mathcal{R};\Theta)$ it makes sense to say that the analytic operator function $(\mathcal{H},\mathcal{D}_*;\Theta_T)$ defined by (3.2) is the *maximal function* of the bicontraction $T = (T_0, T_1)$ (see also [13]).

Remark 3.3. Recall from [6] that $\Delta[F_T] = \Theta_T(0)^*\Theta_T(0)$ is the Szegő operator of the semispectral measure F_T. From Theorem 1 of [8] we have $\Delta[F_T] = 0$ if $\mathcal{R} = \{0\}$, or, equivalently, T is a bidisc coisometry on $\mathcal{H}$.

Remark 3.4. If $T_i \in C_{\cdot 0}$, $i = 0,1$, then, as it was shown in formula (3.2), we have

$$\Theta_T(z)h = P[Ah](z), \qquad z \in \mathbb{D}^2,\ h \in \mathcal{H}.$$

On the other hand, from Proposition 2.9 it follows $T_i \in C_{\cdot 0}$, $i = 0,1$ iff $P_{\mathcal{R}}\mathcal{H} = \mathcal{R}_i = \{0\}$, $i = 0,1$, or equivalently, $F_T = F_\Theta$. In this case (see Theorem 6 in [5] or Lemma 2 in [2]), F_T is mutually absolutely continuous with respect to the Lebesgue measure m_2 on $\mathbb{T}^2$. Moreover, we have:

Proposition 3.5. *For a bicontraction $[\mathcal{H}, T]$ having a $*$-regular isometric dilation, the following assertions are equivalent:*

(i) *The $*$-regular semispectral measure F_T of T is boundedly dominated by m_2 on $\mathbb{T}^2$;*
(ii) *The maximal function $(\mathcal{H}, \mathcal{D}_*; \Theta_T)$ is bounded on $\mathbb{D}^2$;*
(iii) *The spectra of $T_i, i = 0,1$, are contained in $\mathbb{D}$.*

Proof. The equivalence (i) $\Leftrightarrow$ (ii) follows from Proposition 2 from [13]. Next, from (i) we deduce that the semispectral measures of $T_i, i = 0,1$, are boundedly dominated by the Lebesgue measure on $\mathbb{T}$, which means (iii) (see [12]). Therefore (i) $\Rightarrow$ (iii). Now supposing (iii), we have $r(T_i) < 1$, $i = 0,1$, where $r(\cdot)$ is the spectral radius in $\mathcal{B}(\mathcal{H})$. For $n \in \mathbb{Z}^2$ set $T(n) = T^{n_+}T^{*n_-}$ and as usual we put $T_i^{(p)} = T_i^p$ if $p \geq 0$ and $T_i^{(p)} = T_i^{*(-p)}$ if $p < 0$. Then for $n = (p,q) \in \mathbb{Z}^2$, $||T(n)|| \leq ||T_0^{(p)}|| \cdot ||T_1^{(q)}||$, so we get

$$\sum_{n\in\mathbb{Z}^2} ||T(n)|| \leq \sum_{p=-\infty}^{\infty} ||T_0^{(p)}|| \sum_{q=-\infty}^{\infty} ||T_1^{(q)}|| < \infty.$$

Thus we can define the function $Q_T : \mathbb{T}^2 \longrightarrow \mathcal{B}(\mathcal{H})$ by

$$Q_T(t) = \sum_{n\in\mathbb{Z}^2} \bar{t}^n T(n), \qquad t \in \mathbb{T}^2, \tag{3.3}$$

the series being uniformly convergent (in norm) on $\mathbb{T}^2$. Moreover, Q_T is a norm-continuous function on $\mathbb{T}^2$ and we have

$$\int_{\mathbb{T}^2} t^n (Q_T(t)h, h)\, dm_2(t) = (T(n)h, h) = \int_{\mathbb{T}^2} t^n d(F_T(t)h, h)$$

for $n \in \mathbb{Z}^2$ and $h \in \mathcal{H}$. It follows that

$$(F_T(\sigma)h, h) \leq \sup_{t\in\mathbb{T}^2} ||Q_T(t)|| \cdot ||h||^2 m_2(\sigma), \qquad \sigma \in \mathrm{Bor}\,(\mathbb{T}^2),\ h \in \mathcal{H},$$

so F_T is boundedly dominated by m_2 on $\mathbb{T}^2$. □

Proposition 3.5 and Theorem 5 from [8] give:

Corollary 3.6. *Let $[\mathcal{H},T]$ be a bicontraction having $*$-regular isometric dilation. Then the maximal function $(\mathcal{H},\mathcal{D}_*;\Theta_T)$ is bounded and has a bounded inverse on $\mathbb{D}^2$ iff there exists a constant $c \geq 1$ such that*

$$\frac{1}{c}m_2(\sigma)||h||^2 \leq (F_T(\sigma)h,h) \leq cm_2(\sigma)||h||^2$$

for all $\sigma \in \operatorname{Bor}(\mathbb{T}^2)$ and $h \in \mathcal{H}$. In this case we have $\dim \mathcal{D}_* = \dim \mathcal{H}$.

By the conditions of Proposition 3.5 we have shown in fact that the $*$-regular semispectral measure of T is

$$dF_T(t) = Q_T(t)dm_2(t), \qquad t \in \mathbb{T}^2,$$

where the function Q_T is defined in (3.3). Note that because of $r(T_i) < 1, i = 0,1$, the series which defines the function Q_T is uniformly convergent on $\overline{\mathbb{D}}^2$, so Q_T has a (norm) continuous extension to $\overline{\mathbb{D}}^2$.

Recall from [3] that if the Taylor spectrum $\sigma(T^*)$ of T^* is contained in $\mathbb{D}^2$, then by using $I := (I,I)$ and the multiindex notation, the *operator Poisson kernel* of T^* is the function $P(T^*;\,\cdot\,) : \mathbb{D}^2 \longrightarrow \mathcal{B}(\mathcal{H})$ given by

$$P(T^*;z) = (I - zT)^{-1}\Delta_{T^*}(I - \overline{z}T^*)^{-1}.$$

It is clear that

$$P(T^*;z) = \Theta_T(\overline{z})^*\Theta_T(\overline{z}), \qquad z \in \overline{\mathbb{D}}^2.$$

Also, if we denote by E the spectral measure of the regular unitary dilation of T^* and by $P(z,t)$ the Poisson kernel on $\overline{\mathbb{D}}^2 \times \mathbb{T}^2$, then for every $f \in C(\mathbb{T}^2)$ and $h \in \mathcal{H}$ we have

$$\begin{aligned}
\int_{\mathbb{T}^2} f(t)||\Theta_T(\overline{t})h||^2 dm_2(t) &= \lim_{r\to 1^-}\int_{\mathbb{T}^2} f(t)||\Theta_{rT}(\overline{t})h||^2 dm_2(t)\\
&= \lim_{r\to 1^-}\int_{\mathbb{T}^2} f(t)(P(rT^*;t)h,h)dm_2(t)\\
&= \lim_{r\to 1^-}\int_{\mathbb{T}^2} f(t)\Big[\int_{\mathbb{T}^2} P(rw,t)d(E(w)h,h)\Big]dm_2(t)\\
&= \lim_{r\to 1^-}\int_{\mathbb{T}^2}\Big[\int_{\mathbb{T}^2} f(t)P(rw,t)dm_2(t)\Big]d(E(w)h,h)\\
&= \lim_{r\to 1^-}\int_{\mathbb{T}^2} P[f](rw)d(E(w)h,h)\\
&= \int_{\mathbb{T}^2} f(w)d(F_{T^*}(w)h,h),
\end{aligned}$$

where the last equality results from the fact that the (continuous) extension of the Poisson transform $P[f]$ coincides with f on $\mathbb{T}^2$ (see [9]). Thus we get:

Proposition 3.7. *Let $[\mathcal{H}, T]$ be a bicontraction having a regular isometric dilation and $\sigma(T_i) \subset \mathbb{D}$, $i = 0, 1$. Then the regular semispectral measure of T is*

$$dF_T(t) = \Theta_{T^*}(\bar{t})^*\Theta_{T^*}(\bar{t})dm_2(t) = P(T;t)dm_2(t), \qquad t \in \mathbb{T}^2.$$

Finally, if for a contraction B on $\mathcal{H}$ we set $D_B = (I - B^*B)^{\frac{1}{2}}$ the defect operator $\Theta_B(z) = D_{B^*}(I - zB^*)^{-1}$, $z \in \mathbb{D}$, the maximal function and $P(B;z) = \Theta_{B^*}(\bar{z})^*\Theta_{B^*}(\bar{z})$, $z \in \overline{\mathbb{D}}$, the Poisson kernel of B, from Proposition 3.7 we obtain:

Corollary 3.8. *Let T_0 and T_1 be doubly commuting contractions on $\mathcal{H}$, with $\sigma(T_i) \subset \mathbb{D}$, $i = 0, 1$. Then the regular (and $*$-regular) semispectral measure of $T = (T_0, T_1)$ is*

$$dF_T(t) = \Theta_{T_0^*}(\bar{t})^*\Theta_{T_0^*}(\bar{t})\Theta_{T_1^*}(\bar{t})^*\Theta_{T_1^*}(\bar{t})dm_2(t) = P(T_0, t_0)P(T_1, t_1)dm_2(t),$$

for $t = (t_0, t_1) \in \mathbb{T}^2$.

Remark 3.9. If we regard $\lambda \in \mathbb{D}^2$ as a bicontraction on $\mathbb{C}$, then its regular semispectral measure is the absolutely continuous representing measure (with respect to the bidisc algebra $A(\mathbb{D}^2)$) of λ, having its Radon-Nikodym derivative given by the Poisson kernel corresponding to λ :

$$dm_\lambda = P(\lambda, \cdot\,)dm_2.$$

Acknowledgements

The work of the first named author on this research was finished during a visit at the Universität des Saarlandes, Saarbrücken, as an Alexander von Humboldt-Fellow. He wishes to thank Professor Ernst Albrecht for hospitality.

References

[1] ANDO, T.: On a pair commutative contractions; Acta Sci. Math. 24 (1963), 88–90.

[2] ATHAVALE, A.: Relating the normal extension and the regular unitary dilation of a subnormal tuple of contractions; preprint.

[3] CURTO, R.E., VASILESCU, F.H.: Standard operator models in the polydisc; Indiana Univ. Math. J. 42 (1993), 791–810.

[4] GAŞPAR, D., SUCIU, N.: On the structure of isometric semigroups; Spectral Theory of Linear Operators and Related Topics, Birkhäuser Verlag, Basel 1984, 125–139.

[5] GAŞPAR, D., SUCIU, N.: On Wold decompositions of isometric semigroups; Approximation Theory and Functional Analysis, Birkhäuser Verlag, Basel 1984, 99–108.

[6] GAŞPAR, D., SUCIU, N.: Analyticity of the semispectral measures on the bitorus and the Szegö operator; J. Operator Theory 24 (1990), 3–16.

[7] GAŞPAR, D., SUCIU, N.: On the intertwining of regular dilations; Ann. Polon. Math. LXVI (1997), 105–121.

[8] GAŞPAR, D., SUCIU, N., VALUŞESCU, I.: Semispectral measures on the bitorus and Wold-type functions on the bidisc; Rev. Roumaine Math. Pures Appl. 38:3 (1993), 253–266.

[9] RUDIN, W.: Function theory in polydiscs; Benjamin Inc., New York Amsterdam 1969.

[10] SŁOCINSKI, M.: Isometric dilations of doubly commuting contractions and related models; Bull. Pol. Acad. Sci. Math. 25:12 (1977), 1233–1242.

[11] SŁOCINSKI, M.: On the Wold type decomposition of a pair of commuting isometries; Ann. Polon. Math. 37 (1980), 255–262.

[12] SCHREIBER, M.: Absolutely continuous operators; Duke Math. J. 29 (1962), 175–190.

[13] SUCIU, N., VALUŞESCU, I.: The maximal function of doubly commuting contractions; Operator Theory: Adv. Appl. 2 Birkhäuser Verlag, Basel 1982, 295–309.

[14] SZŐKEFALVY-NAGY, B., FOIAŞ, C.: Harmonic analysis of operators on Hilbert spaces; North Holland, Amsterdam 1970.

[15] VASILESCU, F.H.: On operator-valued Poisson kernel; J. Funct. Anal. 110:1 (1992), 47–72.

Department of Mathematics
West University of Timişoara
Bul. V. Pârvan 4
1900 Timişoara
Romania
gaspar@tim1.uvt.ro

Department of Mathematics
West University of Timişoara
Bul. V. Pârvan 4
1900 Timişoara
Romania

1991 Mathematics Subject Classification: Primary 47A20; Secondary 47A60

Operator Theory:
Advances and Applications, Vol. 103

On generalized interpolation and shift invariant maximal semidefinite subspaces

A. Gheondea

We consider the bitangential Nevanlinna-Pick problem for meromorphic matrix functions with upper bounded total pole multiplicity. We follow the approach of J.A. Ball and J.W. Helton to view this problem as a shift invariant maximal semi-definite subspace problem in a space with indefinite inner product. The abstract problem to be solved is a slightly modified version of the generalized interpolation problem for matrix valued functions, considered by Ball and Helton. Using a generalization of a theorem of I.S. Iokhvidov on the existence of invariant maximal semi-definite subspaces, the meromorphic case of the generalized interpolation problem is solved, and it is shown that this contains the bitangential Nevanlinna-Pick problem with poles.

1. Introduction

This paper concerns generalized interpolation problems for meromorphic matrix valued functions with upper bounded total pole multiplicity. This kind of problem appeared implicitly in the work of V.M. Adamyan, D.Z. Arov and M.G. Kreĭn [1] and in an explicit way in the work of M.G. Kreĭn and H. Langer [12]. The approach that we follow comes from a paper of J.A. Ball and J.W. Helton [2], using an older idea of D. Sarason [13], in which they put the bitangential Nevanlinna-Pick problem for meromorphic matrix valued functions into the framework of a generalized interpolation problem. The approach in [2] is based on an elegant reformulation of this problem, via the angular operator method, as a shift invariant maximal semi-definite subspace problem. For the existence and the "parametrization" of the set of solutions they use an indefinite variant of the Beurling-Lax theorem. Some later closer look at that indefinite variant of the theorem of Beurling-Lax type showed that only the rational case is covered, see the comments and corrections in [3].

We slightly modify the generalized interpolation problem as formulated in [2] and then, following the above mentioned approach, we prove that the test of solvability, in terms of the number of negative squares of the defect of an associated angular operator, is true in the meromorphic context, cf. Theorem 3.9. The idea of the proof uses, of course, much of the geometry of inner product spaces as angular operators and pseudo-regularity and, since this is a subtle point where serious obstructions prevent us from obtaining the full generality as stated in [2], we explain with details all the necessary ingredients in Section 2. For the existence of shift invariant maximal semi-definite subspaces in a certain indefinite inner product space we use an improved version, cf. S. Treil and A. Volberg [15], of a theorem of I.S. Iokvidov obtained via a fixed point theorem of Fan-Glicksberg, cf. [5], [10].

One good quality of the proof of Theorem 3.9 we see is that it does not assume the invertibility condition on the "generalized" Pick matrix, a condition which seems to appear frequently in other approaches of similar problems. Unfortunately, we do not have a proof for the test of the solvability of the generalized interpolation problem in its full generality as formulated by Ball and Helton in [2].

2. Preliminaries

2.1. Subspaces of a Kreĭn space. A *Kreĭn space* is, by definition, a complex vector space $\mathcal{K}$ endowed with an (indefinite) inner product $[\cdot,\cdot]$ such that $\mathcal{K} = \mathcal{K}^- \dot{+} \mathcal{K}^+$, $[\mathcal{K}^-, \mathcal{K}^+] = 0$ and the subspaces $(\mathcal{K}^-, -[\cdot,\cdot])$ and $(\mathcal{K}^+, [\cdot,\cdot])$ are Hilbert spaces. Such a decomposition is denoted by $\mathcal{K} = \mathcal{K}^-[+]\mathcal{K}^+$, and it is called a *fundamental decomposition.* Associated with it is a *fundamental symmetry* $J : \mathcal{K} \to \mathcal{K}$, $J(x^+ + x^-) = x^+ - x^-$ for all $x^\pm \in \mathcal{K}^\pm$, and a corresponding positive definite inner product $\langle\cdot,\cdot\rangle_J$,

$$\langle x, y\rangle_J = [Jx, y], \qquad x, y \in \mathcal{K}.$$

The *strong topology* is defined by the norm $\|x\| = \langle x, x\rangle_J^{1/2}$ and this topology does not depend on the fundamental symmetry. The *negative/positive signatures* of the Kreĭn space $\mathcal{K}$ are the cardinal numbers $\kappa^\pm(\mathcal{K}) = \dim(\mathcal{K}^\pm)$ and they do not depend on the chosen fundamental decomposition. In this paper *we will consider only separable Kreĭn spaces* and hence all the signatures will be either positive integer numbers or the symbol ∞.

Let $\mathcal{M}$ be a *subspace* of the Kreĭn space $\mathcal{K}$, that is, $\mathcal{M}$ is a closed linear manifold in $\mathcal{K}$. Then, the orthogonal subspace associated with $\mathcal{M}$ is by definition $\mathcal{M}^\perp = \{y \in \mathcal{K} \mid [x, y] = 0,\ x \in \mathcal{M}\}$. The subspace $\mathcal{M}^0 = \mathcal{M} \cap \mathcal{M}^\perp$ is called the *isotropic subspace* of $\mathcal{M}$ and has the property that the inner product $[\cdot,\cdot]$ vanishes on it.

For two subspaces $\mathcal{A}$ and $\mathcal{B}$ of a Kreĭn space, the notation $\mathcal{A}[+]\mathcal{B}$ is used for the algebraic sum $\mathcal{A} + \mathcal{B}$ if and only if this sum is closed, $\mathcal{A} \perp \mathcal{B}$, and $\mathcal{A} \cap \mathcal{B} = 0$.

A subspace $\mathcal{M}$ is called *nonpositive* (*negative, uniformly negative*) if $[x, x] \leq 0$ for all $x \in \mathcal{M}$ (if $[x, x] < 0$ for all $x \in \mathcal{M} \setminus \{0\}$, if there exists $\delta > 0$ such that $[x, x] \leq \delta\langle x, x\rangle_J$ for all $x \in \mathcal{M}$, respectively). A nonpositive (negative, uniformly negative) subspace $\mathcal{M}$ can always be extended to a maximal nonpositive (maximal negative, maximal uniformly negative, respectively) subspace.

Similarly one defines *nonnegative, positive* and *uniformly positive* subpaces and the corresponding maximal extension property holds. Moreover, maximal uniformly positive/negative subspaces coincide exactly with fundamental decompositions in the sense that a subspace $\mathcal{M}$ is maximal uniformly positive/negative if and only there exists a fundamental decomposition $\mathcal{K} = \mathcal{K}^-[+]\mathcal{K}^+$ such that $\mathcal{K}^+ = \mathcal{M}$, $\mathcal{K}^- = \mathcal{M}$, respectively.

A very useful notion in dealing with semidefinite subspaces is that of an *angular operator*. Let $\mathcal{M}$ be a nonpositive subspace of the Kreĭn space $\mathcal{K}$ in which a fundamental decomposition $\mathcal{K} = \mathcal{K}^-[+]\mathcal{K}^+$ is fixed. If J is the corresponding fundamental symmetry then $J = J^+ - J^-$ where $J^\pm$ is the projection of $\mathcal{K}$ onto $\mathcal{K}^\pm$ along $\mathcal{K}^\mp$. Since $\mathcal{M}$ is nonpositive, we have

$$0 \geq [x,x] = \langle J^+x, J^+x\rangle_J - \langle J^-x, J^-x\rangle_J, \qquad x \in \mathcal{M},$$

and hence $J^-\mathcal{M}$ is a closed subspace of the Hilbert space $(\mathcal{K}^-, \langle\cdot,\cdot\rangle)$ and the operator $K_\mathcal{M} : J^-\mathcal{M} \to \mathcal{K}^+$ defined by $K_\mathcal{M}J^-x = J^+x$ for all $x \in \mathcal{M}$ is correctly defined and contractive. The subpace $\mathcal{M}$ is recovered from $K_\mathcal{M}$ as graph, that is,

$$\mathcal{M} = \{x + K_\mathcal{M}x \mid x \in J^-\mathcal{M}\}.$$

Let $\mathcal{M}$ be a subspace of a Kreĭn space $\mathcal{K}$. Then there exist a positive subspace $\mathcal{M}^+$ and a negative subspace $\mathcal{M}^-$ such that

$$\mathcal{M} = \mathcal{M}^-[+]\mathcal{M}^0[+]\mathcal{M}^+.$$

Such a decomposition is also called a *fundamental decomposition* of the subspace $\mathcal{M}$. The *negative/positive signature* of $\mathcal{M}$ is by definition $\kappa^\mp(\mathcal{M}) = \dim(\mathcal{M}^\mp)$. As an application of the Open Mapping Principle it can be proved that these notions do not depend on the chosen fundamental decomposition.

We start with a first application of the angular operator method.

Lemma 2.1. *Given a nonpositive subspace $\mathcal{M}$ of $\mathcal{K}$, the codimension of $\mathcal{M}$ in any maximal nonpositive extension $\widetilde{\mathcal{M}}$ coincides with the negative signature of $\mathcal{M}^\perp$.*

Proof. Fix a fundamental decomposition $\mathcal{K} = \mathcal{K}^-[+]\mathcal{K}^-$ and let J be the corresponding fundamental symmetry. Since $\mathcal{M}$ is nonpositive we consider its angular operator $T : J^-\mathcal{M} \to \mathcal{K}^+$, such that T is a contraction, that is, $\langle Tx, Tx\rangle \leq \langle x, x\rangle$ for all $x \in J^-\mathcal{M}$, where $\langle\cdot,\cdot\rangle$ is the positive definite inner product on $\mathcal{K}$ associated with the fixed fundamental symmetry J, and

$$\mathcal{M} = \{Tx + x \mid x \in J^-\mathcal{M}\}. \tag{2.1}$$

We first prove that

$$\mathcal{M}^\perp = \{y + T^*y \mid y \in \mathcal{K}^+\}[+](\mathcal{K}^- \ominus J^-\mathcal{M}). \tag{2.2}$$

Indeed, let $y = y_+ + y_-$ be an arbitrary vector in $\mathcal{K}$, with $y_\pm \in \mathcal{K}^\pm$ and let $Tx + x$ be an arbitrary vector in $\mathcal{M}$, with $x \in J^-\mathcal{M}$. Then $y \in \mathcal{M}^\perp$ if and only if

$$\langle Tx, y_+\rangle = \langle x, y_-\rangle, \qquad x \in J^-\mathcal{M}. \tag{2.3}$$

Then the inclusion $\supseteq$ in (2.2) follows immediately. To prove the converse inclusion we represent the negative component of the arbitrary vector $y \in \mathcal{M}^\perp$ as $y_- = y_{-1} + y_{-2}$ where $y_{-1} \in J^-\mathcal{M}$ and $y_{-2} \in \mathcal{K}^- \ominus J^-\mathcal{M}$. Then (2.3) becomes

$$\langle x, T^*y_+\rangle = \langle x, y_{-1}\rangle, \qquad x \in J^-\mathcal{M},$$

and hence $y_{-1} = T^*y_+$ and y_{-2} is arbitrary in $\mathcal{K}^- \ominus J^-\mathcal{M}$. This proves (2.2).

Further, let $\widetilde{\mathcal{M}}$ be an arbitrary maximal nonpositive subspace of $\mathcal{K}$ such that $\mathcal{M} \subseteq \widetilde{\mathcal{M}}$ and let $\widetilde{T} : \mathcal{K}^- \to \mathcal{K}^+$ be the angular operator of $\widetilde{\mathcal{M}}$. Note that due to the maximality of the subspace $\widetilde{\mathcal{M}}$ the angular operator $\widetilde{T}$ is defined everywhere on $\mathcal{K}^-$. Then, with respect to the decomposition $\mathcal{K}^- = J^-\mathcal{M} \oplus (\mathcal{K}^- \ominus J^-\mathcal{M})$, we have

$$\widetilde{T} = [T \quad X],$$

where $X \in \mathcal{L}(\mathcal{K}^- \ominus J^-\mathcal{M}, \mathcal{K}^+)$. Then we consider the subspace

$$\mathcal{N} = \{Xx + x \mid x \in \mathcal{K}^- \ominus J^-\mathcal{M}\},$$

and from here it follows that $\widetilde{\mathcal{M}} = \mathcal{M} \dot{+} \mathcal{N}$. Therefore, $\operatorname{codim}_{\widetilde{\mathcal{M}}} \mathcal{M} = \dim(\mathcal{K}^- \ominus J^-\mathcal{M})$. Since from (2.2) we have $\kappa^-(\mathcal{M}^\perp) = \dim(\mathcal{K}^- \ominus J^-\mathcal{M})$ it follows that $\operatorname{codim}_{\widetilde{\mathcal{M}}} \mathcal{M} = \kappa^-(\mathcal{M}^\perp)$. □

In general, given a subspace $\mathcal{M}$ of the Kreĭn space $\mathcal{K}$ it is not true that $\mathcal{K} = \mathcal{M} \dot{+} \mathcal{M}^\perp$. One obstruction is the presence of the isotropic part $\mathcal{M}^0$, more precisely, it holds $(\mathcal{M}^0)^\perp = \operatorname{clos}(\mathcal{M} + \mathcal{M}^\perp)$. But even if the subspace $\mathcal{M}$ is *nondegenerate*, that is, $\mathcal{M}^0 = 0$, all we have in this case is that $\mathcal{M} \dot{+} \mathcal{M}^\perp$ is dense in $\mathcal{M}$. A subspace $\mathcal{M}$ is called *regular* (or a *Kreĭn subspace*) if $\mathcal{K} = \mathcal{M}[+]\mathcal{M}^\perp$. An equivalent characterization of the regularity of a subspace $\mathcal{M}$ is the existence of a decomposition $\mathcal{M} = \mathcal{M}_-[+]\mathcal{M}_+$, where $\mathcal{M}_\pm$ is a uniformly positive/negative subspace.

Also, given a subspace $\mathcal{M}$ of a Kreĭn space $\mathcal{K}$, it is not true, in general, that the algebraic sum $\mathcal{M} + \mathcal{M}^\perp$ is closed. The subspace $\mathcal{M}$ is called *pseudo-regular* if the algebraic sum $\mathcal{M} + \mathcal{M}^\perp$ is closed. The following proposition can be found in [8].

Proposition 2.2. *Let $\mathcal{M}$ be a subspace of the Kreĭn space $\mathcal{K}$. The following assertions are equivalent:*

(a) *$\mathcal{M}$ is pseudo-regular.*

(b) *$\mathcal{M} + \mathcal{M}^\perp = \mathcal{M}^{0\perp}$.*

(c) *There exists a regular subspace $\mathcal{R}$ such that $\mathcal{M} = \mathcal{M}^0[+]\mathcal{R}$.*

(d) *If $\mathcal{L}$ is a subspace of $\mathcal{K}$ such that $\mathcal{M} = \mathcal{M}^0[+]\mathcal{L}$ then $\mathcal{L}$ is regular.*

(e) *There exists a regular subspace $\mathcal{S} \supseteq \mathcal{M}$ such that $\mathcal{S} \cap \mathcal{M}^\perp = \mathcal{M}^0$.*

Morever, the subspaces $\mathcal{R}$ from (c) *and $\mathcal{S}$ from* (e) *are uniquely determined, modulo an isometric isomorphism.*

Lemma 2.3.[2] *Let $\mathcal{M}$ be a pseudo-regular subspace of the Kreĭn space $\mathcal{K}$, and let $\mathcal{L} \subseteq \mathcal{M}$ be an $\mathcal{M}$-maximal nonpositive subspace. Then $\kappa^-(\mathcal{L}^\perp) = \kappa^-(\mathcal{M}^\perp)$.*

Proof. If $\mathcal{M}$ is pseudo-regular, then so is $\mathcal{M}^\perp$, and hence by Proposition 2.2 we have

$$\mathcal{M}^\perp = \mathcal{M}'_-[+]\mathcal{M}^0[+]\mathcal{M}'_+,$$

where $\mathcal{M}'_\pm$ is a uniformly positive/negative subspace.

Let $\mathcal{L} \subseteq \mathcal{M}$ be an $\mathcal{M}$-maximal nonpositive subspace. Then $\mathcal{M} \cap \mathcal{L}^{\perp}$ is a nonnegative subspace.
By Lemma 2.1, in order to prove $\kappa^{-}(\mathcal{L}^{\perp}) = \kappa^{-}(\mathcal{M}^{\perp})$ it is sufficient to prove that $\mathcal{L}\dot{+}\mathcal{N}$ is $\mathcal{K}$-maximal nonpositive, for some nonnpositive subspace $\mathcal{N}$ with $\dim(\mathcal{N}) = \kappa^{-}(\mathcal{M}^{\perp})$. In the following we will prove that $\mathcal{L}[+]\mathcal{M}'_{-}$ is $\mathcal{K}$-maximal nonpositive.
Indeed, $\mathcal{L} \perp \mathcal{M}'_{-}$ since $\mathcal{L} \subseteq \mathcal{M}$ and $\mathcal{M}'_{-} \subseteq \mathcal{M}^{\perp}$. Since $\mathcal{M}'_{-}$ is uniformly negative, it follows that $\mathcal{L} + \mathcal{M}'_{-}$ is closed and nonpositive. Therefore, again by Lemma 2.1, in order to prove that $\mathcal{L}[+]\mathcal{M}'_{-}$ is $\mathcal{K}$-maximal nonpositive, it is necessary and sufficient to prove that its orthogonal $\mathcal{L}^{\perp} \cap \mathcal{M}'^{\perp}_{-}$ is nonnegative. To this end let us first note that since $\mathcal{M}$ is $\mathcal{M}$-maximal nonpositive, we have $\mathcal{M}^{0} \subseteq \mathcal{L}$ and hence $\mathcal{L}^{\perp} \subseteq \mathcal{M}^{0\perp} = \mathcal{M} + \mathcal{M}^{\perp}$. Therefore, taking into account that $\mathcal{M}^{\perp} \subseteq \mathcal{L}^{\perp}$ we have

$$\mathcal{L}^{\perp} = \mathcal{L}^{\perp} \cap (\mathcal{M} + \mathcal{M}^{\perp}) = (\mathcal{L}^{\perp} \cap \mathcal{M}) + \mathcal{M}^{\perp}. \tag{2.4}$$

On the other hand, it is easy to see that

$$\mathcal{M}'^{\perp}_{-} \cap (\mathcal{M} + \mathcal{M}^{\perp}) = \mathcal{M}[+]\mathcal{M}'_{+}. \tag{2.5}$$

Since $\mathcal{M}^{0} \subseteq \mathcal{L} + \mathcal{M}'_{-}$ it follows $(\mathcal{L} + \mathcal{M}'_{-})^{\perp} \subseteq \mathcal{M}^{0\perp} = \mathcal{M} + \mathcal{M}^{\perp}$. Therefore, using (2.4) and (2.5) we get

$$(\mathcal{L} + \mathcal{M}'_{-})^{\perp} = \mathcal{L}^{\perp} \cap \mathcal{M}'^{\perp}_{-} \cap (\mathcal{M} + \mathcal{M}^{\perp}) = (\mathcal{L}^{\perp} \cap \mathcal{M})[+]\mathcal{M}'_{-}.$$

Since the subspace $\mathcal{L}^{\perp} \cap \mathcal{M}$ is nonnegative and the subspace $\mathcal{M}'_{+}$ is uniformly positive, it follows that the subspace $(\mathcal{L} + \mathcal{M}'_{-})^{\perp}$ is nonnegative. □

2.2 Generalized angular operators. Let $\mathcal{H}$ be a subspace of a Kreĭn space $(\mathcal{K}, [\cdot, \cdot])$ and for some fixed admissible positive definite inner product $\langle \cdot, \cdot \rangle$ on $\mathcal{K}$ let $G \in \mathcal{L}(\mathcal{H})$ be the *Gram operator* of $\mathcal{H}$, that is, $G \in \mathcal{L}(\mathcal{H})$ is selfadjoint with respect to the positive definite inner product $\langle \cdot, \cdot \rangle$ and $[Gx, y] = \langle x, y \rangle$ holds for all $x, y \in \mathcal{H}$. Consider the Jordan decomposition $G = G_{+} - G_{-}$ of G and let $\mathcal{H}_{+}$ denote the spectral subspace corresponding to the positive semiaxis $(0, +\infty)$ and $\mathcal{H}_{-}$ be the spectral subspace corresponding to the nonpositive semiaxis $(-\infty, 0]$. Clearly, we have the decomposition

$$\mathcal{H} = \mathcal{H}_{+} \oplus \mathcal{H}_{-} \tag{2.6}$$

and, if $x = x_{+} + x_{-}$ and $y = y_{+} + y_{-}$ are the corresponding representations of arbitrary vectors $x, y \in \mathcal{H}$, then

$$[x, y] = \langle G_{+}x_{+}, x_{-} \rangle - \langle G_{-}x_{-}, y_{-} \rangle.$$

In the following we will use the notions of positivity, negativity, neutrality, etc. with respect to the indefinite inner product space $(\mathcal{H}, [\cdot, \cdot])$ and fix the decomposition (2.6).

Let $\mathcal{M}$ be a nonpositive subspace of $\mathcal{H}$, that is, a closed linear manifold such that $[x,x] \leq 0$ for all $x \in \mathcal{M}$. With respect to the decomposition (2.6) this means

$$\langle G_+ x_+, x_+\rangle \leq \langle G_- x_-, x_-\rangle, \qquad x = x_+ + x_- \in \mathcal{H}. \tag{2.7}$$

As in the case of Kreĭn spaces this enables us to introduce an angular operator. Let $P_\pm$ denote the projection on $\mathcal{H}_\pm$ with respect to the decomposition (2.6). Clearly $P_\pm$ are orthogonal projections in the Hilbert space $\mathcal{H}$, in particular their norms are ≤ 1. Let us define an operator $K_{\mathcal{M}} : P_-\mathcal{M} \to \mathcal{H}_+$ by

$$K_{\mathcal{M}} : P_- x \mapsto P_+ x, \qquad x \in \mathcal{M}. \tag{2.8}$$

By (2.7) and taking into account that G_+ is injective on $\mathcal{H}_+$, this definition is correct and

$$\mathcal{M} = \{x + K_{\mathcal{M}} x \mid x \in P_-\mathcal{M}\}. \tag{2.9}$$

Since $\mathcal{M}$ is closed, this implies that the operator $K_{\mathcal{M}}$ is closed. The operator $K_{\mathcal{M}}$ is called the *generalized angular operator* of the nonpositive subspace $\mathcal{M}$. Note that, in this general setting, there is no reason to conclude that $P_+\mathcal{M}$ is a (closed) subspace. This anomaly is remedied if an extra condition is imposed.

Remark 2.4. The condition that in the Jordan decomposition of the Gram operator G the operator G_+ has closed range is equivalent with the condition that the spectrum of G has a gap $(0,\epsilon)$. In particular, this shows that this condition is independent on which admissible positive definite inner product $\langle\cdot,\cdot\rangle$ we consider on the subspace $\mathcal{H}$, since, by changing it with another Gram operator, say B, we have $B = C^*GC$ for some boundedly invertible $C \in \mathcal{L}(\mathcal{H})$ (the inner products on the incoming Hilbert space $\mathcal{H}$ and the outgoing Hilbert space $\mathcal{H}$ are different) and this transformation preserves the topology of the spectrum.

A basic step in this approach is the possibility of handling generalized angular operators in subspaces in a similar fashion as the angular operators in Kreĭn spaces. A detailed proof of the following proposition of Treil [14], cf. [15], can be found also in [9].

Proposition 2.5. *With the previous notation, assume that the operator G_+ has closed range. Then:*

(1) *$\mathcal{M}$ is a nonpositive subspace of $\mathcal{H}$ if and only if $P_-\mathcal{M}$ is closed, $K_{\mathcal{M}}$ is bounded and the following inequality holds:*

$$K_{\mathcal{M}}^* G_+ K_{\mathcal{M}} \leq P_{P_-\mathcal{M}} G_- | P_-\mathcal{M}. \tag{2.10}$$

(2) *Let $\mathcal{M}$ and $\mathcal{N}$ be nonpositive subspaces. Then $\mathcal{M} \subseteq \mathcal{N}$ if and only if $K_{\mathcal{M}} \subseteq K_{\mathcal{N}}$, that is, $P_-\mathcal{M} \subseteq P_-\mathcal{N}$ and $K_{\mathcal{M}} x = K_{\mathcal{N}} x$ for all $x \in P_-\mathcal{M}$.*

(3) *For any nonpositive subspace $\mathcal{M}$ there exists a maximal nonpositive subspace $\widetilde{\mathcal{M}}$ such that $\mathcal{M} \subseteq \widetilde{\mathcal{M}}$.*

(4) *A nonpositive subspace* $\mathcal{M}$ *is maximal if and only if* $P_-\mathcal{M} = \mathcal{H}_-$.
(5) *If* $\operatorname{rank} G_+ = \kappa < \infty$ *then a nonnegative subsapace* $\mathcal{M}$ *is maximal nonpositive if and only if* $\operatorname{codim} \mathcal{M} = \kappa$.

Corollary 2.6. *If the operator* G_+ *has closed range, then the set*

$$\mathcal{X} = \{K_{\mathcal{M}} \mid \mathcal{M} \text{ maximal nonpositive subspace}\}$$

is convex and compact with respect to the weak operator topology on $\mathcal{L}(\mathcal{H}_-, \mathcal{H}_+)$.

2.3. Invariant maximal nonpositive subspaces. We now recall a theorem on the existence of invariant maximal semidefinite subspaces in a subspace of a Kreĭn space obtained by S. Treil and A. Volberg [15], which is slightly more general than a theorem of Ky Fan [6] and I.S. Iokhvidov [11]. Its proof is based on Proposition 2.5, its Corollary 2.6, and a fixed point theorem of Ky Fan [5] and I.L. Glicksberg [10].

Theorem 2.7. *With the notation as before, assume that the operator* G_+ *has closed range and let* $V \in \mathcal{L}(\mathcal{H})$ *be an operator subject to the following conditions:*
(i) *For any nonpositive vector* $x \in \mathcal{H}$ *the vector* Vx *is also nonpositive.*
(ii) *The operator* $G_-^{1/2} P_{\mathcal{H}_-} V P_{\mathcal{H}_+}$ *is compact.*
Then there exists a maximal nonpositive subspace $\mathcal{L} \subseteq \mathcal{H}$ *which is invariant under the operator* V.

3. Generalized interpolation

3.1. A generalized interpolation problem. Let m, n be nonnegative integer numbers and denote by $M_{m,n}$ the set of $m \times n$ matrices with complex entries. We denote by $H^\infty(M_{m,n})$ the space of all functions $F : \mathbb{D} \to M_{m,n}$ which are analytic and uniformly bounded in $\mathbb{D} = \{z \in \mathbb{C} \mid |z| < 1\}$,

$$\|F\|_\infty = \sup_{z \in \mathbb{D}} \|F(z)\| < \infty.$$

By $BH^\infty(M_{m,n})$ we denote the unit ball of $H^\infty(M_{m,n})$, that is,

$$BH^\infty(M_{m,n}) = \{F \in H^\infty(M_{m,n}) \mid \|F\| \leq 1\}.$$

Similarly one introduces the spaces $L^\infty(M_{m,n})$ and its unit ball $BL^\infty(M_{m,n})$. We also consider the Hardy space $H^2(\mathbb{C}^n)$ and the space $L^2(\mathbb{C}^n)$ of square integrable $\mathbb{C}^n$-valued functions. For any function $\phi \in L^\infty(M_{m,n})$ a multiplication operator $\cdot : L^2(\mathbb{C}^m) \to L^2(\mathbb{C}^n)$ is defined. If the function $\phi \in H^\infty(M_{m,n})$ then the corresponding multiplication operator maps $H^2(\mathbb{C}^m)$ into $H^2(\mathbb{C}^n)$.

A *phase function* is, by definition, a function $\phi \in L^\infty(M_n)$ such that $\phi(e^{it})$ is isometric for a.e. $t \in [0, 2\pi]$. If ϕ is a phase function then $\phi \in BL^\infty(M_n)$ and the corresponding multiplication operator is isometric.

Let $\alpha \in \mathbb{D}$. We consider the Möbius transformation

$$b_\alpha(z) = \frac{|\alpha|}{\alpha} \frac{\alpha - z}{1 - \overline{\alpha} z}, \qquad z \in \mathbb{D},$$

which maps the unit disk conformally into itself. A *Blaschke-Potapov cell* of order $q \leq n$ is by definition a square matrix of order n

$$B_\alpha(z) = \begin{bmatrix} I_{n-r} & 0 & 0 \\ 0 & b_\alpha(z) I_q & 0 \\ 0 & 0 & I_{n-s} \end{bmatrix},$$

where $n = r + q + s$. A *Blaschke-Potapov product of finite order* is by definition a finite product of analytic functions, each one being unitary equivalent with a Blaschke-Potapov cell. The order of a Blaschke-Potapov product is the sum of the orders of all its factors. Since the function b_α maps $\partial\mathbb{D}$ into itself, a Blaschke-Potapov product is always of norm one and the corresponding multiplication operator is isometric $H^2(\mathbb{C}^n) \to H^2(\mathbb{C}^n)$.

Let l be a nonnegative integer. $H_l^\infty(M_{m,n})$ is the set of functions F which can be represented as $F = G\Psi^{-1}$ where $G, \Psi \in H^\infty(M_{m,n})$ and Ψ is a Blaschke-Potapov product of order $\leq l$.

By $BH_l^\infty(M_{m,n})$ we denote the set of functions F which can be represented as $F = G\Psi^{-1}$ where $G, \Psi \in BH^\infty(M_{m,n})$ and Ψ is a Blaschke-Potapov product of order $\leq l$. If we impose the additional condition that no zero of the function G coincides with some zero of the Blaschke-Potapov product Ψ, then this representation of functions in $BH_l^\infty(M_{m,n})$ is unique, modulo unitary equivalence. Such a factorization is called *right coprime.*

We formulate now the problem of *generalized interpolation* by slightly modifying the generalized interpolation problem in [2]. For $K \in L^\infty(M_{m,n})$ and phase functions $\theta \in BL^\infty(M_m)$ and $\phi \in BL^\infty(M_n)$ there is considered the set

$$C_l(K, \theta, \phi) = (K + \theta H_l^\infty(M_{m,n})\phi) \cap BL^\infty(M_{m,n}), \tag{3.1}$$

and it is required to determine its elements.

3.2 The Ball-Helton approach. Since ϕ is a phase function, the same is true for ϕ^* and hence the corresponding multiplication operator is isometric, in particular $\phi H^2(\mathbb{C}^n)$ is a closed subspace of $L^2(\mathbb{C}^n)$. Thus we can consider the Hilbert space $L^2(\mathbb{C}^m) \oplus \phi^* H^2(\mathbb{C}^n)$ regarded as a subspace of the Kreĭn space $L^2(\mathbb{C}^m \oplus \mathbb{C}^n) = L^2(\mathbb{C}^m) \oplus L^2(\mathbb{C}^n)$, with the indefinite inner product $[\cdot,\cdot]$ induced by the indefinite inner product on $\mathbb{C}^m \oplus \mathbb{C}^n$, where the first component $\mathbb{C}^m$ is uniformly positive definite and the second component $\mathbb{C}^n$ is unifomly negative definite. In this section the ambiental Kreĭn space will be

$$\mathcal{K} = L^2(\mathbb{C}^m) \oplus \phi^* H^2(\mathbb{C}^n), \tag{3.2}$$

with the induced structure as before, more precisely the indefinite inner product $[\cdot,\cdot]$ is defined by

$$(3.3)\ [f_1 \oplus \phi^* g_1, f_2 \oplus \phi^* g_2] = \frac{1}{2\pi}\int_0^{2\pi} \langle f_1(e^{it}), f_2(e^{it}\rangle dt - \frac{1}{2\pi}\int_0^{2\pi} \langle g_1(e^{it}), g_2(e^{it})\rangle dt.$$

On the Kreĭn space $L^2(\mathbb{C}^m \oplus \mathbb{C}^n)$ the bilateral shift operator $S = M_{e^{it}}$ is defined, and it is unitary with respect to both positive inner product and indefinite inner product. Clearly, $\mathcal{K}$ is invariant under S.

By slightly modifying the approach in [2], we denote by $\mathcal{M}_{K,\theta,\phi}$ the subspace of $\mathcal{K}$ generated by the graphs of multiplication operators with all the functions of type $K + \theta h\phi$, $h \in H^\infty(M_{m,n})$, more precisely,

$$(3.4)\ \ \mathcal{M}_{K,\theta,\phi} = \operatorname{clos}\left\{\begin{bmatrix} (K+\theta h\phi)\phi^* g \\ \phi^* g \end{bmatrix} \mid h \in H^\infty(M_{m,n}),\ g \in H^2(\mathbb{C}^n)\right\}.$$

Lemma 3.1. *The subspace $\mathcal{M}_{K,\theta,\phi}$ is invariant under the shift operator S and*

$$\begin{aligned} \mathcal{M}_{K,\theta,\phi} &= \left\{\begin{bmatrix} K\phi^* g + \theta f \\ \phi^* g \end{bmatrix} \mid f \in H^2(\mathbb{C}^m),\ g \in H^2(\mathbb{C}^n)\right\} \\ &= \left\{\begin{bmatrix} K\phi^* g \\ \phi^* g \end{bmatrix} \mid g \in H^2(\mathbb{C}^n)\right\} \dot{+} \left\{\begin{bmatrix} \theta f \\ 0 \end{bmatrix} \mid f \in H^2(\mathbb{C}^m)\right\}. \end{aligned}$$

Proof. Let $h \in H^\infty(M_{m,n})$ and $g \in H^2(\mathbb{C}^n)$ be arbitrary. Since $\phi\phi^* = I_n$, we have

$$(K + \theta h\phi)\phi^* g = K\phi^* g + \theta h g,$$

and taking into account that $H^\infty(M_{m,n})H^2(\mathbb{C}^n) = H^2(\mathbb{C}^m)$ we denote $f = hg$. Then the subspace $\mathcal{M}_{K,\theta,\phi}$ is generated by functions

$$\begin{bmatrix} K\phi^* g + \theta f \\ \phi^* g \end{bmatrix}, \qquad f \in H^2(\mathbb{C}^m),\ g \in H^2(\mathbb{C}^n).$$

On the other hand, we note that the subspace

$$\mathcal{M}_1 = \left\{\begin{bmatrix} K\phi^* g \\ \phi^* g \end{bmatrix} \mid g \in H^2(\mathbb{C}^n)\right\}$$

is closed since it is the graph of a bounded operator. Clearly, the subspace

$$\mathcal{M}_2 = \left\{\begin{bmatrix} \theta f \\ 0 \end{bmatrix} \mid f \in H^2(\mathbb{C}^m)\right\},$$

is uniformly positive. Then it easily follows that $\mathcal{M}_1 \cap \mathcal{M}_2 = 0$ and the sum $\mathcal{M}_1 \dot{+} \mathcal{M}_2$ is closed.

The subspace $\mathcal{M}_{K,\theta,\phi}$ is invariant under the shift S since it is generated by graphs of multiplication operators. □

The orthogonal companion of the subspace $\mathcal{M}_{K,\theta,\phi}$ can be calculated. For this reason consider the operator $\Gamma_{\theta,\phi}(K)$ defined as follows:

$$L^2(\mathbb{C}^m) \ominus \theta H^2(\mathbb{C}^m) \ni f \mapsto \Gamma_{\theta,\phi}(K)^* f = P_{\phi^* H^2(\mathbb{C}^n)} K^* f \in H^2(\mathbb{C}^n). \tag{3.5}$$

Lemma 3.2. *The orthogonal subspace of $\mathcal{M}_{K,\theta,\phi}$ in the Kreĭn space $\mathcal{K}$ (see (3.2)) has the angular operator $\Gamma_{\theta,\phi}(K)$, more precisely,*

$$\mathcal{K} \cap \mathcal{M}_{K,\theta,\phi}^{\perp} = \{a \oplus \Gamma_{\theta,\phi}(K)^* a \mid a \in L^2(\mathbb{C}^m) \ominus \theta H^2(\mathbb{C}^m)\}.$$

Proof. By Lemma 3.1, a function $a \oplus b$ belongs to $\mathcal{K} \cap \mathcal{M}_{K,\theta,\phi}^{\perp}$ if and only if

$$\langle K\phi^* g + \theta f, a\rangle = \langle \phi^* g, b\rangle, \qquad f \in H^2(\mathbb{C}^m),\ g \in H^2(\mathbb{C}^n).$$

Letting first $g = 0$ and $f \in H^2(\mathbb{C}^m)$ arbitrary we get $a \in L^2(\mathbb{C}^m) \ominus \theta H^2(\mathbb{C}^m)$. Therefore $b = P_{\phi^* H^2(\mathbb{C}^n)} K^* a$. This proves that the subspace $\mathcal{K} \cap \mathcal{M}_{K,\theta,\phi}^{\perp}$ has the angular operator $\Gamma_{\theta,\phi}(K)$. □

In the following, for a given selfadjoint operator A in a Hilbert space $\mathcal{H}$, we denote by $\kappa^-(A)$ the dimension of the spectral subspace of A corresponding to the negative semi-axis. As a consequence of Lemma 3.2 we get:

Corollary 3.3. $\kappa^-(\mathcal{M}_{K,\theta,\phi}^{\perp}) = \kappa^-(I - \Gamma_{\theta,\phi}(K)\Gamma_{\theta,\phi}(K)^*)$.

We recall a theorem whose proof, for the scalar case, can be found in [7]. The matrix valued version follows in a similar way but uses the theorem on the structure of the vector valued inner functions and the theorem of Beurling-Lax characterizing the shift invariant subspaces in $H^2(\mathbb{C}^n)$.

Theorem 3.4. *A subspace $\mathcal{S} \in H^2(\mathbb{C}^n)$ is shift invariant and of codimension $l < \infty$ if and only if $\mathcal{S} = \Psi H^2(\mathbb{C}^n)$ where $\Psi \in BH^\infty(M_n)$ is a Blaschke-Potapov product of order l.*

The main idea in the approach of Ball and Helton is the modelling of the generalized interpolation problem by means of shift invariant subspaces in $\mathcal{M}_{K,\theta,\phi}$.

Lemma 3.5. *With the notation as before, the angular operator induces a bijective correspondence between the set $C_l(K,\theta,\phi)$ and the set of all shift-invariant nonpositive subspaces $\mathcal{L}$ of $\mathcal{M}_{K,\theta,\phi}$ such that $\kappa^-(\mathcal{K} \cap \mathcal{L}^{\perp}) \leq l$.*

Proof. Let $F \in C_l(K,\theta,\phi)$, that is, $F = K + \theta G \Psi^{-1}\phi$ for some $G \in H^\infty(M_{m,n})$ and $\Psi \in BH^\infty(M_n)$ is a Blaschke-Potapov product of degree at most l. Taking into account that $\phi\phi^* = I_n$, we consider the subspace

$$\mathcal{L} = \begin{bmatrix} F \\ I \end{bmatrix} \phi^* \Psi H^2(\mathbb{C}^n) = \left\{ \begin{bmatrix} K\phi^*\Psi g + \theta G g \\ \phi^*\Psi g \end{bmatrix} \mid g \in H^2(\mathbb{C}^n) \right\} \subseteq \mathcal{M}_{K,\theta,\phi}.$$

From $\|F\|_\infty \le 1$ it follows that $\mathcal{L}$ is a nonpositive subspace. Since Ψ is a Blaschke-Potapov product of order $\le l$, it follows that the codimension of $\Psi H^2(\mathbb{C}^n)$ in $H^2(\mathbb{C}^n)$ is $\le l$ and hence, the codimension of $\phi^*\Psi H^2(\mathbb{C}^n)$ in $\phi^* H^2(\mathbb{C}^n)$ is $\le l$. Then the codimension of $\mathcal{L}$ in some $\mathcal{K}$-maximal nonpositive subspace is at most l. Since $\mathcal{L}$ is the graph of a multiplication operator, it follows that $\mathcal{L}$ is invariant under the shift S.

Conversely, let $\mathcal{L}$ be a shift-invariant nonpositive subspace of $\mathcal{M}_{K,\theta,\phi}$ of codimension at most l in some $\mathcal{K}$-maximal nonpositive subspace. Then $\mathcal{L}$ is the graph of some contraction $F : \phi^* H^2(\mathbb{C}^n) \to L^2(\mathbb{C}^m)$,

$$\mathcal{L} = \{Fx \oplus x \mid x \in P_{\phi^* H^2(\mathbb{C}^n)}\mathcal{L}\}.$$

Writing down the invariance of $\mathcal{L}$ under the shift operator S we get that the subspace $P_{\phi^* H^2(\mathbb{C}^n)}\mathcal{L}$ is shift invariant and hence, by the Beurling-Lax Theorem there exists an inner function Ψ such that $P_{\phi^* H^2(\mathbb{C}^n)}\mathcal{L} = \phi^*\Psi H^2(\mathbb{C}^n)$, and

$$SF\phi^*\Psi g = FS\phi^*\Psi g, \qquad g \in H^2(\mathbb{C}^n). \tag{3.6}$$

Since the subspace $\mathcal{L}$ has codimension at most l in some $\mathcal{K}$-maximal nonpositive subspace, then (see Subsection 2.1) the codimension of $\phi^*\Psi H^2(\mathbb{C}^n)$ in $\phi^* H^2(\mathbb{C}^n)$ is at most l and hence by Theorem 3.4 it follows that Ψ is a Blaschke-Potapov factor of order at most l.

Further, taking into account that the commutant of the shift operator consists of multiplication operators it follows that $\mathcal{L}$ is the graph of some multiplication operator, that is,

$$\mathcal{L} = \begin{bmatrix} F \\ I \end{bmatrix} \phi^*\Psi H^2(\mathbb{C}^n),$$

where $F \in BH^\infty(M_{m,n})$.

Now we take into account that $\mathcal{L}$ is a subspace of $\mathcal{M}_{K,\theta,\phi}$ and, by means of Lemma 3.1, we get that $(F - K)\phi^*\Psi = \theta G$ for some $G \in H^\infty(M_{m,n})$, that is, $F = K + \theta G \Psi^{-1}\phi$. Since $\|F\|_\infty \le 1$, this means that $F \in C_l(K, \theta, \phi)$. □

Corollary 3.6. *If the set $C_l(K,\theta,\phi)$ has solutions, then $\kappa^-(\mathcal{M}^\perp_{K,\theta,\phi}) \le l$.*

Proof. Indeed, by Lemma 3.5, if $C_l(K,\theta,\phi)$ is non-void, there exists a nonpositive subspace $\mathcal{L}$ of $\mathcal{M}_{K,\theta,\phi}$ which has codimension at most l in some $\mathcal{K}$-maximal nonpositive subspace. According to Lemma 2.1 the codimension of $\mathcal{L}$ in some $\mathcal{K}$-maximal nonpositive subspace is $\kappa^-(\mathcal{L}^\perp)$. Since $\mathcal{M}^\perp_{K,\theta,\phi} \subseteq \mathcal{L}^\perp$, this implies $\kappa^-(\mathcal{M}^\perp_{K,\theta,\phi}) \le \kappa^-(\mathcal{L}^\perp) \le l$. □

3.3. The case when K is analytic and θ and ϕ are Blaschke-Potapov products of finite order. In the following we assume that K is analytic and ϕ and θ are Blaschke-Potapov products of finite order. A basic observation that can be made in this case is the following:

Lemma 3.7. *The subspace $\mathcal{M}_{K,\theta,\phi}$ is pseudo-regular.*

Proof. Indeed, we use the second representation in Lemma 3.1 to decompose the subspace $\mathcal{M}_{K,\theta,\phi}$ into a direct orthogonal sum

$$\begin{aligned}\mathcal{M}_{K,\theta,\phi} &= \left\{\begin{bmatrix} P_{L^2(\mathbb{C}^m)\ominus\theta H^2(\mathbb{C}^m)}K\phi^* g \\ \phi^* g\end{bmatrix} \mid g \in H^2(\mathbb{C}^n)\right\} \\ &[+]\left\{\begin{bmatrix}\theta f \\ 0\end{bmatrix} \mid f \in H^2(\mathbb{C}^m)\right\}.\end{aligned} \tag{3.7}$$

Clearly, the second subspace in the right hand side of (3.7) is uniformly positive, so it remains to prove that the first space is pseudo-regular. To see this, we consider first the decomposition

$$H^2(\mathbb{C}^n) = \phi H^2(\mathbb{C}^n) \oplus (H^2(\mathbb{C}^2) \ominus \phi H^2(\mathbb{C}^2)). \tag{3.8}$$

Since ϕ is a Blaschke-Potapov product of finite order, the second subspace on the right hand side is finite dimensional. Taking into account that $\phi^*\phi = I_m$ and that K is analytic, it follows that

$$\begin{aligned}&\left\{\begin{bmatrix} P_{L^2(\mathbb{C}^m)\ominus\theta H^2(\mathbb{C}^m)}K\phi^* g \\ \phi^* g\end{bmatrix} \mid g \in \phi H^2(\mathbb{C}^n)\right\} \\ &= \left\{\begin{bmatrix} P_{H^2(\mathbb{C}^m)\ominus\theta H^2(\mathbb{C}^m)}K g \\ g\end{bmatrix} \mid g \in H^2(\mathbb{C}^n)\right\}.\end{aligned} \tag{3.9}$$

Therefore, since the subspace $H^2(\mathbb{C}^m)\ominus\theta H^2(\mathbb{C}^m)$ is finite dimensional, the angular operator of the subspace in (3.9) is of finite rank and hence the subspace is pseudo-regular. □

Corollary 3.8. *If $\kappa^-(\mathcal{M}_{K,\theta,\phi}^\perp) = l$, then the angular operator induces a bijective correspondence between the set $C_l(K,\theta,\phi)$ and the set of all shift-invariant $\mathcal{M}_{K,\theta,\phi}$-maximal nonpositive subspaces.*

Proof. This is a consequence of Lemma 3.5, Lemma 3.7 and Lemma 2.3. □

The main result is a necessary and sufficient condition for the existence of elements in the set $C_l(K,\theta,\phi)$.

Theorem 3.9. *Suppose that K is in $H^\infty(M_{m,n})$ and the functions $\theta \in BH^\infty(M_m)$ and $\phi \in BH^\infty(M_n)$ are Blaschke-Potapov products of finite orders. Then the set $C_l(K,\theta,\phi)$ is non-void if and only if $\kappa^-(I - \Gamma_{\theta,\phi}(K)\Gamma_{\theta,\phi}(K)^*) \leq l$.*

Proof. Indeed, if the set $C_l(K,\theta,\phi)$ is non-void then, on the base of Lemma 3.5, it follows that $\kappa^-(\mathcal{M}_{K,\theta,\phi}^\perp) \leq l$ and then, taking into account Lemma 3.2 we obtain that $\kappa^-(I - \Gamma_{\theta,\phi}(K)\Gamma_{\theta,\phi}(K)^*) \leq l$.
Conversely, assume that $\kappa^-(I - \Gamma_{\theta,\phi}(K)\Gamma_{\theta,\phi}(K)^*) \leq l$ and hence, by Lemma 3.2, $\kappa^-(\mathcal{M}_{K,\theta,\phi}^\perp) \leq l$. In view of Lemma 3.5 we only have to prove that there exists at

least one shift-invariant $\mathcal{M}_{K,\theta,\phi}$-maximal nonpositive invariant subspace. For this we will use Theorem 2.7.

We consider the shift operator S on the indefinite inner product space $\mathcal{M}_{K,\theta,\phi}$ and we verify that the following hold:

(1) *The operator S maps nonpositive vectors into nonpositive vectors.*

Indeed, this follows trivially since S is isometric on the Kreĭn space $\mathcal{K}$ and, in particular, on its subspace $\mathcal{M}_{K,\theta,\phi}$.

(2) *Let $\mathcal{M} = \mathcal{M}_{K,\theta,\phi}$, $G = P_{\mathcal{M}}J|\mathcal{M}$ its Gram operator and let $G = G_+ - G_-$ be the corresponding Jordan decomposition as in Subsection 2.2. Then G_+ has closed range.*

Indeed, we proved in Lemma 3.7 that the subspace $\mathcal{M}_{K,\theta,\phi}$ is pseudo-regular. With respect to the fundamental symmetry J on the Kreĭn space $\mathcal{K}$ we have $G = P_{\mathcal{M}_{K,\theta,\phi}}J|\mathcal{M}_{K,\theta,\phi}$. Now, in order to prove that G_+ has closed range we have to prove that the corresponding spectral subspace $\operatorname{clos} G_+\mathcal{M}_{K,\theta,\phi}$ is regular. But this is a consequence of Proposition 2.2.

(3) *Let $\mathcal{M} = \mathcal{M}^+[+]\mathcal{M}^-$ be the corresponding spectral decomposition associated with the Jordan decomposition $G = G_+ - G_-$, as in Subsection 2.2. Then the operator $P_{\mathcal{M}^+}S|\mathcal{M}^-$ is of finite rank, in particular, it is compact.*

Indeed, we go back to the proof of Lemma 3.7 and note that from (3.7), (3.8), and (3.9) we get the following decomposition:

$$\begin{aligned}\mathcal{M}_{K,\theta,\phi} &= \left(\left\{\begin{bmatrix} P_{H^2(\mathbb{C}^m)\ominus\theta H^2(\mathbb{C}^m)}Kh \\ h\end{bmatrix} \mid h \in H^2(\mathbb{C}^n)\right\}\right.\\ &\quad \dot{+}\left.\left\{\begin{bmatrix} P_{L^2(\mathbb{C}^m)\ominus\theta H^2(\mathbb{C}^m)}K\phi^*g \\ \phi^*g\end{bmatrix} \mid g \in H^2(\mathbb{C}^n)\ominus\phi H^2(\mathbb{C}^n)\right\}\right)\\ &\quad [+]\left\{\begin{bmatrix}\theta f \\ 0\end{bmatrix} \mid f \in H^2(\mathbb{C}^m)\right\}.\end{aligned}$$

Taking into account that the third subspace is contained in $\mathcal{M}_+$ we obtain

$$\operatorname{rank}(P_{\mathcal{M}^+}S|\mathcal{M}^-) \le \dim(H^2(\mathbb{C}^m)\ominus\theta H^2(\mathbb{C}^m)) + \dim(H^2(\mathbb{C}^n)\ominus\phi H^2(\mathbb{C}^n)) < \infty.$$

and the claim is proved. We can now apply Theorem 2.7 to conclude that there exists a shift-invariant $\mathcal{M}_{K,\theta,\phi}$-maximal nonpositive subspace and hence, in view of Corollary 3.8, the set $C_l(K,\theta,\phi)$ is non-void. □

4. The bitangential Nevanlinna-Pick problem

In the remaining part of this paper we recall the approach of Ball-Helton to the bitangential Nevanlinna-Pick problem in order to show that it fits with the assumptions of Theorem 3.9.

Let m, n be nonnegative integers, $\mathbf{z} = \{z_j\}_{j=1}^{N} \cup \{z'_j\}_{j=1}^{N'}$ be a set of complex numbers in the unit disk $\mathbb{D}$, $\mathbf{p} = \{p_j\}_{j=1}^{N} \cup \{p'_j\}_{j=1}^{N'}$ be a set of points in $\mathbb{C}^m$ and $\mathbf{q} = \{q_j\}_{j=1}^{N} \cup \{q'_j\}_{j=1}^{N'}$ be a set of points in $\mathbb{C}^n$.

With the notation as in Section 3, the set we are interested in is the following:

$$NP_l(\mathbf{z},\mathbf{p},\mathbf{q}) = \{F \in BH_l^\infty(M_{m,n}) \mid F(z_j)^* p_j = q_j,\ j = 1,\dots,N,$$
$$F(z'_j)\, q'_j = p'_j,\ \ j = 1,\dots,N'\}.$$

The problem of determining the elements of the set $NP_l(\mathbf{z},\mathbf{p},\mathbf{q})$ is called the *bitangential Nevanlinna-Pick problem.*

As in [2] the set $NP_l(\mathbf{z},\mathbf{p},\mathbf{q})$ can be realized as a set of type $C_l(K,\theta,\phi)$ as in (3.1).

Lemma 4.1. *There exist Blaschke-Potapov functions $\theta \in BH^\infty(M_m)$ and $\phi \in BH^\infty(M_n)$ of finite order such that*

$$NP_l(\mathbf{z},\mathbf{p},\mathbf{q}) = C_l(K,\theta,\phi),$$

where $K \in H^\infty(M_{m,n})$ is any function satisfying the interpolation conditions $F(z_j)^* p_j = q_j$, $j = \overline{1,N}$ and $F(z'_j)q'_j = p'_j$, $j = \overline{1,N'}$.

Proof. Indeed, consider the space

$$\mathcal{S} = \{f \in H^2(\mathbb{C}^m) \mid \langle p_j, f(z_j)\rangle_{\mathbb{C}^m} = 0,\ j = \overline{1,N}\}.$$

Then $\mathcal{S}$ is a closed subspace of $H^2(\mathbb{C}^m)$, invariant under the shift S and of finite codimension. According to Theorem 3.4, there exists a Blaschke-Potapov product of finite order $\theta \in BH^\infty(\mathbb{C}^m)$ such that $\mathcal{S} = \theta H^\infty(\mathbb{C}^m)$. Let

$$\mathcal{S}' = \{g \in H^2(\mathbb{C}^n) \mid \langle q'_j, g(z_j)\rangle_{\mathbb{C}^n} = 0,\ j = \overline{1,N'}\}.$$

Similarly, $\mathcal{S}'$ is a closed subspace of $H^2(\mathbb{C}^n)$, invariant under the shift S and of finite codimension. According to Theorem 3.4, there exists a Blaschke-Potapov product of finite order $\psi \in BH^\infty(\mathbb{C}^n)$ such that $\mathcal{S}' = \psi H^\infty(\mathbb{C}^m)$. We take $\phi \in BH^\infty(\mathbb{C}^n)$ defined by $\phi(e^{it}) = \psi^*(e^{-it})$, for all $t \in [0, 2\pi)$.

We now make the remark that it is always possible to find a function $K \in H^\infty(M_{m,n})$ satisfying the interpolation conditions $F(z_j)^* p_j = q_j$, $j = \overline{1,N}$ and $F(z'_j)q'_j = p'_j$, $j = \overline{1,N'}$. Indeed, let $\{F_j\}_{j=1}^{N} \in M_{m,n}$ and $\{F'_j\}_{j=1}^{N'} \in M_{m,n}$ be such that $F_j^* p_j = q_j$, $j = \overline{1,N}$ and $F'_j q'_j = p'_j$, $j = \overline{1,N'}$. Then we can use Lagrange interpolation corresponding to complex points $\{z_j\}_{j=1}^{N} \cup \{z'_j\}_{j=1}^{N'}$ and values $\{F_j\}_{j=1}^{N} \cup \{F'_j\}_{j=1}^{N'}$ to determine such a function K which is a complex polynomial with matrix coefficients, in particular $K \in H^\infty(M_{m,n})$ (if $\{z_j\}_{j=1}^{N} \cap \{z'_j\}_{j=1}^{N'} \neq \emptyset$ then we have to impose the additional conditions $F_j = F'_k$ for all pairs $z_j = z'_k$, and this is possible only if appropriate consistency conditions are imposed on the data).

Let us now consider a function $F \in C_l(K,\theta,\phi)$, that is, $F \in BH^\infty(M_{m,n})$ and $F = K + \theta H \Psi^{-1}\phi$ for some $H \in H^\infty(M_{m,n})$ and some function $\Psi \in BH^\infty(M_n)$ which is a Blaschke-Potapov product of order at most l. Since K satisfies the interpolation conditions, to prove that F satisfies the interpolation conditions we have to show that

$$\begin{aligned} (\theta H \Psi^{-1}\phi)^*(z_j)p_j &= 0, \qquad j=\overline{1,N}, \\ (\theta H \Psi^{-1}\phi)\,(z_j')q_j' &= 0, \qquad j=\overline{1,N'}. \end{aligned}$$

These assertions can be proved easily using the properties of the functions θ, ϕ. Conversely, let $F \in NP_l(\mathbf{z},\mathbf{p},\mathbf{q})$ be arbitrary. Since K satisfies the interpolation conditions, it follows that the function $F-K$ satisfies the conditions

$$\begin{aligned} (F-K)^*(z_j)p_j &= 0, \qquad j=\overline{1,N}, \\ (F-K)(z_j')\,q_j' &= 0, \qquad j=\overline{1,N'}. \end{aligned}$$

Using the properties of the functions θ and ϕ it follows from here that $F-K=\theta G\phi$, and taking into account that $F \in BH^\infty(M_{m,n})$ we get that $G \in BH^\infty(M_{m,n})$. □

As a consequence of Lemma 4.1 and Theorem 3.9 a necessary and sufficient condition of solvability for the bitangential Nevanlinna-Pick problem for meromorphic functions can be obtained.

Theorem 4.2. *Let the functions K,θ and ϕ be associated with the data $(\mathbf{z},\mathbf{p},\mathbf{q})$ as in Lemma 4.1. There exist solutions of the bitangential Nevannlinna-Pick problem $NP_l(\mathbf{z},\mathbf{p},\mathbf{q})$ if and only if $\kappa^-(I-\Gamma_{\theta,\phi}(K)\Gamma_{\theta,\phi}(K)^*) \le l$.*

To relate this criterion with the criterion of solvability of the classical Nevanlinna-Pick problem in terms of the Pick matrix, as in [2], pp. 126–127, one can prove:

Proposition 4.3. *Assume that we are given the data $\mathbf{z},\mathbf{p},\mathbf{q}$ and $N'=0$. Then the one-sided tangential Nevanlinna-Pick problem $NP_l(\mathbf{z},\mathbf{p},\mathbf{q})$ has at least one solution if and only if the Pick matrix $\Lambda_{\mathbf{z},\mathbf{p},\mathbf{q}}$ defined by*

$$\Lambda_{\mathbf{z},\mathbf{p},\mathbf{q}} = \left[\frac{\langle p_j,p_i\rangle_{\mathbb{C}^m} - \langle q_j,q_i\rangle_{\mathbb{C}^n}}{1-\overline{z}_j z_i}\right]_{i,j=1}^N$$

has at most l negative eigenvalues, counted with their multiplicities.

Acknowledgements

This research was partially supported by the Romanian Academy, grant GAR 3720/1995.

References

[1] ADAMYAN, V.M., AROV, D.Z., KREĬN, M.G.: Infinite block Hankel matrices and some related continuation problems; Izv. Akad. Nauk Armyan SSR Ser. Mat. 6 (1971), 87–112.

[2] BALL, J.A., HELTON, J.W.: A Beurling-Lax theorem for the Lie group $U(m,n)$ which contains most classical interpolation theory; J. Operator Theory 9 (1983), 107–142.

[3] BALL, J.A., HELTON, J.W.: Interpolation problems of Pick-Nevanlinna and Loewner types for meromorphic matrix functions: parametrization of the set of all solutions; Integral Equations Operator Theory 9 (1986), 155–203.

[4] BOGNÁR, J.: Indefinite Inner Product Spaces; Springer Verlag, Berlin 1974.

[5] FAN, K.: A generalization of Tychonov's fixed point theorem; Math. Ann. 142 (1961), 305–510.

[6] FAN, K.: Invariant subspaces of certain linear operators; Bull. Amer. Math. Soc. 69 (1963), 773–777.

[7] FOIAŞ, C., FRAZHO, A.: The Commutant Lifting Approach to Interpolation Problems; Birkhäuser Verlag, Basel 1990.

[8] GHEONDEA, A.: On the geometry of pseudo-regular subspaces of a Kreĭn space; Operator Theory: Adv. Appl. 14, Birkhäuser Verlag, Basel 1984, 141–156.

[9] GHEONDEA, A.: Contractive intertwining dilations for quasi-contractions; Z. Anal. Anwendungen 16 (1996), 31–44.

[10] GLICKSBERG, I.L.: A further generalization of the Kakutani fixed point theorem, with applications to Nash equilibrium points; Proc. Amer. Math. Soc. 3 (1952), 170–174.

[11] IOKHVIDOV, I.S.: On a lemma of Ky Fan generalizing the fixed-point principle of A.N. Tikhonov; Dokl. Akad. Nauk. SSSR 159 (1964), 501–504.

[12] KREĬN, M.G., LANGER, H.: Über einige Fortsetzungsprobleme, die eng mit der Theorie hermitescher Operatoren im Raume Π_κ zusammenhängen. I. Einige Funktionenklassen und ihre Darstellungen; Math. Nachr. 77 (1977), 187–236.

[13] SARASON, D.: Generalized interpolation in H^∞; Trans. Amer. Math. Soc. 127 (1967), 179–203.

[14] TREIL, S.: Adamyan-Arov-Kreĭn theorem: vector valued version; Zap. Nauchn. Sem. LOMI 141 (1985), 56–81.

[15] TREIL, S., VOLBERG, A.: A fixed point approach to Nehari's problem and its applications; Operator Theory: Adv. Appl. 71, Birkhäuser Verlag, Basel 1994, 165–186.

Institutul de Matematică
C.P. 1-764
70700 Bucureşti
România
gheondea@imar.ro

1991 Mathematics Subject Classification: Primary 47A57; Secondary 30E05, 47A53, 47B50

Submitted: May 21, 1996

Operator Theory:
Advances and Applications, Vol. 103

The sum of matrix Nevanlinna functions and self-adjoint extensions in exit spaces

S. Hassi, M. Kaltenbäck and H.S.V. de Snoo

For a closed symmetric operator or relation, Kreĭn's formula describes all its self-adjoint extensions in terms of certain holomorphic parameters. Our interest is in self-adjoint extensions of a symmetric relation which extends itself an orthogonal sum of two symmetric relations. The corresponding class of parameters in Kreĭn's formula is identified. This leads to a description of (minimal) self-adjoint extensions in a fixed exit space.

1. Introduction

In this paper we discuss Kreĭn's formula for a closed symmetric relation or operator S with finite equal defect numbers in a Hilbert space $\mathfrak{H}$: it provides a one-to-one correspondence between (the compressed resolvents of) all self-adjoint extensions H of S and all relations $E(\ell)$ in $\mathbb{C}^n$, $\ell \in \mathbb{C} \setminus \mathbb{R}$, with $E(\ell)^* = E(\bar{\ell})$ and which are holomorphic and maximal dissipative in $\mathbb{C}^+$. The canonical self-adjoint extensions, i.e. the self-adjoint extensions which take place in $\mathfrak{H}$, correspond to the constant self-adjoint relations $E(\ell) = E$.

We consider a situation where a symmetric relation extends the orthogonal sum of two closed symmetric relations. This leads to a description of the self-adjoint extensions corresponding to holomorphic parameters of the form $Q(\ell) + E$, where $Q(\ell)$ is a fixed Q-function and E varies over the set of self-adjoint matrices. In particular, we obtain an operator representation for the sum of two Q-functions. By means of the holomorphic parameters in Kreĭn's formula and the notion of the Schur complement of a Nevanlinna function, it is possible to apply the above situation to study the exit spaces of noncanonical self-adjoint extensions.

Maximal dissipative relations, Nevanlinna pairs of matrix functions and their reduced form are treated in Section 2; they appear in Kreĭn's formula in Section 3, where also symmetric extensions of the original symmetric relation are discussed. Selfadjoint extensions for the special class of symmetric relations are introduced in Section 4. As an example we consider a class of self-adjoint boundary value problems for the orthogonal sum of two Sturm-Liouville operators in Section 5. Schur complements of matrix Nevanlinna functions and their operator representation are introduced in Section 6. Finally, in Section 7 we treat self-adjoint extensions in exit spaces by means of the operator representation of the Nevanlinna pair in Kreĭn's formula.

2. Nevanlinna pairs

For a pair $(A(\ell), B(\ell))$ of $n \times n$ matrix functions, defined on a subset of $\mathbb{C}\backslash\mathbb{R}$, which is symmetric with respect to $\mathbb{R}$, the Nevanlinna kernel $\mathsf{K}_{A,B}(\ell, \lambda)$ is given by

$$\mathsf{K}_{A,B}(\ell, \lambda) = \frac{A(\lambda)^* B(\ell) - B(\lambda)^* A(\ell)}{\ell - \bar{\lambda}}.$$

A pair of $n \times n$ matrix functions $(A(\ell), B(\ell))$, holomorphic for $\ell \in \mathbb{C}\backslash\mathbb{R}$, is called a Nevanlinna pair if

$$\text{rank} \begin{pmatrix} A(\ell) \\ B(\ell) \end{pmatrix} = n, \tag{2.1}$$

$$A(\bar{\ell})^* B(\ell) - B(\bar{\ell})^* A(\ell) = 0, \tag{2.2}$$

$$\mathsf{K}_{A,B}(\ell, \lambda) \text{ is nonnegative}, \tag{2.3}$$

see [15, 19]. For $\mu \in \mathbb{C} \setminus \mathbb{R}$ the rank condition (2.1) is equivalent to $B(\ell) + \mu A(\ell)$ invertible for some (and hence for all) $\ell \in \mathbb{C}_\mu$, i.e. the halfplane which contains the point μ, cf. [4, 18]. In this section we state and prove some useful properties of Nevanlinna pairs from the point of view of relations.

Let $(A(\ell), B(\ell))$ be a Nevanlinna pair and define the linear relation (multivalued operator) $H_{A,B}(\ell)$ in $\mathbb{C}^n$ by

$$H_{A,B}(\ell) = \{ \{A(\ell)x, B(\ell)x\} : x \in \mathbb{C}^n \}. \tag{2.4}$$

Then $H_{A,B}(\ell)$ is a holomorphic relation [8]. The conditions (2.1), (2.2), and (2.3) imply that $H_{A,B}(\ell)$ is maximal dissipative in $\ell \in \mathbb{C}^+$, i.e. $\text{Im}\,[g, f] \geq 0$ for all $\{f, g\} \in H_{A,B}(\ell)$, and $H_{A,B}(\ell)$, $\ell \in \mathbb{C}^+$, is maximal with respect to this property, while $H_{A,B}(\ell) = H_{A,B}(\bar{\ell})^*$, $\ell \in \mathbb{C}^-$. Conversely, for each holomorphic relation $H(\ell)$ in $\mathbb{C}^n$, for which $H(\ell)^* = H(\bar{\ell})$ and which is maximal dissipative for $\ell \in \mathbb{C}^+$, there exists a Nevanlinna pair $(A(\ell), B(\ell))$ such that $H(\ell) = H_{A,B}(\ell)$, cf. [1, 2]. If $A(\ell)$ and $B(\ell)$ are identified with their graphs, then $H_{A,B}(\ell) = B(\ell)A(\ell)^{-1}$ in the sense of the product of relations [6]. The multivalued part of $H_{A,B}(\ell)$ in (2.4) does not depend on ℓ, cf. [12]. Two Nevanlinna pairs $(A(\ell), B(\ell))$ and $(A'(\ell), B'(\ell))$ are called equivalent if $H_{A,B}(\ell) = H_{A',B'}(\ell)$, in which case $A(\ell)$, $A'(\ell)$, and $B(\ell)$, $B'(\ell)$ have the same range, respectively. Clearly, $(A(\ell), B(\ell))$ and $(A'(\ell), B'(\ell))$ are equivalent if and only if there exists an invertible $n \times n$ matrix function $X(\ell)$, holomorphic on $\mathbb{C} \setminus \mathbb{R}$, such that $A'(\ell) = A(\ell)X(\ell)$ and $B'(\ell) = B(\ell)X(\ell)$.

Proposition 2.1. *Every Nevanlinna pair $(A(\ell), B(\ell))$ is equivalent to a Nevanlinna pair $(P, C(\ell))$, where P is an orthogonal projection and $C(\ell)$ is a matrix Nevanlinna function such that $PC(\ell) = C(\ell)P$, $C(\ell)(I - P) = I - P$.*

Proof. Apply the orthogonal decomposition of maximal dissipative relations into multivalued and operator parts. In $\text{dom}\, H_{A,B}(\ell) = \text{ran}\, A(\ell)$, $H_{A,B}(\ell)$ acts as a Nevanlinna matrix function $C_s(\ell)$. Define $C(\ell) = \text{diag}\,(C_s(\ell), I)$, then $C(\ell)$ is a Nevanlinna function and $H_{A,B}(\ell) = H_{P,C}(\ell)$, which implies that $(A(\ell), B(\ell))$ and $(P, C(\ell))$ are equivalent. Moreover, it is clear that $PC(\ell) = C(\ell)P$ and $C(\ell)(I-P) = I-P$. □

Assume that X is invertible and that $E(\ell)$ is a Nevanlinna function. Then, with $(A(\ell), B(\ell))$, also

$$(X^*A(\ell), X^{-1}B(\ell)), \tag{2.5}$$

$$(A(\ell), B(\ell) + E(\ell)A(\ell)), \tag{2.6}$$

$$(-B(\ell), A(\ell)), \tag{2.7}$$

are Nevanlinna pairs. In particular, it follows from (2.7) that the null space of a Nevanlinna function does not depend on ℓ, cf. [11].

Proposition 2.2. *Let $(A(\ell), B(\ell))$ be a pair of holomorphic $n \times n$ matrix functions, which satisfies* (2.2) *and* (2.3). *Then $(A(\ell), B(\ell))$ is a Nevanlinna pair if and only if $Q(\ell)A(\ell) + B(\ell)$ is invertible for some (and hence for every) strict $n \times n$ matrix Nevanlinna function $Q(\ell)$.*

Proof. Assume that $(A(\ell), B(\ell))$ satisfies (2.1), in addition to (2.2) and (2.3). Then $Q(\ell)A(\ell)+B(\ell)$ is invertible for every strict $n\times n$ matrix Nevanlinna function $Q(\ell)$, due to Proposition 2.1.
Conversely, let $(A(\ell), B(\ell))$ satisfy (2.2) and (2.3) and let $Q(\ell)A(\ell) + B(\ell)$ be invertible for some strict $n \times n$ matrix Nevanlinna function $Q(\ell)$. Let $\mu \in \mathbb{C} \setminus \mathbb{R}$, then $Q(\mu) = D + \mu G$ for a self-adjoint $n \times n$ matrix D and a positive invertible matrix G. Define $A_1(\ell) = G^{\frac{1}{2}}A(\ell)$ and $B_1(\ell) = G^{-\frac{1}{2}}(B(\ell) + DA(\ell))$. This is a Nevanlinna pair, since clearly (2.2) and (2.3) follow from the corresponding properties of the pair $(A(\ell), B(\ell))$. Furthermore, $B_1(\ell) + \ell A_1(\ell)$ is invertible for $\ell = \mu$, since $B_1(\mu)+\mu A_1(\mu) = G^{-\frac{1}{2}}(B(\mu)+Q(\mu)A(\mu))$. As $A(\ell) = G^{-\frac{1}{2}}A_1(\ell)$ and $B(\ell) = G^{\frac{1}{2}}(B_1(\ell) - G^{-\frac{1}{2}}DG^{-\frac{1}{2}}A_1(\ell))$, it follows from (2.5) and (2.6) with $E(\ell)$ constant, that $(A(\ell), B(\ell))$ is a Nevanlinna pair. □

Corollary 2.3. *Let $(A(\ell), B(\ell))$ be a Nevanlinna pair and let $Q(\ell)$ be a strict Nevanlinna function on $\mathbb{C} \setminus \mathbb{R}$. Then*

$$(Q(\ell) + H_{A,B}(\ell))^{-1} = A(\ell)(Q(\ell)A(\ell) + B(\ell))^{-1}, \tag{2.8}$$

and, in particular,

$$A(\ell)(Q(\ell)A(\ell) + B(\ell))^{-1} = (Q(\bar{\ell})A(\bar{\ell}) + B(\bar{\ell}))^{-*}A(\bar{\ell})^*. \tag{2.9}$$

Proof. Assume that $\{f, g\} \in A(\ell)(Q(\ell)A(\ell) + B(\ell))^{-1}$. Then there exists an element h such that $\{h, f\} \in Q(\ell)A(\ell) + B(\ell)$ and $g = A(\ell)h$. Hence $\{g, f\} = \{A(\ell)h, Q(\ell)A(\ell)h+B(\ell)h\} \in Q(\ell)+B(\ell)A(\ell)^{-1}$. Since $B(\ell)A(\ell)^{-1} = H_{A,B}(\ell)$, it follows that $A(\ell)(Q(\ell)A(\ell)+B(\ell))^{-1} \subset (Q(\ell)+H_{A,B}(\ell))^{-1}$. The reverse inclusion is shown similarly. □

Proposition 2.4. *A Nevanlinna pair $(A(\ell), B(\ell))$ is equivalent to a pair (U, V), which satisfies*

$$\operatorname{rank} \begin{pmatrix} U \\ V \end{pmatrix} = n, \qquad U^*V = V^*U, \tag{2.10}$$

if and only if

$$\mathsf{K}_{A,B}(\ell, \lambda) = 0. \tag{2.11}$$

Proof. Let (U, V) be a pair, which satisfies (2.10). If $(A(\ell), B(\ell))$ is a Nevanlinna pair equivalent to (U, V), then (2.11) is automatically satisfied.
Conversely, let $(A(\ell), B(\ell))$ be a Nevanlinna pair which also satisfies (2.11). According to Proposition 2.1, $(A(\ell), B(\ell))$ is equivalent to $(P, C(\ell))$, where $C(\ell)$ is a Nevanlinna function such that $\operatorname{Im} C(\ell) = 0$. Therefore, $C(\ell)$ is constant by the maximum principle. □

Let the Nevanlinna pair (U, V) satisfy (2.10). Then the linear relation $H_{U,V}$ in $\mathbb{C}^n$ defined by

$$H_{U,V} = \{ \{Ux, Vx\} : x \in \mathbb{C}^n \} \tag{2.12}$$

is self-adjoint. Conversely, for each self-adjoint relation H in $\mathbb{C}^n$, there exists a Nevanlinna pair (U, V) such that $H = H_{U,V}$. Equivalence and reduction of such Nevanlinna pairs can be treated in a similar way: two Nevanlinna pairs (U, V) and (U', V') are called equivalent if $H_{U,V} = H_{U',V'}$; in this case U, U', and V, V' have the same range, respectively. Clearly, (U, V) and (U', V') are equivalent if and only if there exists an invertible $n \times n$ matrix X such that $U' = UX$ and $V' = VX$. A Nevanlinna pair (U, V) is equivalent to a Nevanlinna pair (P, W), where P is an orthogonal projection and W is a self-adjoint matrix such that $PW = WP$ and $W(I - P) = I - P$.

3. Kreĭn's formula

Let H be a self-adjoint relation in a Hilbert space $\mathfrak{K}$. Let $\mathfrak{H}$ be a closed subspace of $\mathfrak{K}$ and let $P_{\mathfrak{H}}$ be the orthogonal projection onto $\mathfrak{H}$. The compressed resolvent $R(\ell)$, $\ell \in \mathbb{C} \setminus \mathbb{R}$, of H is a function with values in $\mathbf{L}(\mathfrak{H})$, the algebra of the bounded linear operators on the Hilbert space $\mathfrak{H}$, defined by $R(\ell) = P_{\mathfrak{H}}(H - \ell)^{-1} |_{\mathfrak{H}}$, $\ell \in \mathbb{C} \setminus \mathbb{R}$. The relation H is minimal with respect to $\mathfrak{H}$ if for some $\mu \in \mathbb{C} \setminus \mathbb{R}$,

$$\mathfrak{K} = \overline{\operatorname{span}} \{ (I + (\ell - \mu)(H - \ell)^{-1})f : f \in \mathfrak{H}, \ \ell \in \mathbb{C} \setminus \mathbb{R} \}.$$

Minimal self-adjoint relations with the same compressed resolvents are isomorphically equivalent. A relation is called completely non-self-adjoint if it does not contain nontrivial reducing parts which are self-adjoint. The following assertions are easily seen to be equivalent:

(i) The relation H is minimal with respect to $\mathfrak{H}$.
(ii) $\mathfrak{K} \ominus \mathfrak{H}$ does not contain nontrivial reducing subspaces for H.
(iii) $H \cap (\mathfrak{K} \ominus \mathfrak{H})^2$ is completely non-self-adjoint.

Let S be a closed symmetric linear relation in $\mathfrak{H}$ with defect numbers (n,n) and let A be a fixed self-adjoint extension of S in $\mathfrak{H}$. For $\mu \in \mathbb{C} \setminus \mathbb{R}$ let the $1 \times n$ matrix of vector valued entries $\chi(\mu) = (\chi_1(\mu), \ldots, \chi_n(\mu))$ be a basis for $\ker(S^* - \mu)$. Then

$$\chi(\ell) = \left(I + (\ell - \mu)(A - \ell)^{-1}\right)\chi(\mu), \qquad \ell \in \mathbb{C} \setminus \mathbb{R}, \tag{3.1}$$

defines a $1 \times n$ holomorphic basis for $\ker(S^* - \ell)$ and

$$S = \{\, \{f,g\} \in A : \, [g - \ell f, \chi(\bar{\ell})] = 0 \,\}. \tag{3.2}$$

Note that $\mathfrak{H} = \overline{\operatorname{span}}\{\, \chi(\ell) : \, \ell \in \mathbb{C} \setminus \mathbb{R} \}$ if and only if S is completely non-self-adjoint (in which case S is an operator). An $n \times n$ matrix function $Q(\ell)$, $\ell \in \mathbb{C} \setminus \mathbb{R}$, is a Q-function of S and A, uniquely determined modulo a self-adjoint $n \times n$ matrix, if it satisfies

$$\frac{Q(\ell) - Q(\lambda)^*}{\ell - \bar{\lambda}} = [\chi(\ell), \chi(\lambda)], \qquad \ell, \lambda \in \mathbb{C} \setminus \mathbb{R}. \tag{3.3}$$

A Q-function belongs to the class $\mathbf{N}$ of $n \times n$ matrix Nevanlinna functions, i.e. it is defined and holomorphic in $\ell \in \mathbb{C} \setminus \mathbb{R}$, $Q(\ell)^* = Q(\bar{\ell})$, and $\operatorname{Im} Q(\ell)/\operatorname{Im} \ell \geq 0$ for $\ell \in \mathbb{C} \setminus \mathbb{R}$. Moreover, any Q-function $Q(\ell)$ is a strict Nevanlinna function, i.e. $\operatorname{Im} Q(\ell)$ is invertible for all $\ell \in \mathbb{C} \setminus \mathbb{R}$. Conversely, for any strict $n \times n$ matrix Nevanlinna function $Q(\ell)$ there exists a Hilbert space $\mathfrak{H}_Q$, a closed symmetric, completely non-self-adjoint operator S_Q with defect numbers (n,n) and a self-adjoint extension A_Q, such that $Q(\ell)$ is the Q-function of S_Q and A_Q, cf. [17]. Such a representation is called minimal and it is uniquely determined up to unitary equivalence. The following result is well-known [17].

Theorem 3.1. *Let S be a closed symmetric linear relation with defect numbers (n,n) and let A be a fixed self-adjoint extension of S in $\mathfrak{H}$. There is a one-to-one correspondence between the (compressed resolvents of) self-adjoint extensions of S and the (equivalence classes of) Nevanlinna pairs $(A(\ell), B(\ell))$ of $n \times n$ matrix functions, via*

$$R(\ell) = (A - \ell)^{-1} - \chi(\ell)A(\ell)(Q(\ell)A(\ell) + B(\ell))^{-1}[\,\cdot\,, \chi(\bar{\ell})]. \tag{3.4}$$

The resolvents of canonical self-adjoint extensions of S correspond to the Nevanlinna pairs of the form (2.10).

Suppose that the Nevanlinna pair $(A(\ell), B(\ell))$ is equivalent to the Nevanlinna pair $(P, C(\ell))$ as in Proposition 2.1 with $\operatorname{rank} A(\ell) = \operatorname{rank} P = k$, $0 \leq k \leq n$. Then S_P, defined by

$$S_P = \{\, \{f,g\} \in A : \, [g - \ell f, \chi(\bar{\ell})P] = 0 \,\},$$

is a closed symmetric relation. Its defect space $\ker(S_P^* - \ell)$ is spanned by $\chi_P(\ell) = \chi(\ell)P$, and is therefore k-dimensional. The function $Q_P(\ell) = PQ(\ell)P$ (as a $k \times k$ matrix function in $\operatorname{ran} P$) is the Q-function of S_P and A. With P and $C(\ell)$ define $C_P(\ell) = PC(\ell)P$, so that $C_P(\ell)$ is a Nevanlinna function in a k-dimensional space. The Kreĭn formula (3.4) reduces to

$$R(\ell) = (A - \ell)^{-1} - \chi_P(\ell)(Q_P(\ell) + C_P(\ell))^{-1}[\,\cdot\,, \chi_P(\bar{\ell})], \tag{3.5}$$

see also [16, pp. 223, 224]. Note that if $S \subset S_1 \subset A$, then to any self-adjoint extension of S_1, there exists a Nevanlinna pair $(P, C(\ell))$ such that $S_1 = S_P$. For a related approach see [5].

Define the Štraus family of relations $T(\ell)$ by $T(\ell) = \ell + R(\ell)^{-1}$, i.e.

$$T(\ell) = \{\,\{R(\ell)h, (I + \ell R(\ell))h\} : h \in \mathfrak{H}\,\}, \qquad \ell \in \mathbb{C} \setminus \mathbb{R}. \tag{3.6}$$

Clearly, $T(\ell)$ is holomorphic [8], $T(\ell)^* = T(\bar{\ell})$, and $\operatorname{ran}(T(\ell) - \ell) = \mathfrak{H}$, cf. [12].

Corollary 3.2. *Let $(A(\ell), B(\ell))$ be a Nevanlinna pair, which is equivalent to $(P, C(\ell))$. The Štraus relations $T(\ell)$ associated with* (3.4) *satisfy*

$$T(\ell) \cap A = S_P, \qquad \ell \in \mathbb{C} \setminus \mathbb{R}.$$

Proof. Let $\{f, g\} \in T(\ell) \cap A$, so that $R(\ell)(g - \ell f) = f$ and $(A - \ell)^{-1}(g - \ell f) = f$. Then it follows from (3.5) that $[g - \ell f, \chi_P(\bar{\ell})] = 0$, so that $\{f, g\} \in S_P$. For the converse inclusion reverse the steps. □

Let H be a minimal self-adjoint extension of S in a Hilbert space $\mathfrak{K} \supset \mathfrak{H}$, associated with $R(\ell)$ and $T(\ell)$ in (3.6). Then $H \cap A = S_P$. To see this, let $\{f, g\} \in T(\ell) \cap A$. As $\{f, g\} \in T(\ell)$, there exists $\varphi \in \mathfrak{K} \ominus \mathfrak{H}$ such that $\{f, g\} - \{\varphi, \ell\varphi\} \in H$. Now $\{f, g\} \in A \subset \mathfrak{H}^2$ shows that $\varphi = 0$ and $\{f, g\} \in H$. Hence $T(\ell) \cap A \subset H \cap A$. The reverse inclusion is clear, so that $H \cap A = T(\ell) \cap A = S_P$.

4. The sum of Q-functions

Let $\mathfrak{H}_1$ and $\mathfrak{H}_2$ be Hilbert spaces, let S_1 and S_2 be closed symmetric linear relations in them, each with equal defect numbers (n, n), and let A_1 and A_2 be self-adjoint extensions of S_1 and S_2. In $\mathfrak{H} = \mathfrak{H}_1 \ominus \mathfrak{H}_2$, the orthogonal sum $S = S_1 \oplus S_2$ is a closed symmetric relation with defect numbers $(2n, 2n)$ and the orthogonal sum $A = A_1 \oplus A_2$ is a self-adjoint extension of $S_1 \oplus S_2$. Clearly, the matrix function

$$X(\ell) = \begin{pmatrix} \chi_1(\ell) & 0 \\ 0 & \chi_2(\ell) \end{pmatrix} \tag{4.1}$$

satisfies

$$\frac{X(\ell) - X(\lambda)}{\ell - \lambda} = (A - \ell)^{-1} X(\lambda) \tag{4.2}$$

and spans $\ker((S_1 \oplus S_2)^* - \ell)$. Define the $2n \times 2n$ matrix function $Q(\ell)$ by

$$Q(\ell) = \operatorname{diag}(Q_1(\ell), Q_2(\ell)). \tag{4.3}$$

Then

$$\frac{Q(\ell) - Q(\lambda)^*}{\ell - \bar{\lambda}} = [X(\ell), X(\lambda)]. \tag{4.4}$$

Hence $Q(\ell)$ is the Q-function of $S = S_1 \oplus S_2$ and $A = A_1 \oplus A_2$. Define the relation $\widetilde{S}$ in $\mathfrak{H}_1 \oplus \mathfrak{H}_2$ by

$$\widetilde{S} = \left\{ \{f, g\} \in A : \left[g - \ell f, \begin{pmatrix} \chi_1(\bar{\ell}) \\ \chi_2(\bar{\ell}) \end{pmatrix} \right] = 0 \right\}. \tag{4.5}$$

Then $\widetilde{S} \subset A$ is a closed symmetric extension of S.

Proposition 4.1. *The closed symmetric relation $\widetilde{S}$ in (4.5) does not depend on $\ell \in \mathbb{C} \setminus \mathbb{R}$ and has defect numbers (n, n). In fact,*

$$\ker(\widetilde{S}^* - \ell) = \operatorname{span}\left\{ \begin{pmatrix} \chi_1(\ell) \\ \chi_2(\ell) \end{pmatrix} \right\}, \qquad \ell \in \mathbb{C} \setminus \mathbb{R}. \tag{4.6}$$

Moreover, the $n \times n$ matrix Nevanlinna function $Q_1(\ell) + Q_2(\ell)$ is the Q-function of $\widetilde{S}$ and A.

Proof. It follows from (4.2) that

$$\begin{pmatrix} \chi_1(\ell) \\ \chi_2(\ell) \end{pmatrix} = \left(I - (\ell - \lambda)(A - \ell)^{-1}\right) \begin{pmatrix} \chi_1(\lambda) \\ \chi_2(\lambda) \end{pmatrix}.$$

This shows that $\widetilde{S}$ in (4.5) does not depend on $\ell \in \mathbb{C} \setminus \mathbb{R}$ and that (4.6) holds. By multiplication from the left by $(I\ I)$ and from the right by $(I\ I)^\top$, it follows directly from (4.4) and (4.1), that

$$\frac{(Q_1(\ell) + Q_2(\ell)) - (Q_1(\lambda) + Q_2(\lambda))^*}{\ell - \bar{\lambda}} = \left[\begin{pmatrix} \chi_1(\ell) \\ \chi_2(\ell) \end{pmatrix}, \begin{pmatrix} \chi_1(\lambda) \\ \chi_2(\lambda) \end{pmatrix} \right],$$

so that $Q_1(\ell) + Q_2(\ell)$ is the Q-function of $\widetilde{S}$ and A. □

The operator representation of $Q_1(\ell) + Q_2(\ell)$ is in general not minimal as $\widetilde{S}$ need not be completely non-self-adjoint. In the language of reproducing kernel Hilbert spaces the maximal self-adjoint part of $\widetilde{S}$ corresponds to the overlapping space of the two model spaces for $Q_1(\ell)$ and $Q_2(\ell)$, cf. [3]. A nontrivial example occurs in the case of the sum of a Q-function with itself.

Since the self-adjoint extensions H of $\widetilde{S}$ are canonical extensions of $S = S_1 \oplus S_2$ in $\mathfrak{H}_1 \oplus \mathfrak{H}_2$, they correspond to equivalence classes of Nevanlinna pairs of $2n \times 2n$ matrices (U, V) via Kreĭn's formula:

$$(H - \ell)^{-1} = (A - \ell)^{-1} - X(\ell)U(Q(\ell)U + V)^{-1}[\,\cdot\,, X(\bar{\ell})]. \tag{4.7}$$

By Corollary 3.2 the condition $H \cap A = \widetilde{S}$ is equivalent to

$$U = \begin{pmatrix} I & 0 \\ I & 0 \end{pmatrix}. \tag{4.8}$$

The Nevanlinna pairs (U, V) with U as in (4.8) are now of the form

$$U = \begin{pmatrix} I & 0 \\ I & 0 \end{pmatrix}, \qquad V = \begin{pmatrix} V_{11} & I \\ V_{21} & -I \end{pmatrix}, \tag{4.9}$$

with $E = V_{11} + V_{21}$ self-adjoint. Multiplication on the right by

$$\begin{pmatrix} I & 0 \\ V_{21} & I \end{pmatrix} \quad \text{or} \quad \begin{pmatrix} I & 0 \\ -V_{11} & I \end{pmatrix}$$

shows that V in (4.9) can be replaced by

$$V = \begin{pmatrix} E & I \\ 0 & -I \end{pmatrix} \quad \text{or} \quad V = \begin{pmatrix} 0 & I \\ E & -I \end{pmatrix}. \tag{4.10}$$

Hence the self-adjoint extensions H in $\mathfrak{H}_1 \ominus \mathfrak{H}_2$ of $\widetilde{S}$ in (4.5) for which $H \cap A = \widetilde{S}$ are in one-to-one correspondence with all $n \times n$ self-adjoint matrices E, via

$$(H - \ell)^{-1} = (A - \ell)^{-1} - \begin{pmatrix} \chi_1(\ell) \\ \chi_2(\ell) \end{pmatrix} (Q_1(\ell) + Q_2(\ell) + E)^{-1} \left[\cdot, \begin{pmatrix} \chi_1(\bar{\ell}) \\ \chi_2(\bar{\ell}) \end{pmatrix} \right]. \tag{4.11}$$

Proposition 4.2. *Let H be a self-adjoint extension of $\widetilde{S}$. Then the following statements are equivalent:*

(i) $H \cap A = \widetilde{S}$;

(ii) $H \cap \mathfrak{H}_1^2 = S_1$;

(iii) $H \cap \mathfrak{H}_2^2 = S_2$.

Proof. Let H be a self-adjoint extension of $\widetilde{S}$ for which $H \cap A = \widetilde{S}$. By writing

$$H = \{ \{(H - \ell)^{-1} h, (I + \ell(H - \ell)^{-1}) h\} : h \in \mathfrak{H}_1 \oplus \mathfrak{H}_2 \},$$

and using the special form of $(H - \ell)^{-1}$ in (4.11) it follows that $H \cap \mathfrak{H}_1^2 = S_1$ and $H \cap \mathfrak{H}_2^2 = S_2$.

Now let H be a self-adjoint extension of $\widetilde{S}$ for which $H \cap A$ is a proper extension of $\widetilde{S}$. Hence there is a Nevanlinna pair of $n \times n$ matrices (P, W), where P is an orthogonal projection with $\operatorname{rank} P < n$ and W is a self-adjoint matrix with $PW = WP$ and $W(I - P) = I - P$, such that

$$(H - \ell)^{-1} = (A - \ell)^{-1} - \begin{pmatrix} \chi_1(\ell) \\ \chi_2(\ell) \end{pmatrix} P((Q_1(\ell) + Q_2(\ell))P + W)^{-1} \left[\cdot, \begin{pmatrix} \chi_1(\bar{\ell}) \\ \chi_2(\bar{\ell}) \end{pmatrix} \right].$$

Then H is a self-adjoint extension of the symmetric extension $\widetilde{T}$ of $\widetilde{S}$, defined by

$$\widetilde{T} = \left\{ \{f,g\} \in A : \left[g - \ell f, \begin{pmatrix} \chi_1(\bar{\ell}) \\ \chi_2(\bar{\ell}) \end{pmatrix} P \right] = 0 \right\}.$$

This implies that

$$S_1 = \widetilde{S} \cap \mathfrak{H}_1^2 \subset \widetilde{T} \cap \mathfrak{H}_1^2 \subset H \cap \mathfrak{H}_1^2, \qquad S_2 = \widetilde{S} \cap \mathfrak{H}_2^2 \subset \widetilde{T} \cap \mathfrak{H}_2^2 \subset H \cap \mathfrak{H}_2^2.$$

Clearly, $\widetilde{T} \cap \mathfrak{H}_1^2$ and $\widetilde{T} \cap \mathfrak{H}_2^2$ are proper symmetric extensions of S_1 and S_2, respectively, and therefore, $H \cap \mathfrak{H}_1^2$ and $H \cap \mathfrak{H}_2^2$ are proper symmetric extensions of S_1 and S_2, respectively. □

Theorem 4.3. *Let S_1 and S_2 be closed symmetric relations in $\mathfrak{H}_1$ and $\mathfrak{H}_2$ with defect numbers (n,n). Let $\widetilde{S}$ be the symmetric extension of $S_1 \oplus S_2$ defined by (4.5). Then there is a one-to-one correspondence between the (compressed resolvents of) self-adjoint extensions H of $\widetilde{S}$ with $H \cap A = \widetilde{S}$ and the self-adjoint $n \times n$ matrices $E = E^*$, via*

$$(4.12) \quad P_{\mathfrak{H}_1}(H-\ell)^{-1}|_{\mathfrak{H}_1} = (A_1 - \ell)^{-1} - \chi_1(\ell)(Q_1(\ell) + Q_2(\ell) + E)^{-1}[\,\cdot\,, \chi_1(\bar{\ell})].$$

Proof. Let H be a self-adjoint extension of $\widetilde{S}$ with $H \cap A = \widetilde{S}$. Then there is an $n \times n$ matrix E as in (4.11) and the compression of $(H-\ell)^{-1}$ to $\mathfrak{H}_1$ is given by (4.12). Conversely, any self-adjoint $n \times n$ matrix $E = E^*$ gives a self-adjoint extension H with $H \cap A = \widetilde{S}$ in (4.11). □

By way of illustration we consider the case where $\mathfrak{H}_2$ is a one-dimensional Hilbert space. Assume that S_1 is a closed symmetric relation with defect numbers $(1,1)$ and that S_2 is the zero operator in $\mathbb{C}$. Let A_2 be the self-adjoint extension of S_2 given by span $\{0,1\}$ (all other self-adjoint extensions of S_2 are operators), so that $(A_2 - \ell)^{-1} = 0$. Then $\chi_2(\ell) = 1$ and $Q_2(\ell) = \ell$ gives the corresponding Q-function. The other self-adjoint extensions H_2 of S_2 are in one-to-one correspondence with $\tau \in \mathbb{R}$ via

$$(H_2 - \ell)^{-1} = -\frac{1}{\ell - \tau} \quad \Longleftrightarrow \quad H_2 = \operatorname{span}\{\{1,\tau\}\}.$$

Then $\widetilde{S}$ is given by

$$\widetilde{S} = \left\{ \left\{ \begin{pmatrix} f_1 \\ 0 \end{pmatrix}, \begin{pmatrix} g_1 \\ c \end{pmatrix} \right\} : \{f_1, g_1\} \in A_1,\ c \in \mathbb{C},\ [g_1 - \ell f_1, \chi_1(\bar{\ell})] + c = 0 \right\}.$$

The one-to-one correspondence between the (compressed resolvents of) self-adjoint extensions H of $\widetilde{S}$ with $H \cap A = \widetilde{S}$ and $E \in \mathbb{R}$ is given via

$$P_{\mathfrak{H}_1}(H-\ell)^{-1}|_{\mathfrak{H}_1} = (A_1 - \ell)^{-1} - \chi_1(\ell)(Q_1(\ell) + \ell + E)^{-1}[\,\cdot\,, \chi_1(\bar{\ell})].$$

By Proposition 4.1, $Q_1(\ell) + \ell$ is the Q-function of $\widetilde{S}$ and $A_1 \oplus A_2$. The above operator representation of $Q_1(\ell) + \ell$ need not be minimal. In fact, by means of the model space in [13, Section 2] it is easy to check that if S_1 is completely non-self-adjoint, then $\widetilde{S}$ is completely non-self-adjoint if and only if $Q_1(\ell)$ itself does not have a linear part.

5. The orthogonal sum of Sturm-Liouville operators

We consider the orthogonal sum of Sturm-Liouville operators on $(0,\infty)$ and on $(-\infty,0)$, which are in the limit point case at their singular endpoints ∞ and $-\infty$. This results in a boundary value problem at 0 for a differential operator on $\mathbb{R}$ with an interior singularity. For general background we refer to [7], [14], and [21].

In $\mathfrak{H}_1 = L^2(0,\infty)$ we consider the Sturm-Liouville operator $-D^2+q_1$, where q_1 is a locally integrable real-valued function on $(0,\infty)$, such that ∞ is in the limit point case. Let $u_1(\cdot,\ell)$ and $u_2(\cdot,\ell)$ be solutions of $-u''(x,\ell)+q_1(x)u(x,\ell)=\ell u(x,\ell)$, $x>0$, determined by the initial conditions

$$\begin{pmatrix} u_1(0+,\ell) & u_2(0+,\ell) \\ u_1'(0+,\ell) & u_2'(0+,\ell) \end{pmatrix} = \begin{pmatrix} 1 & 0 \\ 0 & 1 \end{pmatrix}.$$

There is precisely one Nevanlinna function $m_1(\ell)$, the so-called Titchmarsh-Weyl coefficient, such that

$$\chi_1(\cdot,\ell) = u_1(\cdot,\ell)m_1(\ell) - u_2(\cdot,\ell) \in \mathfrak{H}_1.$$

The maximal differential operator T_1 is defined by

$$T_1 = \{\,\{f,-f''+q_1f\} \in \mathfrak{H}_1^2 : f,f' \text{ absolutely continuous}\,\}.$$

Its adjoint $S_1 = T_1^*$ is a closed symmetric restriction of T_1 with defect numbers $(1,1)$, given by

$$S_1 = \{\,\{f,-f''+q_1f\} \in T_1 : f,f' \text{ absolutely continuous}, \ f(0+)=f'(0+)=0\}$$

and $\ker(S_1^*-\ell) = \operatorname{span}\{\chi_1(\ell)\}$. The canonical self-adjoint extensions of S_1 are in one-to-one correspondence with $\tau \in \mathbb{R}\cup\{\infty\}$ via

$$H_1(\tau) = \{\,\{f,-f''+q_1f\} \in T_1 : f(0+) = \tau f'(0+)\,\}, \qquad \tau \in \mathbb{R},$$

and

$$H_1(\infty) = \{\,\{f,-f''+q_1f\} \in T_1 : f'(0+) = 0\,\}.$$

Clearly, the resolvent operator of $H_1(\infty)$ has the integral representation

$$(H_1(\infty)-\ell)^{-1}f(x) = u_1(x,\ell)\int_x^\infty f(t)\chi_1(t,\ell)\,dt + \chi_1(x,\ell)\int_0^x f(t)u_1(t,\ell)\,dt,$$

when $f \in \mathfrak{H}_1$ vanishes in a neighborhood of ∞. The Kreĭn formula

$$(H_1(\tau)-\ell)^{-1} = (H_1(\infty)-\ell)^{-1} - \chi_1(\ell)\frac{1}{m_1(\ell)+\tau}\,[\,\cdot\,,\chi_1(\bar{\ell})]$$

corresponds with these self-adjoint extensions. The Titchmarsh-Weyl coefficient $m_1(\ell)$ is the Q-function for $H_1(\infty)$.

Similarly in $\mathfrak{H}_2 = L^2(-\infty,0)$ we consider the operator $-D^2+q_2$, where q_2 is a locally integrable real-valued function on $(-\infty,0)$, such that $-\infty$ is in the limit point case. Let $u_1(\,\cdot\,,\ell)$ and $u_2(\,\cdot\,,\ell)$ be solutions of $-u''(x,\ell)+q_2(x)u(x,\ell) = \ell u(x,\ell)$, $x<0$, determined by the initial conditions

$$\begin{pmatrix} u_1(0-,\ell) & u_2(0-,\ell) \\ u_1'(0-,\ell) & u_2'(0-,\ell) \end{pmatrix} = \begin{pmatrix} 1 & 0 \\ 0 & 1 \end{pmatrix}.$$

There is precisely one Nevanlinna function $m_2(\ell)$, such that

$$\chi_2(\,\cdot\,,\ell) = -u_1(\,\cdot\,,\ell)m_2(\ell) - u_2(\,\cdot\,,\ell) \in \mathfrak{H}_2.$$

The maximal differential operator T_2 is defined by

$$T_2 = \{\,\{f,-f''+q_2 f\} \in \mathfrak{H}_2^2 :\, f, f' \text{ absolutely continuous}\,\}.$$

Its adjoint $S_2 = T_2^*$ is a closed symmetric restriction of T_2 with defect numbers $(1,1)$, given by

$$S_2 = \{\,\{f,-f''+q_2 f\} \in T_2 :\, f, f' \text{ absolutely continuous, } f(0-) = f'(0-) = 0\,\}$$

and $\ker(S_2^*-\ell) = \operatorname{span}\{\chi_2(\ell)\}$. The canonical self-adjoint extensions of S_2 are in one-to-one correspondence with $\tau \in \mathbb{R}\cup\{\infty\}$ via

$$H_2(\tau) = \{\,\{f,-f''+q_2 f\} \in T_2 : f(0-) = -\tau f'(0-)\,\}, \qquad \tau \in \mathbb{R},$$

and

$$H_2(\infty) = \{\,\{f,-f''+q_2 f\} \in T_1 : f'(0-) = 0\,\}.$$

The resolvent operator of $H_2(\infty)$ has the integral representation

$$(H_2(\infty)-\ell)^{-1}f(x) = -u_1(x,\ell)\int_{-\infty}^{x} f(t)\chi_2(t,\ell)\,dt - \chi_2(x,\ell)\int_{x}^{0} f(t)u_1(t,\ell)\,dt,$$

when $f \in \mathfrak{H}_2$ vanishes in a neighborhood of $-\infty$. The Kreĭn formula

$$(H_2(\tau)-\ell)^{-1} = (H_2(\infty)-\ell)^{-1} - \chi_2(\ell)\frac{1}{m_2(\ell)+\tau}\,[\,\cdot\,,\chi_2(\bar{\ell})]$$

corresponds with these self-adjoint extensions. The Titchmarsh-Weyl coefficient $m_2(\ell)$ is the Q-function for $H_2(\infty)$.

Proposition 5.1. *The self-adjoint extensions H of $S_1 \oplus S_2$ in $\mathfrak{H}_1 \oplus \mathfrak{H}_2$ are in one-to-one correspondence with the Nevanlinna pairs (U,V) of 2×2 matrices via*

$$\operatorname{dom} H = \left\{ \begin{pmatrix} f_1 \\ f_2 \end{pmatrix} \in \operatorname{dom} T_1 \oplus \operatorname{dom} T_2 : U^* \begin{pmatrix} f_1(0+) \\ -f_2(0-) \end{pmatrix} = V^* \begin{pmatrix} f_1'(0+) \\ f_2'(0-) \end{pmatrix} \right\}.$$

The Kreĭn formula (4.7) corresponds to these self-adjoint extensions when we take $A_1 = H_1(\infty)$ and $A_2 = H_2(\infty)$, so that $Q_1(\ell) = m_1(\ell)$ and $Q_2(\ell) = m_2(\ell)$. We formulate the analog of (4.11) and Theorem 4.3 in terms of boundary conditions.

Corollary 5.2. *The self-adjoint extensions H of $S_1 \oplus S_2$ in $\mathfrak{H}_1 \oplus \mathfrak{H}_2$ which satisfy $H \cap A = \widetilde{S}$ are in one-to-one correspondence with $E \in \mathbb{R}$ via*

$$\operatorname{dom} H = \left\{ \begin{pmatrix} f_1 \\ f_2 \end{pmatrix} \in \operatorname{dom} T_1 \oplus \operatorname{dom} T_2 : \right. \\ \left. f_1(0+) - f_2(0-) = E f_1'(0+),\ f_1'(0+) = f_2'(0-) \right\}.$$

The corresponding Štraus extension $T(\ell)$, $\ell \in \mathbb{C} \setminus \mathbb{R}$, of S_1 in $\mathfrak{H}_1$, is given by the boundary condition

$$\operatorname{dom} T(\ell) = \{ f_1 \in \operatorname{dom} T_1 : f_1(0+) = (m_2(\ell) + E) f_1'(0+) \}.$$

The operator $T_1 \oplus T_2$ has been interpreted as a differential operator in $\mathfrak{H}_1 \oplus \mathfrak{H}_2 = L^2(\mathbb{R})$ with an interior singularity 0, whose adjoint is given by $S_1 \oplus S_2$. The self-adjoint extensions of $S_1 \oplus S_2$ are described by boundary conditions in terms of the boundary values

$$\left\{ \begin{pmatrix} f_1(0+) \\ -f_2(0-) \end{pmatrix}, \begin{pmatrix} f_1'(0+) \\ f_2'(0-) \end{pmatrix} \right\}.$$

In particular we have studied the self-adjoint extensions of the operator $\widetilde{S}$:

$$\operatorname{dom} \widetilde{S} = \left\{ \begin{pmatrix} f_1 \\ f_2 \end{pmatrix} \in \operatorname{dom} T_1 \oplus \operatorname{dom} T_2 : \right. \\ \left. f_1(0+) = f_2(0-),\ f_1'(0+) = f_2'(0-) = 0 \right\}.$$

When the interior singularity is more singular the same methods can be applied if we can use the "generalized" boundary values in the sense of Rellich. Our example is a prototype for the use of Weyl functions in the sense of Derkach and Malamud in a more abstract situation, cf. [9, 10].

6. Schur complements of Q-functions

Let S be a closed symmetric linear relation in a Hilbert space $\mathfrak{H}$ with defect numbers (n,n). Let A be a canonical self-adjoint extension of S. Let $Q(\ell)$ be the corresponding $n \times n$ matrix Q-function. Decompose $\mathbb{C}^n$ as the orthogonal sum $\mathbb{C}^n = \mathfrak{G}_1 \oplus \mathfrak{G}_2$ with $\dim \mathfrak{G}_i = n_i$, $i = 1, 2$, and decompose $Q(\ell)$ accordingly:

$$Q(\ell) = \begin{pmatrix} Q_{11}(\ell) & Q_{12}(\ell) \\ Q_{21}(\ell) & Q_{22}(\ell) \end{pmatrix}.$$

The Schur complement of $Q(\ell)$ with respect to this decomposition is the $n_1 \times n_1$ matrix function defined by

$$\widehat{Q}(\ell) = Q_{11}(\ell) - Q_{12}(\ell) Q_{22}(\ell)^{-1} Q_{21}(\ell). \tag{6.1}$$

The above decomposition of $\mathbb{C}^n$ leads to a symmetric extension $\widehat{S}$ and a self-adjoint extension $\widehat{A}$ of $\widehat{S}$: $S \subset \widehat{S} \subset \widehat{A}$. Indeed, the relation $\widehat{A}$, defined by

$$(\widehat{A} - \ell)^{-1} = (A - \ell)^{-1} - \chi(\ell) \begin{pmatrix} 0 & 0 \\ 0 & Q_{22}(\ell)^{-1} \end{pmatrix} [\,\cdot\,, \chi(\bar{\ell})], \tag{6.2}$$

is a canonical self-adjoint extension of S. This follows from Kreĭn's formula with the Nevanlinna pair given by

$$U = \begin{pmatrix} 0 & 0 \\ 0 & I \end{pmatrix}, \qquad V = \begin{pmatrix} I & 0 \\ 0 & 0 \end{pmatrix}.$$

Define the $1 \times n_1$ matrix function $\widehat{\chi}(\ell)$ by

$$\widehat{\chi}(\ell) = \chi(\ell) \begin{pmatrix} I \\ -Q_{22}(\ell)^{-1} Q_{21}(\ell) \end{pmatrix}. \tag{6.3}$$

Clearly, $\widehat{\chi}(\ell)$ spans an n_1–dimensional subspace for each $\ell \in \mathbb{C} \setminus \mathbb{R}$. The closed symmetric restriction $\widehat{S}$ of $\widehat{A}$ is defined by

$$\widehat{S} = \{\, \{f, g\} \in \widehat{A} : \, [g - \ell f, \widehat{\chi}(\bar{\ell})] = 0 \,\}. \tag{6.4}$$

Proposition 6.1. *The closed symmetric relation $\widehat{S}$ in (6.4) does not depend on $\ell \in \mathbb{C} \setminus \mathbb{R}$ and has defect numbers (n_1, n_1). In fact,*

$$\ker (\widehat{S}^* - \ell) = \operatorname{span} \{\widehat{\chi}(\ell)\}, \qquad \ell \in \mathbb{C} \setminus \mathbb{R}. \tag{6.5}$$

Moreover, the Schur complement $\widehat{Q}(\ell)$ is the Q-function of $\widehat{S}$ and $\widehat{A}$.

Proof. It follows from (6.2) that

$$(\widehat{A} - \ell)^{-1} \widehat{\chi}(\lambda) = (A - \ell)^{-1} \widehat{\chi}(\lambda) - \chi(\ell) \begin{pmatrix} 0 & 0 \\ 0 & Q_{22}(\ell)^{-1} \end{pmatrix} [\widehat{\chi}(\lambda), \chi(\bar{\ell})],$$

and by (6.3) the right side is equal to

$$\left((A - \ell)^{-1} \chi(\lambda) - \chi(\ell) \begin{pmatrix} 0 & 0 \\ 0 & Q_{22}(\ell)^{-1} \end{pmatrix} [\chi(\lambda), \chi(\bar{\ell})] \right) \begin{pmatrix} I \\ -Q_{22}(\lambda)^{-1} Q_{21}(\lambda) \end{pmatrix}.$$

Since $Q(\ell)$ is the Q-function of A and S, the first factor is equal to

$$\frac{\chi(\ell)-\chi(\lambda)}{\ell-\lambda} - \chi(\ell)\begin{pmatrix}0 & 0\\ 0 & Q_{22}(\ell)^{-1}\end{pmatrix}\frac{Q(\ell)-Q(\lambda)}{\ell-\lambda}$$

$$=\frac{1}{\ell-\lambda}\left(\chi(\ell)\left(I-\begin{pmatrix}0 & 0\\ 0 & Q_{22}(\ell)^{-1}\end{pmatrix}(Q(\ell)-Q(\lambda))\right)-\chi(\lambda)\right).$$

The identity

$$\begin{pmatrix}I\\ -Q_{22}(\ell)^{-1}Q_{21}(\ell)\end{pmatrix}=\left(I-\begin{pmatrix}0 & 0\\ 0 & Q_{22}(\ell)^{-1}\end{pmatrix}(Q(\ell)-Q(\lambda))\right)\begin{pmatrix}I\\ -Q_{22}(\lambda)^{-1}Q_{21}(\lambda)\end{pmatrix}$$

then leads to

$$(\widehat{A}-\ell)^{-1}\widehat{\chi}(\lambda)=\frac{\widehat{\chi}(\ell)-\widehat{\chi}(\lambda)}{\ell-\lambda}.$$

This implies that $\widehat{S}$ in (6.4) does not depend on ℓ and that (6.5) holds. Next, multiply the identity (3.3) from the right by $\begin{pmatrix}I\\ -Q_{22}(\ell)^{-1}Q_{21}(\ell)\end{pmatrix}$ and from the left by $\begin{pmatrix}I\\ -Q_{22}(\lambda)^{-1}Q_{21}(\lambda)\end{pmatrix}^*$. It follows that

$$\frac{\widehat{Q}(\ell)-\widehat{Q}(\lambda)^*}{\ell-\bar{\lambda}}=[\widehat{\chi}(\ell),\widehat{\chi}(\lambda)],$$

so that $\widehat{Q}(\ell)$ is the Q-function of $\widehat{S}$ and $\widehat{A}$. □

7. Nevanlinna functions and exit spaces

Let S be a closed symmetric linear relation with defect numbers (n,n) and let A be a canonical self-adjoint extension of S. Let $(A(\ell),B(\ell))$ be a Nevanlinna pair of $n\times n$ matrix functions. If $\mathsf{K}_{A,B}(\ell,\lambda)\neq 0$, then (3.4) gives the compressed resolvent of a self-adjoint extension of S in an exit space. We show how to construct this exit space and the self-adjoint extension in terms of the Nevanlinna pair $(A(\ell),B(\ell))$.

By Proposition 2.1 the pair $(A(\ell),B(\ell))$ is equivalent to the Nevanlinna pair $(P,C(\ell))$ and the Kreĭn formula (3.4) then reduces to the form (3.5). The closed symmetric extension S_P of S has defect numbers (k,k), $k=\operatorname{rank}P$, and is a restriction of the self-adjoint extension A of S. Moreover, we may interpret $C_P(\ell)=PC(\ell)P$ as a $k\times k$ matrix Nevanlinna function. Hence, by eventually allowing a symmetric extension of S as the basic symmetric relation, we may restrict ourselves to Nevanlinna pairs $(A(\ell),B(\ell))$, which are equivalent to $(I,C(\ell))$ with $C(\ell)$ an $n\times n$ matrix Nevanlinna function. Let $C(\ell)$ be an $n\times n$ matrix Nevanlinna function, then by Theorem 3.1, the function $R(\ell)$ defined by

(7.1) $$R(\ell)=(A-\ell)^{-1}-\chi(\ell)(Q(\ell)+C(\ell))^{-1}[\cdot,\chi(\bar{\ell})]$$

is a compressed resolvent of S. The previous results in this note are now applied to associate with $R(\ell)$ a self-adjoint extension H of S in an exit space. First consider the special case that $C(\ell)$ is a strict $n \times n$ matrix Nevanlinna function. Let $\mathfrak{H}_C$ be a Hilbert space, S_C be a closed completely non-self-adjoint, symmetric operator in $\mathfrak{H}_C$ with defect numbers (n,n) and let A_C be a self-adjoint extension of S_C, such that the Q-function of S_C and A_C is $C(\ell)$, cf. [17]. Define the symmetric extension $\widetilde{S}$ of $S \oplus S_C$ in $\mathfrak{H} \oplus \mathfrak{H}_C$ by

$$\widetilde{S} = \left\{ \{f,g\} \in A \oplus A_C : \left[g - \ell f, \begin{pmatrix} \chi(\ell) \\ \chi_C(\ell) \end{pmatrix} \right] = 0 \right\}.$$

It follows from Theorem 4.3 that the self-adjoint extensions H of $S \oplus S_C$ in $\mathfrak{H} \oplus \mathfrak{H}_C$ with the property $H \cap (A \oplus A_C) = \widetilde{S}$ are in one-to-one correspondence with the $n \times n$ self-adjoint matrices E. The compression of the resolvent $(H-\ell)^{-1}$ to $\mathfrak{H}$ is given by

$$P_{\mathfrak{H}}(H-\ell)\,|_{\mathfrak{H}} = (A-\ell)^{-1} - \chi(\ell)(Q(\ell)+C(\ell)+E)^{-1}[\,\cdot\,,\chi(\bar{\ell})]. \tag{7.2}$$

For $E = 0$ this compressed resolvent coincides with $R(\ell)$ in (7.1).

Now consider (7.1) when the $n \times n$ matrix function $C(\ell)$ is not necessarily strict. Then there is a decomposition of the space $\mathbb{C}^n$ such that

$$C(\ell) = \begin{pmatrix} G(\ell) & M \\ M^* & N \end{pmatrix}, \tag{7.3}$$

where $G(\ell)$ is a strict $k \times k$ matrix Nevanlinna function and the matrix N is self-adjoint. The strict Nevanlinna function $G(\ell)$ gives rise to an exit space $\mathfrak{H}_G$, a closed completely non-self-adjoint symmetric operator S_G with defect numbers (k,k), $k < n$, and a self-adjoint extension A_G. Since the defect numbers of S_G and S are different, we proceed by decomposing the function $Q(\ell)$ as $C(\ell)$ above:

$$Q(\ell) = \begin{pmatrix} Q_{11}(\ell) & Q_{12}(\ell) \\ Q_{21}(\ell) & Q_{22}(\ell) \end{pmatrix},$$

and let $\widehat{Q}(\ell) = Q_{11}(\ell) - (Q_{12}(\ell)+M)(Q_{22}(\ell)+N)^{-1}(Q_{21}(\ell)+M^*)$ be the Schur complement of

$$\begin{pmatrix} Q_{11}(\ell) & Q_{12}(\ell)+M \\ Q_{21}(\ell)+M^* & Q_{22}(\ell)+N \end{pmatrix}. \tag{7.4}$$

Define the self-adjoint extension $\widehat{A}$ of S generated by $\widehat{Q}(\ell)$ as in (6.2), and define the function $\widehat{\chi}(\ell)$ by

$$\widehat{\chi}(\ell) = \chi(\ell) \begin{pmatrix} I \\ -(Q_{22}(\ell)+N)^{-1}(Q_{21}(\ell)+M^*) \end{pmatrix}.$$

It is a basis for $\ker(\widehat{S}^* - \ell)$, where $\widehat{S}$ is defined as in (6.4). Hence the defect numbers of $\widehat{S}$ match those of S_G. Define the symmetric extension $\widetilde{S}$ of $\widehat{S} \oplus S_G$ by

$$\widetilde{S} = \left\{ \{f,g\} \in \widehat{A} \oplus A_G : \left[g - \ell f, \begin{pmatrix} \widehat{\chi}(\ell) \\ \chi_G(\ell) \end{pmatrix} \right] = 0 \right\}.$$

We can now interpret (7.1) as the compression of a self-adjoint extension of $\widetilde{S}$ relative to the exit space $\mathfrak{H}_G$.

Theorem 7.1. *The self-adjoint extensions H of S with an exit in $\mathfrak{H}_G$ and with the property $H \cap (\widehat{A} \oplus A_G) = \widetilde{S}$ are in one-to-one correspondence with the $k \times k$ self-adjoint matrices E. The compression of the resolvent $(H-\ell)^{-1}$ to $\mathfrak{H}$ is given by*

$$(7.5) \quad P_{\mathfrak{H}}(H-\ell)^{-1}\,|_{\mathfrak{H}} = (\widehat{A}-\ell)^{-1} - \widehat{\chi}(\ell)\left(\widehat{Q}(\ell)+G(\ell)+E\right)^{-1}[\,\cdot\,,\widehat{\chi}(\bar{\ell})].$$

For $E=0$ this compressed resolvent coincides with $R(\ell)$ in (7.1).

Proof. Rewrite $(Q(\ell)+C(\ell))^{-1}$ in (7.1) in terms of $\widehat{Q}(\ell)$ of (7.4):

$$\begin{aligned}
&(Q(\ell)+C(\ell))^{-1} - \begin{pmatrix} 0 & 0 \\ 0 & (Q_{22}(\ell)+N)^{-1} \end{pmatrix} \\
&= \begin{pmatrix} I \\ -(Q_{22}(\ell)+N)^{-1}(Q_{21}(\ell)+M^*) \end{pmatrix} (\widehat{Q}(\ell)+G(\ell))^{-1} \\
&\quad \cdot \begin{pmatrix} I \\ -(Q_{22}(\bar{\ell})+N)^{-1}(Q_{21}(\bar{\ell})+M^*) \end{pmatrix}^* .
\end{aligned}$$

Hence it follows from the analog of (6.2) that (7.1) has the form

$$P_{\mathfrak{H}}(H-\ell)^{-1}\,|_{\mathfrak{H}} = (\widehat{A}-\ell)^{-1} - \widehat{\chi}(\ell)(\widehat{Q}(\ell)+G(\ell))^{-1}[\,\cdot\,,\widehat{\chi}(\bar{\ell})].$$

Then apply Proposition 6.1 and Theorem 4.3 to obtain (7.5). □

It follows from Proposition 4.2, the minimality of the model spaces, and (i) in Section 3, that the self-adjoint extensions described in Theorem 7.1 are minimal with respect to $\mathfrak{H}$. Note that they all satisfy $H \cap \mathfrak{H}_G^2 = S_G$. By replacing $C(\ell)$ in (7.1) with $PC(\ell)P$ (or $G(\ell)$ with $PG(\ell)P$) one obtains a symmetric extension $\widehat{S}_G$ of S_G in $\mathfrak{H}_G$. This decreases the size of the Schur complement in (7.3) and produces a symmetric extension of $\widehat{S}$ in $\mathfrak{H}$. Theorem 7.1 provides now a family of self-adjoint extensions $\widehat{H}$ of S with an exit in $\mathfrak{H}_G$ such that $\widehat{H} \cap \mathfrak{H}_G^2 = \widehat{S}_G \supset S_G$. If the function $C(\ell)$ is strict, then clearly $\widehat{S} = S$ and (7.5) coincides with (7.2). See also [20] for some applications of self-adjoint extensions in a fixed exit space.

Acknowledgements

We thank Professor Heinz Langer for initial discussions about this project. S. Hassi was supported by the EC-programme "Human Capital and Mobility", M. Kaltenbäck by "Fonds zur Förderung der wissenschaftlichen Forschung" of Austria, Project P 09832-MAT.

References

[1] ALPAY, D., BRUINSMA, P., DIJKSMA, A., DE SNOO, H.S.V.: Interpolation problems, extensions of symmetric operators and reproducing kernel spaces I; Operator Theory: Adv. Appl. 50 (1991), 35–82.

[2] ALPAY, D., BRUINSMA, P., DIJKSMA, A., DE SNOO, H.S.V.: Interpolation problems, extensions of symmetric operators and reproducing kernel spaces II; Integral Equations Operator Theory 14 (1991), 465–500, 15 (1992), 378–388.

[3] ANDO, T.: Reproducing kernel spaces and quadratic inequalities; Lecture notes, Hokkaido University, Sapporo 1987.

[4] BRUINSMA, P.: Interpolation problems for Schur and Nevanlinna pairs; Doctoral dissertation, Groningen 1991.

[5] BRUINSMA, P.: Degenerate interpolation problems for Nevanlinna pairs; Indag. Math., N.S. 2(2) (1991), 179–200.

[6] CODDINGTON, E.A.: Extension theory of formally normal and symmetric subspaces; Mem. Amer. Math. Soc. 134, Providence, R.I. 1973.

[7] CODDINGTON, E.A., LEVINSON, N.: Theory of ordinary differential equations; McGraw-Hill, New York Toronto London 1955.

[8] DERKACH, V.A.: On Weyl function and generalized resolvents of a hermitian operator in a Kreĭn space; Integral Equations Operator Theory 23 (1995), 387–415.

[9] DERKACH, V.A., MALAMUD, M.M.: Generalized resolvents and the boundary value problems for hermitian operators with gaps; J. Funct. Anal. 95 (1991), 1–95.

[10] DERKACH, V.A., MALAMUD, M.M.: The extension theory of hermitian operators and the moment problem; J. Math. Sciences 73 (1995), 141–242.

[11] DIJKSMA, A., LANGER, H., DE SNOO, H.S.V.: Eigenvalues and pole functions of Hamiltonian systems and eigenvalue depending boundary conditions; Math. Nachr. 161 (1993), 107–154.

[12] DIJKSMA, A., DE SNOO, H.S.V.: Selfadjoint extensions of symmetric subspaces; Pacific J. Math. 54 (1974), 71–99.

[13] HASSI, S., LANGER, H., DE SNOO, H.S.V.: Selfadjoint extensions for a class of symmetric operators with defect numbers $(1,1)$; 15th OT Conference Proceedings (1995), 115–145.

[14] HILLE, E.: Lectures on ordinary differential equations, Addison-Wesley, Reading, Menlo Park London Don Mills 1969.

[15] KATSNELSON, V.: Methods of J-theory in continuous interpolation problems of analysis; Part I, VINITI, Kharkov 1982 (in Russian), English translation by T. Ando, Sapporo 1985.

[16] KREĬN, M.G., LANGER, H.: Defect subspaces and generalized resolvents of an hermitian operator in the space Π_κ; Funct. Anal. Appl. 5 (1971), 136–146, 217–228.

[17] LANGER, H., TEXTORIUS, B.: On generalized resolvents and Q-functions of symmetric linear relations (subspaces) in Hilbert space; Pacific J. Math. 72 (1977), 135–165.

[18] LANGER, H., TEXTORIUS, B.: *L*-resolvent matrices of symmetric linear relations with equal defect numbers; applications to canonical differential equations; Integral Equations Operator Theory 5 (1982), 208–243.

[19] NUDELMAN, A.A.: On a generalization of classical interpolation problems; Dokl. Akad. Nauk SSSR 256 (1981), 790–793 (in Russian), English translation: Sov. Math. Dokl. 23 (1981), 125–128.

[20] PAVLOV, B.S.: The theory of extensions and explicitly-soluble models; Russian Math. Surveys 42:6 (1987), 127–168.

[21] TITCHMARSH, E.C.: Eigenfunction expansions associated with second-order differential equations, Part One, second edition, Oxford University Press, Oxford 1962.

S. Hassi
Department of Statistics
University of Helsinki
PL 54, 00014 Helsinki
Finland
hassi@cc.helsinki.fi

H.S.V. de Snoo
Department of Mathematics
University of Groningen
Postbus 800, 9700 AV Groningen
Nederland
desnoo@math.rug.nl

M. Kaltenbäck
Institut für Analysis, Technische Mathematik und Versicherungsmathematik
Technische Universität Wien
Wiedner Hauptstrasse 8-10/114, A-1040 Wien
Österreich
mbaeck@geometrie.tuwien.ac.at

1991 Mathematics Subject Classification: Primary 47A20, 47B25; Secondary 34B24, 46E22, 47A57

Submitted: May 31, 1996

Operator Theory:
Advances and Applications, Vol. 103

Properties of "derived" Hankel matrices

G. HEINIG

Subject of the paper are matrices $[a_{ij}]$ the entries of which satisfy a difference equation $a_{i+2,j} - 2a_{i+1,j+1} + a_{i,j+2} = 0$ or an analogous equation of higher order. Such matrices will be called *derived Hankel matrices.* The main results are a generalization of Kronecker's Theorem, Vandermonde factorization of infinite finite-rank derived Hankel matrices, description of their range, rank and signature, and inversion of finite triangular derived Hankel matrices. A definition of "derived Bezoutians" is also included.

1. Introduction

Hankel matrices generated by rational functions have several applications, for example in linear systems theory and root localization problems. In [H2] they were used in order to describe the inverse of a Sylvester transformation $\mathcal{S}(X) = AX - XB$. For the description of the group inverse of $\mathcal{S}$ in the same paper a new class of matrices was introduced. Since in the definition of this class besides rational functions their derivatives are included the matrices of the new class were called *derived Hankel matrices.* The aim of the present paper is to study some properties of derived Hankel matrices.

Note that derived Hankel matrices (actually this concerns rather derived Toeplitz matrices) appear very naturally in connection with the power series approach and numerical methods like difference methods for the solution of linear differential equations.

The entries a_{ij} of an infinite Hankel matrix satisfy the difference equation $a_{i+1,j} - a_{i,j+1} = 0$ whereas the entries of an infinite derived Hankel matrix (in the sense of [H2]) satisfy the equations

$$a_{i+2,j} - 2a_{i+1,j+1} + a_{i,j+2} = 0, \qquad i,j = 0,1\ldots.$$

They are therefore ω-structured matrices in the sense of [HR2] (see also [H1]) with the "structure matrix"

$$\omega = \begin{bmatrix} 0 & 0 & 1 \\ 0 & -2 & 0 \\ 1 & 0 & 0 \end{bmatrix}.$$

In [H1] it was shown that in the main ω-structured matrices with $\operatorname{rank}\omega = 2$ have the same properties as classical Hankel matrices. For example their inverses are some kind of Bezoutians and certainly fast algorithms for their inversion can be developed. It can be easily checked that all this fails to be true if $\operatorname{rank}\omega > 2$. In this case very little is known concerning the structure of the inverse. For derived Hankel matrices we have rank $\omega = 3$. Therefore the present paper can be considered as a contribution to the general understanding of ω-structured matrices.

Let us sketch the contents of the paper. In Section 2 the classes of m-derived Hankel matrices is introduced which is the class of matrices for which the entries on the antidiagonals are arithmetic progressions of m-th order. For $m = 0$ we get the classical Hankel matrices, for $m = 1$ the derived Hankel matrices. It is shown that Kronecker's Theorem, which claims that finite-rank Hankel matrices are generated by rational functions, can be generalized to m-derived Hankel matrices. Furthermore we present a representation of m-derived Hankel matrices with the help of differential operators.

In Section 3 we show that finite-rank derived Hankel matrices can be represented, like finite-rank Hankel matrices, as a product of a Vandermonde matrix, its transpose and a block diagonal matrix as a middle factor between them. For Hankel matrices the blocks of the middle factor are upper triangular Hankel matrices whereas for derived Hankel matrices these blocks are upper triangular derived Hankel matrices. One can expect that this can also be done for general m-derived Hankel matrices, only the proof of this is connected with some technical difficulties. The Vandermonde representation allows us to describe range, rank and signature of these matrices. It also can be used in order to construct generalized inverses for derived Hankel matrices.

In Section 4 we represent derived Hankel matrices with the help of "generating functions". This representation was the original definition of them in [H2]. In this connection "derived Bezoutians" are introduced. Whereas the description of the signature of Hankel matrices is very important for root localization problems, the consideration of derived Hankel matrices seems to add not very much to the classical root localization theory (as it is described, for example, in the classical paper [KN]).

In Section 5 we consider finite triangular derived Hankel matrices, which appear, in particular, in the middle factor of the Vandermonde factorization but are also of independent interest. The main result is a formula for the inverse matrix involving only triangular Toeplitz matrices and a diagonal matrix. The proof of this formula is nothing else than the translation of the method of variation of the parameter for linear first order differential equations into matrix language.

The inversion of general finite derived Hankel matrices remains an open problem. It is even unclear whether this problem has a reasonable solution, that means whether there exists a formula for the inverse or whether the structure of the inverse can be described in another reasonable way. But the result on Vandermonde factorization leads to the conjecture that a positive answer is possible for the case when the matrix admits an extension to an infinite derived Hankel matrix with the same rank, because then the inverse can be represented as the product of inverse Vandermonde and triangular derived Hankel matrices and is, therefore, highly structured.

2. Representations of m-derived Hankel matrices

We consider infinite matrices $A = [a_{ij}]_0^\infty$ with complex entries. A matrix of this form generates a linear operator acting from the space of all finite sequences into the space of all sequences. We say that A has finite rank r if the range of the corresponding operator has the dimension r. The subspace of all finite-rank matrices A will be denoted by $\mathcal{K}$.

We introduce a linear transformation ∇ in $\mathcal{K}$ in the space of infinite matrices by

$$\nabla A = [\, a_{i+1,j} - a_{i,j+1} \,]_0^\infty , \qquad A = [a_{ij}]_0^\infty .$$

Then ∇ maps $\mathcal{K}$ into $\mathcal{K}$. Clearly, a matrix A belongs to the kernel of ∇ if and only if A is a Hankel matrix $A = [s_{i+j}]_0^\infty$. The matrices in the kernel of ∇^m for $m > 1$ will be called *m-derived Hankel matrices.* 1-derived Hankel matrices will simply be called *derived Hankel matrices.*

The intersection of $\mathcal{K}$ with the kernel of the $(m+1)$-th power of ∇ will be denoted by $\mathcal{H}^{(m)}$. For simplicity we set $\mathcal{H} = \mathcal{H}^{(0)}$ and $\mathcal{H}' = \mathcal{H}^{(1)}$.

Let $f(\lambda)$ be a formal Laurent series of the form

$$f(\lambda) = \sum_{k=0}^{\infty} s_k \lambda^{-k-1}. \tag{2.1}$$

If the series converges in a neighbourhood of infinity we will identify the series with the sum function. We associate f with the Hankel matrix

$$H(f) = [s_{i+j}]_0^\infty .$$

We say that H is *the Hankel matrix generated by f*. The m-derived Hankel matrices generated by f are, by definition, the matrices

$$H^{(m)}(f) = \left[\frac{1}{2^m m!} \, (i - j)^m s_{i+j-m} \right]_0^\infty ,$$

where we put $s_i = 0$ for $i < 0$. We write $H'(f)$ instead of $H^{(1)}(f)$. If $\mathbf{f} = (f_0, \dots, f_m)$ is an $(m+1)$ tuple of Laurent series (2.1), then we set

$$H(\mathbf{f}) = \sum_{k=0}^{m} H^{(k)}(f_k).$$

It can easily be checked that $H^{(m)}(\mathbf{f})$ belongs to the kernel of ∇^{m+1}. That means, it belongs to $\mathcal{H}^{(m)}$ if and only if it has finite rank.

Remark 2.1. If m is even then $H^{(m)}(f)$ is symmetric, if m is odd then $H^{(m)}(f)$ is skew-symmetric.

Lemma 2.2. *For f given by (2.1),*

$$\nabla H^{(m)}(f) = \sum_{k=1}^{m} \frac{1-(-1)^k}{2^k k!} H^{(m-k)}(\lambda^{-k+1} f). \tag{2.2}$$

Proof. According to the definition the entries b_{ij} of $\nabla H^{(m)}(f)$ are given by

$$b_{ij} = \frac{1}{2^m m!} \left((i-j+1)^m - (i-j-1)^m\right) s_{i+j-m+1}. \tag{2.3}$$

Hence

$$\begin{aligned} b_{ij} &= \frac{1}{2^m m!} \sum_{k=1}^{m} \binom{m}{k} (1+(-1)^k)(i-j)^{m-k} s_{i+j+1-m} \\ &= \frac{1-(-1)^k}{2^k k!} \frac{1}{2^{m-k}(m-k)!} (i-j)^{m-k} s_{i+j+k-m-(k-1)}. \end{aligned}$$

This can be written in the form (2.2). □

From Lemma 2.2 we need only the fact that $\nabla H^{(m)}(f)$ is equal to $H^{(m-1)}(f)$ plus an $(m-2)$-derived Hankel matrix. This leads to the following.

Corollary 2.3. *For f of the form (2.1) and $k = 1, 2, \ldots,$*

$$\nabla^k H^{(k)}(f) = H(f).$$

According to Kronecker's Theorem a Hankel matrix has finite rank if and only if it is generated by a (strictly proper) rational function. For m-derived Hankel matrices an analogous theorem is true.

Theorem 2.4. *For any matrix $H \in \mathcal{H}^{(m)}$ there exists a unique $(m+1)$-tuple of strictly proper rational functions $\mathbf{f}$ such that $H = H^{(m)}(\mathbf{f})$.*

Proof. We prove the theorem by induction. For $m = 0$ the assertion is just Kronecker's Theorem. Assume that the assertion is true for $m-1$ and let $H \in \mathcal{H}^{(m)}$. Then, by definition, $\nabla^m H$ belongs to the kernel of ∇ and is therefore a Hankel matrix. Suppose that $\nabla^m H = H(f)$. According to Corollary 2.3 we have $\nabla^m H^{(m)}(f) = H(f)$. Thus $\nabla^m(H - H^{(m)}(f)) = 0$. That means $H = H^{(m)}(f) + H_1$ where $H_1 \in \mathcal{H}^{(m-1)}$. By assumption there exist $f_0, \ldots, f_{m-1}$ such that $H_1 = H(f_0, \ldots, f_{m-1})$. This implies $H = H(f_0, \ldots, f_{m-1}, f)$, which proves the theorem. □

We show now that derived Hankel matrices can be represented with the help of differential operators. For this we introduce the matrix $D = \operatorname{diag}(0, 1, \ldots)$. Then, for $A = [a_{ij}]_0^\infty$, we have

$$D^k A = [i^k a_{ij}]_0^\infty \quad \text{and} \quad AD^k = [j^k a_{ij}]_0^\infty.$$

Since the entries $s_{ij}^{(m)}$ of $H^{(m)}(f)$ are given by

$$s_{ij}^{(m)} = \frac{1}{2^m m!} \sum_{k=0}^{m} (-1)^k \binom{m}{k} i^k j^{m-k} s_{i+j-m},$$

where s_i are the Laurent coefficients of f at ∞, we obtain the following.

Lemma 2.5. *The m-derived Hankel matrix $H^{(m)}(f)$ admits a representation*

$$H^{(m)}(f) = \frac{1}{2^m m!} \sum_{k=0}^{m} (-1)^k \binom{m}{k} D^k H(\lambda^{-m} f) D^{m-k}. \tag{2.4}$$

We translate this into operator language. For this we introduce some spaces. Let $\mathcal{S}$ be the space of all formal Laurent series $\sum_{k=-\infty}^{\infty} x_k \lambda^k$ and $\mathcal{S}_+$ the subspaces of all these series for which $x_k = 0$ for all $k < 0$ and let $\mathcal{F}$, $\mathcal{F}_+$ the subspaces of $\mathcal{S}$, $\mathcal{S}_+$, respectively, of series with a finite number of nonvanishing coefficients. We denote by Q the operator acting from $\mathcal{S}$ to $\mathcal{S}_+$ by

$$Q \sum_{k=-\infty}^{\infty} x_k = \sum_{k=0}^{\infty} x_{-k-1} \lambda^k.$$

Any matrix $A = [a_{ij}]_0^\infty$ generates an operator from $\mathcal{F}_+$ to $\mathcal{S}_+$ via

$$\sum_{k=0}^{\infty} x_k \lambda^k \to \sum_{j=0}^{\infty} y_j \lambda^j, \qquad y_j = \sum_{k=0}^{\infty} a_{jk} x_k,$$

which will be identified with the matrix A.

For a Hankel matrix $H(f)$ we have

$$H(f) x(\lambda) = Q f(\lambda) x(\lambda). \tag{2.5}$$

This can be checked by computing $H(f)\lambda^k$ for $k = 0, 1, \dots$. The matrix D corresponds to the operator in $\mathcal{S}_+$ defined by

$$D x(\lambda) = \lambda x'(\lambda),$$

where the prime denotes the derivative in the power series sense. This relation can also be used to define the operator D in $\mathcal{S}$. With this definition we have

$$DQ + QD + Q = 0. \tag{2.6}$$

This follows from

$$DQx = \sum_{k=0}^{\infty} k x_{-k-1} \lambda^k \quad \text{and} \quad QDx = \sum_{k=0}^{\infty} (-k-1) x_{-k-1} \lambda^k,$$

where $x(\lambda) = \sum_{k=-\infty}^{\infty} x_k \lambda^k$.

In connection with Lemma 2.5 this leads to the following.

Theorem 2.6. *For any r-tuple* $\mathbf{f}$ *of Laurent series* (2.1), *the operator corresponding to the m-derived Hankel matrix* $H(\mathbf{f})$ *is a differential operator of the form*

$$H(\mathbf{f})x(\lambda) = Q\sum_{k=0}^{m} a_k(\lambda)D^k x(\lambda), \tag{2.7}$$

where $a_k(\lambda)$ *are certain rational functions.*

Proof. It suffices to prove the theorem for $H(\mathbf{f}) = H^{(m)}(f)$, where f is a single rational function. From (2.6) we get

$$D^k Q = (-1)^k \sum_{j=0}^{k} \binom{k}{j} QD^j. \tag{2.8}$$

Furthermore, we have the Leibniz formula

$$D^j \lambda^{-m} f(\lambda)x(\lambda) = \sum_{i=0}^{j} \binom{j}{i} g_{j-i}(\lambda)D^i x(\lambda),$$

where $g_k(\lambda) = D^k\left(\lambda^{-m}f(\lambda)\right)$. Inserting this and (2.6) into (2.4) and taking (2.5) into account we obtain

$$H^{(m)}(f)x(\lambda) = \frac{1}{2^m m!} Q \sum_{0\le i\le j\le k\le m} \binom{m}{k}\binom{k}{j}\binom{j}{i} g_{j-i}(\lambda)D^{m-k+i}x(\lambda). \tag{2.9}$$

This leads to a representation (2.7) for $H^{(m)}(f)$ and in this way also for $H(\mathbf{f})$. □

For the special case of $m = 1$ formula (2.9) simplifies to the following.

Corollary 2.7. *The operator corresponding to the derived Hankel matrix* $H(f_0, f_1)$ *can be represented in the form*

$$H(f_0, f_1)x(\lambda) = -Q\left(f_1(\lambda)x'(\lambda) + \left(\frac{1}{2}f_1'(\lambda) - f_0(\lambda)\right)x(\lambda)\right).$$

3. Vandermonde factorization

From now on we deal only with the classes $\mathcal{H}$ and $\mathcal{H}'$. Speaking about $p \times p$ upper triangular (derived) Hankel matrices we have in mind matrices A for which AJ_p, where J_p is the counteridentity,

$$J_p = \underbrace{\begin{pmatrix} & & 1 \\ & \cdot^{\cdot^{\cdot}} & \\ 1 & & \end{pmatrix}}_{p},$$

is upper triangular in the usual sense.

To begin with we introduce some notations. For $c \in \mathbb{C}$ and natural k we denote by $\ell_k(c)$ the column $\ell_k(c) = \text{col}\left(\binom{j}{k}c^{j-k}\right)_0^\infty$ and by $V_k(c)$ the matrix

$$V_k(c) = \text{row}\left[\ell_0(c)\,\ell_1(c)\,\ldots\,\ell_k(c)\right].$$

Let now $\mathbf{c}$ be an r-tuple of complex numbers, $\mathbf{c} = (c_i)_1^r$, and $\mathbf{k}$ an r-tuple of positive integers, $\mathbf{k} = (k_i)_1^r$. Then we denote by $V(\mathbf{c}, \mathbf{k})$ the matrix

$$V(\mathbf{c}, \mathbf{k}) = \text{row}\left[V(c_1, k_1)\,V(c_2, k_2)\,\ldots\,V(c_r, k_r)\right].$$

The matrix $V(\mathbf{c}, \mathbf{k})$ will be called *Vandermonde matrix corresponding to* $(\mathbf{c}, \mathbf{k})$.

It is well-known that finite-rank Hankel matrices admit a factorization with Vandermonde matrix factors. For methodical reason we derive this result.

Lemma 3.1. *For a complex number c, the Hankel matrix $H((\lambda - c)^{-p})$ admits a factorization*

$$H((\lambda - c)^{-p}) = V(c, p) J_p V(c, p)^T.$$

Proof. From the Laurent expansion

$$(\lambda - c)^{-p} = \sum_{j=0}^{\infty} \binom{j}{p-1} c^{j-p+1} \lambda^{-j-1}$$

we conclude the relation

$$H((\lambda - c)^{-p}) = \left[\binom{i+j}{p-1} c^{i+j-p+1}\right]_0^\infty.$$

In view of

(3.1) $$\binom{i+j}{p-1} = \sum_{k=0}^{p-1} \binom{i}{k} \binom{j}{p-1-k}$$

we obtain

$$H((\lambda - c)^{-p}) = \sum_{k=0}^{p-1} \ell_k(c) \ell_{p-k}(c)^T.$$

This is equivalent to the assertion. □

Now let f be any strictly proper rational function. Using the partial fraction decomposition of f we get from Lemma 3.1 the following.

Theorem 3.2. *The Hankel matrix $H(f)$ admits a factorization*

$$H(f) = V(\mathbf{c}, \mathbf{p}) M V(\mathbf{c}, \mathbf{p})^T,$$

where $\mathbf{c} = (c_i)_1^r$, $\mathbf{p} = (p_i)_1^r$, $M = \text{diag}\,(H_1, \ldots, H_r)$ and H_i are $p_i \times p_i$ nonsingular upper triangular Hankel matrices. The numbers c_i and p_i are uniquely determined by f and are the poles of f and their multiplicities.

Note that a very simple proof of this theorem can also be given with the help of realization theory (see [FH]).

An immediate consequence of Theorem 3.2 is the following range and rank description of Hankel matrices.

Corollary 3.3. *The range of $H(f)$ is equal to the span of the vectors $\ell_k(c_i)$ where $k = 0, \dots, p_i - 1$, $i = 1, \dots, r$. The rank of $H(f)$ equals the McMillan degree of f.*

We are going to present similar Vandermonde factorizations for matrices from the class $\mathcal{H}'$. The role of the matrix J_p in Lemma 3.1 will be played by the matrix

$$J'_{p+1} = \frac{1}{2}\begin{pmatrix} \mathbf{0} & & & -p \\ & & -p+2 & \\ & \cdot^{\cdot^{\cdot}} & & \\ & p-2 & & \\ p & & & \mathbf{0} \end{pmatrix}.$$

Note that J_{p+1} is the $(p+1) \times (p+1)$ principal submatrix of the derived Hankel matrix $H'(\lambda^{-p})$.

Lemma 3.4. *For a complex number c and $p = 1, 2, \dots,$ the matrix $H'((\lambda - c)^{-p})$ admits a factorization*

$$H'((\lambda - c)^{-p}) = V(c, p+1) J'_{p+1} V(c, p+1)^T. \tag{3.2}$$

Proof. According to the definition we have

$$H'((\lambda - c)^{-p}) = \left[\, s'_{ij} \,\right]_0^\infty = \left[\frac{1}{2}(i-j)\binom{i+j-1}{p-1} c^{i+j-p}\right]_0^\infty .$$

From (3.1) we get

$$i\binom{i+j-1}{p-1} = i\sum_{k=0}^{p-1}\binom{i-1}{k}\binom{j}{p-1-k} = \sum_{k=0}^{p} k\binom{i}{k}\binom{j}{p-k}$$

and

$$j\binom{i+j-1}{p-1} = j\sum_{k=0}^{p-1}\binom{i}{k}\binom{j-1}{p-1-k} = \sum_{k=0}^{p}(p-k)\binom{i}{k}\binom{j}{p-k}.$$

Hence

$$s'_{ij} = \sum_{k=0}^{p}(k - p/2)\binom{i}{k}\binom{i}{p-k} c^{j-p-k}.$$

This can be written in the form

$$H'((\lambda - c)^{-p}) = \sum_{k=0}^{p} (k - p/2)\ell_k(c)\ell_{p-k}(c)^T$$

and is equivalent to (3.2). □

From Lemma 3.4 we now get the following.

Theorem 3.5. *Let f be a strictly proper rational function, $\mathbf{c} = (c_1, \dots, c_r)$ the poles of f and $\mathbf{p} = (p_1, \dots, p_r)$ the corresponding multiplicities. Then $H'(f)$ admits a factorization*

$$H'(f) = V(\mathbf{c}, \mathbf{p} + \mathbf{1})MV(\mathbf{c}, \mathbf{p} + \mathbf{1})^T,$$

where $\mathbf{1} = (1, 1, \dots, 1)$, $M = \operatorname{diag}(H'_1, \dots, H'_r)$ and H'_i are $(p_i + 1) \times (p_i + 1)$ skew-symmetric upper triangular derived Hankel matrices.

The following range description is an immediate consequence of Theorem 3.5.

Corollary 3.6. *The range of the matrix $H'(f)$ is equal to the span of the vectors $\ell_k(c_i)$ for $k = 0, \dots, p_i$ and $k \neq p/2$. The rank of $H'(f)$ is given by*

$$\operatorname{rank} H'(f) = \deg f + \pi_- = \deg f' - \pi_+ = \sum_{p_i \text{ even}} p_i + \sum_{p_i \text{ odd}} (p_i + 1),$$

where "deg" denotes the McMillan degree, π_+ is the number of even order and π_- the number of odd order poles of f, and f' is the derivative of f.

Corollary 3.7. *The function f has only simple poles if and only if* $\operatorname{rank} H'(f) = 2 \deg f$.

We now consider general matrices from $\mathcal{H}'$.

Theorem 3.8. *Let f_0 and f_1 be strictly proper rational functions, $\mathbf{c} = (c_1, \dots, c_r)$ an r-tuple consisting of all numbers which are either poles of f_0 or f_1, and $(p_{l1}, \dots, p_{lr})$ the corresponding multiplicities for f_l, $l = 0, 1$. Then $H(f_0, f_1)$ admits a factorization*

$$H(f_0, f_1) = V(\mathbf{c}, \mathbf{p})MV(\mathbf{c}, \mathbf{p})^T,$$

where $\mathbf{p} = (p_1, \dots, p_r)$, $p_i = \max\{p_{0i}, p_{1i} + 1\}$, $M = \operatorname{diag}(H'_1, \dots, H'_r)$ and H'_i are $p_i \times p_i$ upper triangular derived Hankel matrices.

Different to the case $H'(f)$, in the general case we will get only an estimation for the range of $H(f_0, f_1)$.

Corollary 3.9. *The range of $H(f_0, f_1)$ is contained in the subspace $L =$* $\operatorname{span}\{\ell_k(c_i) : k = 0, \dots, p_i - 1;\ i = 1, \dots, r\}$. *The codimension of the range of $H(f_0, f_1)$ in L is less or equal to the number of poles of f_1 with even order.*

Corollary 3.10. *If f_0 and f_1 have only simple poles then*

$$\operatorname{rank} H(f_0, f_1) = \deg f_0 + 2\deg f_1 - \rho(f_0, f_1),$$

where $\rho(f_0, f_1)$ denotes the number of common poles of f_0 and f_1. The rank of $H(f_0, f_1)$ equals $2\deg f_1 + \deg f_0$ if and only if f_0 and f_1 have only simple and no common poles.

One of the most important facts concerning Hankel matrices generated by rational functions is that for real f the signature of $H(f)$ is equal to the Cauchy index of the function f along to the real line $\mathbb{R}$. Recall that the Cauchy index is the difference of the number of poles on $\mathbb{R}$ of f for which f jumps from $-\infty$ to ∞ and the number of real poles for which f jumps from ∞ to $-\infty$. The consideration of derived Hankel matrices seems to add not very much to this theorem. Let us explain this.

First we note that the matrix $H \in \mathcal{H}'$ is hermitian if and only if it is of the form $H = H(f_0, \mathrm{i}f_1)$ where f_0 and f_1 are real and $\mathrm{i} = \sqrt{-1}$. Since $H'(\mathrm{i}f)$ is skew-symmetric the signature of $H'(\mathrm{i}f)$ is always equal to 0. That means it contains no information about the function f. In the signature of general hermitian matrices $H(f_0, \mathrm{i}f_1)$ some information about the functions f_0 and f_1 is contained. We formulate a particular result, which follows from Theorem 3.8.

Corollary 3.11. *If f_0 has only simple poles then the signature of $H(f_0, \mathrm{i}f_1)$ equals the difference of the number of real poles of f_0 which are not poles of f_1 and for which f_0 jumps from $-\infty$ to ∞ and the number of these poles for which f_0 jumps from ∞ to $-\infty$.*

Remark 3.12. All statements in this section can also be formulated for the $k \times k$ principal submatrices of the matrices under consideration, provided that k is sufficiently large. For example, rank and signature of $H'(f)$ coincide with those of its $k \times k$ principal submatrix if $k \geq 2\deg f$.

Note that the Vandermonde factorization of derived Hankel matrices can be used to construct generalized inverses of them. Let us shortly explain this. Vandermonde factorization means a representation of the form $H = H(f_0, f_1) = VMV^T$. Assume, for simplicity, that the middle factor M is nonsingular. The matrix V^TV is a generalized Cauchy matrix and its inverse can be constructed easily. Inversion of triangular derived Hankel matrices, which are the blocks of the block diagonal matrix M will be discussed in Section 5. Therefore, it is, in principle, not difficult to find the matrix $X = (V^TV)^{-1}M^{-1}(V^TV)^{-1}$. We form now the matrix $G = VXV^T$. It can be shown that $HGH = H$ and $GHG = G$. That means that G is a symmetric generalized inverse of H.

4. Generating functions and Bezoutians

The $(-)$-*generating function* for an infinite matrix $A = [a_{ij}]_0^\infty$ is, by definition, the formal Laurent series

$$A^-(\lambda,\mu) = \sum_{i,j=0}^{\infty} a_{ij}\lambda^{-i-1}\mu^{-j-1}.$$

If the series converges in a neighbourhood of (∞,∞) then it will be identified with the sum function.

For a Hankel matrix $H(f)$ it is easy to check that

$$H(f)^-(\lambda,\mu) = -\frac{f(\lambda)-f(\mu)}{\lambda-\mu}.$$

For derived Hankel matrices we have the following.

Theorem 4.1. *The $(-)$-generating function of a derived Hankel matrix $H(f_0,f_1)$ is given by*

$$\text{(4.1)} \quad H(f_0,f_1)^-(\lambda,\mu) = \frac{1}{\lambda-\mu}\left(\frac{f_1(\lambda)-f_1(\mu)}{\lambda-\mu} - \frac{1}{2}(f_1'(\lambda)+f_1'(\mu))\right) - \frac{f_0(\lambda)-f_0(\mu)}{\lambda-\mu}.$$

Proof. It is obviously sufficient to prove the theorem for $f_0 = 0$ and $f_1(\lambda) = \lambda^{-p}$. We compute the right-hand side of (4.1) for this choice and obtain for it

$$\frac{\lambda^{-p-1}\mu^{-p-1}}{(\lambda-\mu)^2}\left(\lambda\mu^{p+1} - \lambda^{p+1}\mu + \frac{p}{2}(\mu^{p+1}+\lambda^{p+1})(\lambda-\mu)\right)$$
$$= \frac{\lambda^{-p-1}\mu^{-p-1}}{(\lambda-\mu)^2}\left(\frac{p}{2}\lambda^{p+2} + \frac{p+2}{2}\lambda\mu^{p+1} - \frac{p+2}{2}\lambda^{p+1}\mu - \frac{p}{2}\mu^{p+2}\right).$$

Let $R = [r_{ij}]$ be the matrix with this $(-)$-generating function. Then we see that the entries r_{ij} equal zero for $i+j \neq p$. Suppose that $r_{p-j,j} = \sigma_j$, $j = 0,\dots,p$. Comparing coefficients we obtain

$$\sigma_{j+1} - 2\sigma_j + \sigma_{j-1} = 0, \qquad j = 1,\dots,p-1,$$

and

$$\sigma_0 = -\frac{p}{2}, \qquad \sigma_p = \frac{p}{2}, \qquad \sigma_1 = \frac{-p+2}{2}, \qquad \sigma_{p-1} = \frac{p-2}{2}.$$

From this we conclude that

$$\sigma_j = j - \frac{p}{2} = \frac{j-i}{2}.$$

That means $R = H'(\lambda^{-p})$. □

Hankel matrices generated by rational functions are related to Bezoutians. The advantage of the Bezoutians is that they are finite matrices but reflect many relevant properties of the infinite Hankel matrix. In order to define Bezoutians we use the (+)-generating function of a matrix $A = [a_{ij}]_0^{n-1}$ defined by

$$A^+(\lambda,\mu) = \sum_{i,j=0}^{n-1} a_{ij}\lambda^i\mu^j.$$

The Bezoutian of two polynomials $a(\lambda)$ and $b(\lambda)$ with degree $\leq n$ is, by definition, the matrix $\mathrm{Bez}(a,b)$ with the (+)-generating function

$$\mathrm{Bez}(a,b)^+(a,b) = \frac{a(\lambda)b(\mu) - b(\lambda)a(\mu)}{\lambda - \mu}.$$

The following relation is well-known (see [KN], [HR1]):

$$\mathrm{Bez}(a,b) = -\begin{bmatrix} a_1 & \dots & a_n \\ \vdots & ⋰ & \\ a_n & & \end{bmatrix} H_n(b/a) \begin{bmatrix} a_1 & \dots & a_n \\ \vdots & ⋰ & \\ a_n & & \end{bmatrix},$$

where $a(\lambda) = \sum_{k=0}^n a_k\lambda^k$ and $H_n(f)$ is the $n \times n$ principal submatrix of $H(f)$.

We are looking for a similar representation for derived Hankel matrices. For this we define the *derived Bezoutian* of two polynomials $a(\lambda)$ and $b(\lambda)$ ($n = \deg a > \deg b$) as the $n^2 \times n^2$ matrix given by

$$\mathrm{Bez}'(\mathrm{a,b})^+(\lambda) = \frac{1}{\lambda-\mu}\left(a(\lambda)B(\lambda,\mu)a(\mu) - \frac{1}{2}(g(\lambda)a(\mu)^2 + a(\lambda)^2 g(\mu))\right),$$

where

$$B(\lambda,\mu) = \mathrm{Bez}(a,b)^+(\lambda,\mu), \quad g(\lambda) = a(\lambda)b'(\lambda) - b(\lambda)a'(\lambda).$$

Note that $\mathrm{Bez}'(a,b)$ is always skew-symmetric but in general $\mathrm{Bez}'(a,b) \neq \mathrm{Bez}'(b,a)$.

Theorem 4.2. *The derived Bezoutian* $\mathrm{Bez}'(a,b)$ *is related to the derived Hankel matrix* $H'(b/a)$ *via the relation*

$$\mathrm{Bez}'(a,b) = \begin{bmatrix} c_1 & \dots & c_{2n} \\ \vdots & ⋰ & \\ c_{2n} & & \end{bmatrix} H'_{2n}(b/a) \begin{bmatrix} c_1 & \dots & c_{2n} \\ \vdots & ⋰ & \\ c_{2n} & & \end{bmatrix},$$

where $a(\lambda)^2 = \sum_{k=0}^{2n} c_k\lambda^k$ *and* $H'_{2n}(b/a)$ *is the* $2n \times 2n$ *principal submatrix of* $H'(b/a)$.

Proof. By definition, the $(+)$-generating function of $\mathrm{Bez}'(a,b) = [b_{ij}]_0^{2n-1}$ is equal to the $(-)$-generating function of $H'(b/a) = [s_{ij}]_0^{\infty}$ multiplied by $a(\lambda)^2 a(\mu)^2$.

$$\begin{aligned}\sum_{i,j=0}^{2n-1} b_{ij}\lambda^i\mu^j &= \sum_{p,q=0}^{2n-1}\sum_{r,t=0}^{\infty} c_p s_{rt} c_q \lambda^{p-r-1}\mu^{q-t-1} \\ &= \sum_{i,j=0}^{2n-1}\sum_{r,t=0}^{\infty} c_{i+r+1} s_{rt} c_{j+t+1}\lambda^i\mu^j.\end{aligned}$$

Comparing the coefficients in this equality we obtain the assertion. □

Corollary 4.3. *The matrix* $\mathrm{Bez}'(a,b)$ *is nonsingular if and only if the polynomials* $a(\lambda)$ *and* $b(\lambda)$ *have no roots in common and all roots of* $a(\lambda)$ *are simple.*

5. Triangular derived Hankel matrices

In this section we deal with triangular derived Hankel matrices. Let $\mathcal{H}'_n$ denote the class of all $(n+1)\times(n+1)$ upper triangular derived Hankel matrices. According to Theorem 2.4 for $H_n \in \mathcal{H}'_n$ there exist unique polynomials f_0, f_1 in λ^{-1},

(5.1) $$f_0(\lambda) = s_{00}\lambda^{-1} + \ldots + s_{0n}\lambda^{-n-1}, \qquad f_1(\lambda) = s_{10}\lambda^{-1}\ldots + s_{1,n-1}\lambda^{-n},$$

such that H_n is the $(n+1)\times(n+1)$ principal submatrix of the matrix $H(f_0,f_1)$. We write this in the form

(5.2) $$H_n = H_n(f_0,f_1).$$

For convenience we represent these matrices with the help of triangular Toeplitz matrices. To this end we introduce the following notation. If $a(\lambda) = \sum_{k=0}^{\infty} a_k\lambda^k$ is a formal power series then we denote by $T_n(a)$ the $(n+1)\times(n+1)$ lower triangular Toeplitz matrix $T_n(a) = [\,a_{i-j}\,]_0^n$ (put $a_i = 0$ if $i<0$). Furthermore, we introduce the diagonal matrix $D_n = \mathrm{diag}\,(0,1,\ldots,n)$.

Lemma 5.1. *Let* H_n *be of the form* (5.2) *where* f_0, f_1 *are given by* (5.1). *Then* H_n *admits a representation*

(5.3) $$H_n = J_{n+1}(T_n(a)D_n + T_n(b)),$$

where $a_i = -s_{1,n-i}$ *and* $b_i = s_{0i} + \frac{1}{2}(n-i)s_{1,n-i}$.

The proof is a straightforward verification.

Matrix-vector multiplication by Toeplitz matrices is related to polynomial multiplication. Let, for $x = (x_k)_0^n \in \mathbb{C}^n$, $x(\lambda)$ denote the polynomial $x(\lambda) = x_0 + x_1\lambda + \ldots + x_n\lambda^n$. Then we have, for a polynomial $g(\lambda)$ and $x \in \mathbb{C}^n$,

$$(T_n(g)x)(\lambda) = g(\lambda)x(\lambda) \mod \lambda^{n+1}.$$

Furthermore,

$$(D_n x)(\lambda) = \lambda x'(\lambda),$$

where the prime denotes the derivative. The product rule for differentation $(gx)' = g'x + gx'$ leads to the matrix equality

$$D_n T_n(g) = T_n(\lambda g') + T_n(g) D_n. \tag{5.4}$$

We now consider matrices of the form

$$A = D_n + T_n(c). \tag{5.5}$$

Let $D_n^\dagger$ denote the matrix $D_n^\dagger = \operatorname{diag}(0, 1, \frac{1}{2}, \frac{1}{3}, \ldots, \frac{1}{n})$. Note that $D_n^\dagger$ is the Moore-Penrose inverse of D_n.

Theorem 5.2. *Let A be given by (5.5), and let $g(\lambda) = \exp(D_n^\dagger c)(\lambda)$. Then*

$$A = T_n(g)^{-1}(D_n + c_0 I_{n+1}) T_n(g). \tag{5.6}$$

In particular, A is similar to $D_n + c_0 I_{n+1}$.

Proof. We have

$$(D_n^\dagger c)(\lambda) = \int_0^\lambda \frac{c(\mu) - c_0}{\mu}\, \mathrm{d}\mu.$$

Hence

$$g'(\lambda) = g(\lambda)\frac{c(\lambda) - c_0}{\lambda},$$

which implies

$$T_n(\lambda g'(\lambda)) = T_n(g) T_n(c) - c_0 T_n(g) + T_n(g) D_n.$$

Inserting this relation into (5.4) we get (5.6). □

Note that the proof of Theorem 5.2 is related to the method of variation of the parameter for linear first order differential equations.

Now we present the main result of this section. It follows immediately from Theorem 5.2 using the factorization $H_n = J_{n+1} T_n(a)(D_n + T_n(c))$, where $c(\lambda) = b(\lambda)/a(\lambda)$, provided that $a_0 \neq 0$.

Theorem 5.3. *Let H_n be given by (5.3) and let $a_0 \neq 0$. Furthermore, suppose that*

$$g(\lambda) = \exp \int_0^\lambda b(\mu)/a(\mu) \mathrm{d}\mu,$$

$$u(\lambda) = g(\lambda)^{-1}, \qquad v(\lambda) = g(\lambda)/a(\lambda).$$

Then H_n is nonsingular if and only if $k + b_0/a_0 \neq 0$ for $k = 0, \ldots, n$ and

$$H_n^{-1} = T_n(u) \operatorname{diag}\left((k + b_0/a_0)^{-1}\right)_{k=0}^n T_n(v) J_{n+1}.$$

The situation is a little more involved if the assumption $a_0 \neq 0$ of Theorem 5.3 is not fulfilled. We discuss only the case when $a_0 = 0$ but $a_1 \neq 0$. The general case can be treated similarly. In this case the matrix (5.3) can be represented in the form

$$H_n = J_{n+1}T(\lambda^{-1}a)(S_nD_n + T(c)), \tag{5.7}$$

where S_n denotes the forward shift, $S_n = T_n(\lambda)$, and $c(\lambda) = \lambda b(\lambda)/a(\lambda) = \sum_{k=0}^{\infty} c_k\lambda^k$. Note that

$$c_0 = b_0/a_1 \quad \text{and} \quad c_1 = b_1/a_1 - a_2b_0/a_1^2. \tag{5.8}$$

In view of (5.7) it remains to consider matrices of the form

$$A = S_nD_n + T_n(c).$$

Now, let $g(\lambda)$ be defined by

$$\frac{g'(\lambda)}{g(\lambda)} = \frac{c(\lambda) - c_0 - c_1\lambda}{\lambda^2}. \tag{5.9}$$

Then we obtain from (5.4)

$$T_n(g)^{-1}(c_0I_{n+1} + c_1S_n + S_nD_n)T_n(g) = S_nD_n + T_n(c).$$

In this way the inversion problem for A is reduced to the problem to find the inverse of the 2-diagonal matrix $c_0I_n + c_1S_n + S_nD_n$ where $c_0 \neq 0$. For this an explicit formula exists. Thus we arrived at the following.

Theorem 5.4. *Let H_n be given by* (5.3) *where $a_0 = 0$ and $a_1 \neq 0$. Then H_n is nonsingular if and only if $b_0 \neq 0$ and the inverse can be represented in the form*

$$H_n^{-1} = T_n(u)(c_0I_{n+1} + c_1S_n + S_nD_n)^{-1}T_n(v)J_{n+1},$$

where

$$u(\lambda) = g(\lambda)^{-1}, \quad v(\lambda) = g(\lambda)/\lambda a(\lambda),$$

$g(\lambda)$ is given by (5.9) *and c_0, c_1 by* (5.8).

In case that $a(\lambda)$ has a zero of order m at 0 the middle factor in the inversion formula will be a triangular $(m+1)$-banded derived Toeplitz marix.

References

[EL] Ellis, R.I., Lay, D.C.: Factorization of finite rank Hankel and Toeplitz matrices; Linear Algebra Appl. 173 (1992), 19–38.

[FH] Feldmann, S., Heinig, G.: Vandermonde factorization and canonical representation of block Hankel matrices; Linear Algebra Appl. 241-243 (1996), 247–278.

[H1] HEINIG, G.: On structured matrices, generalized Bezoutians, and generalized Christoffel-Darboux formulas; Operator Theory: Adv. Appl. 50, Birkhäuser Verlag, Basel 1991, 267–281.

[H2] HEINIG, G.: The group inverse of the transformation $\mathcal{S}(X) = AX - XB$; Linear Algebra Appl. 257 (1997), 321–342.

[HR1] HEINIG, G., ROST, K.: Algebraic Methods for Toeplitz-like matrices and operators; Akademie Verlag, Berlin, and Birkhäuser Verlag, Basel 1984.

[HR2] HEINIG, G., ROST, K.: Matrices with displacement structure, generalized Bezoutians, and Möbius transformations; The Gohberg Anniversary Volume I, Operator Theory: Adv. Appl. 40, Birkhäuser Verlag, Basel 1989, 813–834.

[KN] KREIN, M., NAIMARK, M.: The method of symmetric and Hermitian forms in the theory of the separation of the roots of algebraic equations; Linear and Multilinear Algebra 56 (1974), 69–87.

Georg Heinig
Kuwait University
Dept. of Mathematics and Computer Science
P.O.Box 5969
Safat 13060
Kuwait
georg@kuc01.kuniv.edu.kw

1991 Mathematics Subject Classification: Primary 47B35; Secondary 15A09, 15A23, 26C15, 39A70

Submitted: February 27, 1996

Operator Theory:
Advances and Applications, Vol. 103

The probability that a (partial) matrix is positive semidefinite

C.R. JOHNSON and G. NÆVDAL

Assuming that a_{ij} is distributed uniformly in $[-1, 1]$ and $a_{ii} = 1$, we compute the probability that a symmetric matrix $A = [a_{ij}]_{i,j=1}^{n}$ is positive semidefinite. The probability is also computed if A is a Toeplitz matrix. Finally, some results for partial matrices are presented.

1. Introduction

Let A be a real symmetric matrix with ones on the diagonal. What is the probability that such a matrix is positive semidefinite? The probability of course depends on the distribution of the off-diagonal entries. If the entries are independently distributed, with uniform distribution in the interval $[-1, 1]$, it is possible to obtain exact formulas. (Of course we assume $a_{ij} = a_{ji}$.) The basic idea in the computation of this probability is to use the Schur parametrization of positive semidefinite matrices. This technique is even more transparent when we add the extra condition that the matrix is a Toeplitz matrix, so we present this result first. In the final sections we will discuss the case of partial matrices. In this paper we only consider real matrices.

Recall that a *partial matrix* is a rectangular array where some entries are specified, the others unspecified. By assigning values to the unspecified entries we get an ordinary matrix, which is a *completion* of the partial matrix. A partial matrix is *partial positive semidefinite* if every principal submatrix containing only specified entries is positive semidefinite. The positions of the specified entries (unspecified entries, respectively) determine the *pattern* of the partial matrix. Assuming that the partial matrix has ones on the diagonal and $a_{ij} = a_{ji}$, we ask for the probability that the partial matrix is positive semidefinite. One may also ask about the probability that the partial matrix has a positive semidefinite completion. The last question seems in general to be more complicated. For an introduction to matrix completion problems, see [Jo].

The Schur parametrization gives a one to one correspondence between the positive semidefinite matrices with ones on the diagonal and and a set of numbers $s_{ij} \in [-1, 1]$, $i < j$. The numbers s_{ij} are called the Schur parameters. The use of Schur parameters originated in Schur's study of interpolation problems of analytic functions on the unit disc [S]. It was introduced to study positive semidefinite matrices by Constantinescu, [C] (see also [BC, FF, FK]).

Let

$$A = \begin{pmatrix} 1 & a_{12} & \cdots & a_{1n} \\ a_{12} & \ddots & \ddots & \vdots \\ \vdots & \ddots & \ddots & a_{n-1\,n} \\ a_{1n} & \cdots & a_{n-1\,n} & 1 \end{pmatrix}$$

be a positive semidefinite $n \times n$ matrix with ones on the diagonal. Then the Schur parameters, s_{ij}, are uniquely determined, and may be computed recursively as follows. Let

$$s_{i\,i+1} = a_{i\,i+1}, \qquad i = 1, \ldots, n-1, \tag{1.1}$$

$$s_{i\,i+k} = \frac{a_{i\,i+k} - c_{i\,i+k}}{\sqrt{r_{i\,i+k}}\sqrt{l_{i\,i+k}}}, \qquad 2 \le k \le n-i, \tag{1.2}$$

where $s_{i\,i+k} = 0$ if any of the above squareroots is zero, and

$$c_{i\,i+k} = EF^+G \tag{1.3}$$

$$r_{i\,i+k} = \prod_{j=1}^{i+k-1} (1 - s_{ij}^2), \tag{1.4}$$

$$l_{i\,i+k} = \prod_{j=1}^{k-1} (1 - s_{i+j\,i+k}^2), \tag{1.5}$$

with

$$E = \begin{pmatrix} a_{i1} & \cdots & a_{i\,i+k-1} \end{pmatrix},$$

$$F = \begin{pmatrix} 1 & a_{i+1\,2} & \cdots & a_{i+1\,i+k-1} \\ a_{i+1\,2} & \ddots & \ddots & \vdots \\ \vdots & \ddots & \ddots & a_{i+k-2\,i+k-1} \\ a_{i+1\,i+k-1} & \cdots & a_{i+k-2\,i+k-1} & 1 \end{pmatrix},$$

$$G = \begin{pmatrix} a_{i+1\,i+k} \\ \vdots \\ a_{i+k-1\,i+k} \end{pmatrix},$$

where F^+ denotes the Moore-Penrose inverse of F.

If A is a Toeplitz matrix the parametrization is greatly simplified, since in that case there is only one Schur parameter for each superdiagonal.

2. The case of full matrices

In this section we compute the probability that an $n \times n$ matrix with ones on the diagonal is positive semidefinite.

Let S be the event that

$$(2.1)\qquad B = \begin{pmatrix} 1 & a_{12} & \dots & a_{1\,n-1} & ? \\ a_{12} & \ddots & \ddots & & a_{2n} \\ \vdots & \ddots & \ddots & \ddots & \vdots \\ a_{1\,n-1} & & \ddots & \ddots & a_{n-1\,n} \\ ? & a_{2n} & \dots & a_{n-1\,n} & 1 \end{pmatrix}$$

is partial positive semidefinite. Then the probability that the full matrix A is positive semidefinite is

$$(2.2)\qquad \int_S p(a_{1n})\, dF(\{a_{ij} : 1 \leq j - i \leq n-1\})$$

where $p(a_{1n})$ is the probability of filling in the $(1,n)$ entry in B such that A becomes positive semidefinite, and F is the distribution function of the specified entries in B. We are integrating this over the event that B is partial positive semidefinite. This set up holds for a general choice of distributions.

We will now use this general set up to compute the probability of certain matrices being positive semidefinite. We will start with the case when the matrix is a Toeplitz matrix, as the extra structure of these matrices simplifies the calculations. Therefore, let us assume that we have a symmetric Toeplitz matrix with ones on the diagonal, and that the off-diagonal entries a_{i-j} are uniformly distributed in the interval $[-1,1]$. Then the 2×2 symmetric Toeplitz matrix

$$\begin{pmatrix} 1 & a_1 \\ a_1 & 1 \end{pmatrix}$$

is positive semidefinite with probability 1.

To compute the probability that a 3×3 symmetric Toeplitz matrix is positive semidefinite we take advantage of the Schur parametrization of positive semidefinite Toeplitz matrices. The 3×3 Toeplitz matrix

$$\begin{pmatrix} 1 & a_1 & a_2 \\ a_1 & 1 & a_1 \\ a_2 & a_1 & 1 \end{pmatrix}$$

is positive semidefinite if and only if there exist Schur parameters $s_i \in [-1,1]$, $i = 1,2$, such that

$$(2.3)\qquad a_1 = s_1,$$

$$(2.4)\qquad a_2 = s_1^2 + s_2(1 - s_1^2).$$

Taking into account the fact that a_i is uniformly distributed in $[-1,1]$ we get the following when applying (2.2) in this case

$$\begin{aligned} dF(a_1) &= \frac{da_1}{2}, \\ p(a_2) &= 1 - s_1^2. \end{aligned}$$

The last equation follows from the fact that s_1 is fixed when a_1 is given. By changing variables we get

$$\int_{-1}^{1} p(a_2)\frac{da_1}{2} = \int_{-1}^{1} 1 - s_1^2 \frac{ds_1}{2} = \frac{2}{3}.$$

Here we have exploited the fact that the probability that a_2 is chosen such that the matrix is positive semidefinite does only depend on the last term of the right hand side in (2.4).

The general set up for Toeplitz matrices takes advantage of the following Schur parametrization of the positive semidefinite Toeplitz matrices:

$$\begin{aligned} a_1 &= s_1, \\ a_2 &= f_1(s_1) + s_2(1 - s_1^2), \\ a_3 &= f_2(s_1, s_2) + s_3(1 - s_1^2)(1 - s_2)^2, \\ &\vdots \\ a_n &= f_{n-1}(s_1, \ldots, s_{n-1}) + s_n \prod_{i=1}^{n-1}(1 - s_i^2). \end{aligned} \tag{2.5}$$

Then we obtain that the probability that an $n+1 \times n+1$ Toeplitz matrix

$$T_n = \begin{pmatrix} 1 & a_1 & \ldots & a_n \\ a_1 & \ddots & \ddots & \vdots \\ \vdots & \ddots & \ddots & a_1 \\ a_n & \ldots & a_1 & 1 \end{pmatrix}$$

is positive semidefinite if the a_i's are distributed uniformly in the interval $[-1,1]$ is

$$p(T_n) = \int_S p(a_n)\, dF(a_1, \ldots, a_{n-1}) \tag{2.6}$$

where S represents the event that the $n \times n$ Toeplitz matrix

$$T_{n-1} = \begin{pmatrix} 1 & a_1 & \ldots & a_{n-1} \\ a_1 & \ddots & \ddots & \vdots \\ \vdots & \ddots & \ddots & a_1 \\ a_{n-1} & \ldots & a_1 & 1 \end{pmatrix}$$

is positive semidefinite, $p(a_n)$ the probability that a_n is such that a T_n is positive semidefinite given that T_{n-1} is, and $dF(a_1, \ldots, a_{n-1})$ represents the probability distribution of $a_1, \ldots, a_{n-1}$. When the a_i's are uniformly distributed in $[-1, 1]$

$$dF(a_1, \ldots, a_{n-1}) = \frac{da_1}{2}\frac{da_2}{2}\cdots\frac{da_{n-1}}{2}.$$

The probability can be computed changing variables from the given data

$$a_1, \ldots, a_n$$

to the Schur parameters. The relationship is given by (2.5) and the Jacobian determinant for the variable transform is

$$\frac{\partial(a_1, \ldots, a_{n-1})}{\partial(s_1, \ldots, s_{n-1})} = \prod_{i=1}^{n-2}(1 - s_1^2)\cdots(1 - s_i^2).$$

Using Schur parameters it also follows that

$$p(a_n) = \prod_{i=1}^{n-1}(1 - s_i^2).$$

This gives

$$\begin{aligned} p(T_n) &= \int_{-1}^{1}\cdots\int_{-1}^{1}\prod_{i=1}^{n-1}(1 - s_i^2)\frac{\partial(a_1, \ldots, a_{n-1})}{\partial(s_1, \ldots, s_{n-1})}\frac{ds_1}{2}\cdots\frac{ds_{n-1}}{2} \\ &= \frac{1}{2^{n-1}}\int_{-1}^{1}\cdots\int_{-1}^{1}(1 - s_1^2)^{n-1}(1 - s_2^2)^{n-2}\cdots(1 - s_{n-1}^2)ds_1\cdots ds_{n-1} \\ &= \left(\frac{2}{3}\right)^{n-1}\left(\frac{4}{5}\right)^{n-2}\cdots\left(\frac{2n}{2n+1}\right). \end{aligned}$$

In the last of the equalities above it was used that

$$(2.7) \qquad I_{2k+1} = \int_{-1}^{1}\frac{(1 - s^2)^k}{2}ds = \int_0^{\frac{\pi}{2}}(1 - \sin^2\theta)^k\cos\theta d\theta = \frac{2\cdot 4\cdots(2k)}{3\cdot 5\cdots(2k+1)}$$

if k is an integer. The last equality in the equation above is known as a case of Wallis' formulae, see [Ja, p. 581].

Now, we will extend this result to the slightly more complicated case of a general matrix. Then the connection between the entries, a_{ij}, and the Schur parameters, s_{ij}, is

$$(2.8) \qquad a_{i\,i+k} = c_{i\,i+k} + s_{i\,i+k}\prod_{j=1}^{k-1}(1 - s_{i+j\,i+k}^2)^{\frac{1}{2}}\prod_{j=i+1}^{i+k-1}(1 - s_{ij}^2)^{\frac{1}{2}}.$$

This is obtained using (1.2), (1.3), (1.4) and (1.5).

Let us compute the probability that the 3×3 matrix

$$\begin{pmatrix} 1 & a_{12} & a_{13} \\ a_{12} & 1 & a_{23} \\ a_{13} & a_{23} & 1 \end{pmatrix}$$

is positive semidefinite under the assumption that a_{ij} is uniformly distributed in the interval $[-1,1]$. By (2.1) and (2.2) it follows that this probability is

$$\int_S p(a_{13})\, dF(a_{12}, a_{23})$$

where S is the event that the partial matrix

$$\begin{pmatrix} 1 & a_{12} & ? \\ a_{12} & 1 & a_{23} \\ ? & a_{23} & 1 \end{pmatrix}$$

is partial positive semidefinite. This is always true under our assumptions. Here

$$dF(a_{12}, a_{23}) = \frac{da_{12}}{2} \frac{da_{23}}{2}.$$

By a change of variables from the entries of the matrix to its representation in terms of Schur parameters we get

$$\int_S p(a_{13}) dF(a_{12}, a_{23}) = \int_{-1}^{1} \int_{-1}^{1} \sqrt{1 - s_{12}^2} \sqrt{1 - s_{23}^2} \frac{ds_{12}}{2} \frac{ds_{23}}{2} = \left(\frac{\pi}{4}\right)^2.$$

The set of positive semidefinite matrices with ones on the diagonal is known as the set of correlation matrices, and has been studied in several papers. One of the more recent references is the work of M. Laurent and S. Poljak [LP]. They have included a graphical description ([LP, Fig. 1]) of the set

$$\{a_{ij}, 1 \le i < j \le 3 : A = [a_{ij}] \text{ is positive semidefinite and } a_{ii} = 1\},$$

a set which they call the three dimensional elliptope, $\widetilde{\mathcal{L}}_3$. Our previous result thus yields that the volume of this set is $(\pi/4)^2$. For further results about correlation matrices, see the mentioned paper and the references therein.

Now to the general case of computing the probability that the $n \times n$ matrix

$$A = \begin{pmatrix} 1 & a_{12} & \cdots & a_{1n} \\ a_{12} & \ddots & \ddots & \vdots \\ \vdots & \ddots & \ddots & a_{n-1\,n} \\ a_{1n} & \cdots & a_{n-1\,n} & 1 \end{pmatrix}$$

is positive semidefinite. This can be done computing (2.2) under the assumption that the partial matrix B in (2.1) is partial positive semidefinite. Here

$$dF(a_{12},\ldots,a_{n-1\,n}) = \frac{da_{12}}{2}\cdots\frac{da_{n-1\,n}}{2}, \tag{2.9}$$

and S is the event that B is partial positive semidefinite. The probability that a_{1n} is chosen such that A becomes positive semidefinite can be expressed in terms of Schur parameters as

$$p(a_{1\,n}) = \prod_{j=1}^{n-2}(1-s_{1+j\,n}^2)^{\frac{1}{2}}\prod_{j=2}^{n-1}(1-s_{1\,j}^2)^{\frac{1}{2}}. \tag{2.10}$$

This follows from (2.8).

To compute the probability of the existence of a positive semidefinite completion, we again transform from the entries of the matrix to its corresponding Schur parameters. We get a lower triangular Jacobian matrix

$$J = \frac{\partial(a_{12},a_{23},\ldots,a_{n-1\,n},a_{13},\ldots,a_{n-2\,n},a_{1\,n-1},a_{2n})}{\partial(s_{12},s_{23},\ldots,s_{n-1\,n},s_{13},\ldots,s_{n-2\,n},s_{1\,n-1},s_{2n})} \tag{2.11}$$

by ordering the variables starting with the entries closest to the diagonal and moving out from the diagonal.

That the corresponding Jacobian matrix is lower triangular follows from the fact that

$$\frac{\partial a_{i\,i+k}}{\partial s_{\alpha\beta}} = \begin{cases} \prod_{j=1}^{k-1}(1-s_{i+j\,i+k}^2)^{\frac{1}{2}}\prod_{j=i+1}^{i+k-1}(1-s_{i\,j}^2)^{\frac{1}{2}} & \text{if } i=\alpha \text{ and } i+k=\beta, \\ 0 & \text{if any of the inequalities } \beta\le i,\ \beta>i+k,\ \alpha<i \text{ or } \alpha>i+k \text{ are true,} \\ \star & \text{otherwise,} \end{cases} \tag{2.12}$$

which can be obtained using (2.8) and the fact that $c_{i\,i+k}$ in (1.3) is independent of the Schur parameters $s_{\alpha\beta}$ if $\beta-\alpha<k$. Here we have denoted by $\star$ those entries which we do need and therefore do not compute.

Using (2.10), and transforming from the entries of the matrix to the corresponding Schur parameters, we get that the probability of an $n\times n$ symmetric matrix being positive semidefinite, assuming that the entries a_{ij} are uniformly distributed in $[-1,1]$, is

$$\begin{aligned}&\int_S p(a_{1n})\frac{da_{12}}{2}\cdots\frac{da_{n-1\,n}}{2}\\ &=\int_{-1}^{1}\cdots\int_{-1}^{1}\prod_{j=1}^{n-2}(1-s_{1+j\,n}^2)^{\frac{1}{2}}\prod_{j=2}^{n-1}(1-s_{1j}^2)^{\frac{1}{2}}\,J\,\frac{ds_{12}}{2}\cdots\frac{ds_{n-1\,n}}{2}.\end{aligned} \tag{2.13}$$

Taking into account the value of the Jacobian, J, which can be computed from (2.12), the integral in (2.13) can be written as

$$(2.14)\quad \int_{-1}^{1}\cdots\int_{-1}^{1}\prod_{i=1}^{n-1}\prod_{k=1}^{n-i}\left(\prod_{j=1}^{k-1}(1-s_{i+j\,i+k}^2)^{\frac{1}{2}}\prod_{j=i+1}^{i+k-1}(1-s_{ij}^2)^{\frac{1}{2}}\right)\frac{ds_{12}}{2}\cdots\frac{ds_{n-1\,n}}{2}.$$

To compute this integral we need to find the power of all the factors $(1-s_{ij}^2)^{1/2}$. To do this we use the fact that the product in (2.14) is formed by the term (2.10) and the terms from (2.12). Examining these terms, one can see that the factor $(1-s_{ij}^2)^{1/2}$ is occurring once for each α such that $i<\alpha\le n$ and once for each β such that $j<\beta\le n$. From this it follows that the the factor $(1-s_{ij}^2)^{1/2}$ is raised to the power $(n-i)+(n-j)=2n-i-j$ in the product in the integral (2.14).

A convenient way to present the power of each of the factors $(1-s_{ij}^2)^{1/2}$ is to give it as the i,j entry of the $n\times n$ matrix

$$\begin{pmatrix} 0 & n-2 & \dots & 1 & 0 \\ \vdots & \ddots & \ddots & & 1 \\ \vdots & & \ddots & \ddots & \vdots \\ \vdots & & & \ddots & n-2 \\ 0 & \dots & \dots & \dots & 0 \end{pmatrix}.$$

Using this, (2.7), and the fact that for a natural number k,

$$(2.15)\quad \begin{aligned} \int_{-1}^{1}\frac{1}{2}(1-s^2)^{\frac{2k-1}{2}}\,ds &= \int_0^{\frac{\pi}{2}}(1-\sin^2\theta)^{\frac{2k-1}{2}}\cos\theta\,d\theta \\ &= \frac{\pi}{2}\frac{1\cdot 3\cdots(2k-1)}{2\cdot 4\ldots(2k)} \\ &= I_{2k}, \end{aligned}$$

we get that the the expression given in (2.14) evaluates to

$$\prod_{i=2}^{n-1}(I_i)^i$$

where I_i is given by (2.7) if i is odd, and (2.15) if i is even. We can summarise this in the following theorem.

Theorem 2.1. *Let a_{ij} be uniformly distributed in the interval $[-1,1]$. Then the probability that the matrix*

$$A=\begin{pmatrix} 1 & a_{1\,2} & \dots & a_{1\,n} \\ a_{1\,2} & \ddots & \ddots & \vdots \\ \vdots & & \ddots & a_{n-1\,n} \\ a_{1\,n} & \dots & a_{n-1\,n} & 1 \end{pmatrix}$$

is positive semidefinite is

$$\prod_{i=2}^{n-1} (I_i)^i$$

where

$$\begin{aligned} I_{2k} &= \frac{\pi}{2}\frac{1\cdot 3\cdots(2k-1)}{2\cdot 4\ldots(2k)}, \\ I_{2k+1} &= \frac{2\cdot 4\cdots(2k)}{3\cdot 5\cdots(2k+1)}. \end{aligned}$$

3. The case of partial matrices

Here we will work with partial matrices which patterns of specified entries are quasitriangles. We recall the definition of a quasitriangle from [ACC, Definition 1].

Definition 3.1. A set $E \subset \{(i,j) : i \leq j, 1 \leq i \leq n, 1 \leq j \leq n\}$ is called a *quasitriangle* if $j_i = \max\{j : i \leq j \leq n, (i,j) \in E\} \geq i$ for each $1 \leq i \leq n$ and for every (k,l) with $i \leq k \leq l \leq j_i$, $(k,l) \in E$.

To get a suitable formula for the probability that the matrix is partial positive semidefinite we also introduce the numbers

$$i_j = \min\{i : (i,j) \in E\} \tag{3.1}$$

where E is a quasitriangle.

Let A be a partial symmetric matrix specified on a quasitriangle. Since the specified entries are placed in a band it is well known that A always has a positive semidefinite completion if it is partial positive semidefinite. Here we want to compute the probability that this event occurs given that $A = [a_{ij}]$ has the property that $a_{ii} = 1$, $a_{ij} = a_{ji} \in [-1,1]$ uniformly distributed if the entry ij is specified.

In the computation of this probability we may take advantage of the fact that there is a similar representation of partial positive semidefinite matrices as the one presented in the previous section, just disregarding the unspecified entries. If the matrix is partial positive semidefinite and has a quasitriangular pattern, then with probability one it is possible to choose the unspecified entries such that there is a completion $\widetilde{A}$ of A which is positive semidefinite. This means that the probability of A being partial positive semidefinite can be expressed as

$$\int p(a_{1n})\, dF \tag{3.2}$$

where $p(a_{1n}) = 1$ if the $1n$ entry is unspecified (which we may assume) and dF must be a modified version of dF defined in (2.9), taking into account that F only depends on the specified entries of A. For each specified entry a_{ij} we have a formula like (2.8). To compute the probability that A is partial positive semidefinite given that the specified off-diagonal entries are uniformly distributed in $[-1,1]$ is

$$\int_S \frac{da_{ij}}{2} \cdots \frac{da_{kl}}{2} = \int_{-1}^{1} \cdots \int_{-1}^{1} J \frac{ds_{ij}}{2} \cdots \frac{ds_{kl}}{2}$$

where the integration variables run through the specified entries, and J is the corresponding Jacobian. J can be computed using the analogue of equation (2.12) adopted to this setting.

Examining this one can see that the factor $(1 - s_{ij}^2)^{1/2}$ occurs once for each specified entry above a_{ij} and once for each specified entry on the right hand side of a_{ij}. Summing this, and continuing as in the proof of Theorem 2.1, we obtain

Theorem 3.2. *Let E be a quasitriangle, j_i defined as in Definition 3.1 and i_j as in (3.1). Define $I_0 = 1$, and I_{2k} and I_{2k+1} as in Theorem 2.1.*
Then the probability that a partial matrix with pattern E is partial positive semidefinite is

$$\prod_{(i,j)\in E, i<j} I_{\alpha(i,j)} \tag{3.3}$$

where $\alpha = (j_i - j) + (i - i_j)$.

4. The probability of the existence of a positive semidefinite completion

Another problem which may be treated is the following: Suppose we are given a partial positive semidefinite matrix, what is the probability that it has a positive semidefinite completion. A slight variation of this problem is obtained by adding the condition that the matrix also should have the Toeplitz structure.

For both the above mentioned problems the probability depends on the pattern of the specified entries. In [GJSW] necessary and sufficient conditions ensuring that every partial positive semidefinite matrix with a certain pattern has a positive semidefinite completion are given. For the Toeplitz case a similar result is presented in [JLN]. In these cases the probability of the existence of a positive semidefinite completion given that the matrix is partial positive semidefinite is of course one.

Using a result from [N] we can compute the probability that a partial positive semidefinite Toeplitz matrix has a positive semidefinite Toeplitz completion, if the specified entries of the partial Toeplitz matrix consist of two super diagonals, p and n.

Without loss of generality, this is the case when the partial matrix is of the form

$$\begin{pmatrix} 1 & \gamma_1 & \cdots & \cdots & \gamma_{n-1} & c_n \\ \gamma_1 & \ddots & \ddots & & & \gamma_{n-1} \\ \vdots & \ddots & \ddots & \ddots & & \vdots \\ \vdots & & \ddots & \ddots & \ddots & \vdots \\ \gamma_{n-1} & & & \ddots & \ddots & \gamma_1 \\ c_n & \gamma_{n-1} & \cdots & \cdots & \gamma_1 & 1 \end{pmatrix} \tag{4.1}$$

where $\gamma_p = c_p$ for one $p < n$. The other γ_j values are unspecified entries. The partial matrix in (4.1) is partial positive semidefinite if $|c_p| \leq 1$ and $|c_n| \leq 1$. The condition of the existence of a partial positive semidefinite Toeplitz completion in this case is given in [N, Theorem 2.2].

As the computations of the probabilities all follow the same lines, we will only compute the probability in some special cases. Assume $p = 1$ and n is even. If $c_n = \cos\theta$, then there exists a positive semidefinite Toeplitz completion if $c_1 \in \left[-\cos(\frac{\theta}{n}), \cos(\frac{\theta}{n})\right]$. We are still assuming that c_1 and c_n are uniformly distributed in the interval $[-1, 1]$.

Then the probability of the existence of a positive semidefinite Toeplitz completion is

$$\int_0^\pi \frac{\cos(\frac{\theta}{n}) - (-\cos(\frac{\theta}{n}))}{2} \frac{\sin\theta}{2}\, d\theta = n^2 \left(\cos(\frac{\pi}{n}) + 1\right) \left(-2 + 2n^2\right)^{-1}.$$

If n is odd there exists a positive semidefinite Toeplitz completion if $c_1 \in \left[-\cos(\frac{\theta-\pi}{n}), \cos(\frac{\theta}{n})\right]$, and the probability therefore is

$$\int_0^\pi \frac{\cos(\frac{\theta}{n}) + \cos(\frac{\theta-\pi}{n})}{2} \frac{\sin\theta}{2}\, d\theta = n^2 \left(\cos(\frac{\pi}{n}) + 1\right) \left(-2 + 2n^2\right)^{-1},$$

that is exactly the same formula as if n is even.

If $p = 2$ and $(n-1) \bmod 4 = 0$ we get that the probability is

$$\left(n^2 - 4 + \cos(\frac{\pi}{n})n^2\right) \left(2n^2 - 8\right)^{-1},$$

and if $(n-1) \bmod 4 = 0$ we get

$$\left(\left(2 + \cos(\frac{3\pi}{n}) + \cos(\frac{\pi}{n})\right) n^2 - 8\right) \left(4n^2 - 16\right)^{-1}.$$

The remaining computations are left to the interested reader.

References

[ACC] Arsene, G., Ceauşescu, Z., Constantinescu, T.: Schur analysis of some completion problems; Linear Algebra Appl. 109 (1988), 1–35.

[BC] Bakonyi, M., Constantinescu, T.: Schur's Algorithm and Several Applications; Pitman Res. Notes Math. Ser. 261, Longman Scientific & Technical, Essex 1992.

[C] Constantinescu, T.: Schur analysis of positive block-matrices; in: I.C. Gohberg (ed.): I. Schur Methods in Operator Theory and Signal Processing, Operator Theory: Adv. Appl. 18, Birkhäuser Verlag, Basel 1986, 191–206.

[FF] Foias, C., Frazho, A.: The Commutant Lifting Approach to Interpolation Problems; Operator Theory: Adv. Appl. 44, Birkhäuser Verlag, Basel 1990.

[FK] Fritzsche, B., Kirstein, B.: A Schur parametrization of non-negative hermitian and contractive block matrices and the corresponding maximum entropy problems; Optimization 19:5 (1988), 737–755.

[GJSW] Grone, B., Johnson, C.R., Sá, E.M., Wolkowicz, H.: Positive definite completions of partial hermitian matrices; Linear Algebra Appl. 58 (1984), 109–124.

[Ja] James, G.: Modern Engineering Mathematics; Second Edition; Addison-Wesley Longman, Harlow 1996.

[Jo] Johnson, C.R.: Matrix completion problems: a survey; in: C.R. Johnson (ed.): Matrix Theory and Applications, Proc. Symp. Appl. Math. (1990), 171–198.

[JLN] Johnson, C.R., Lundquist, M., Nævdal, G.: Positive definite Toeplitz completions; J. London Math. Soc. (to appear).

[LP] Laurent, M., Poljak, S.: On a positive semidefinite relaxation of the cut polytope; Linear Algebra Appl. 223/224 (1995), 439–461.

[N] Nævdal, G.: On a generalization of the trigonometric moment problem; Linear Algebra Appl. 258 (1997), 1–18.

[S] Schur, I.: Über Potenzreihen, die im Innern des Einheitskreises beschränkt sind; J. Reine Angew. Math. 147 (1917), 205–232.

Department of Mathematics
College of William and Mary
Williamsburg
Virginia 23187-8795
USA
crjohnso@math.wm.edu

Department of Engineering
Stord/Haugesund College
Skaaregt. 103
N-5500 Haugesund
Norway
geir.naevdal@hsh.no

1991 Mathematics Subject Classification: Primary 15A57; Secondary 60B99, 47A20

Submitted: May 31, 1996

Operator Theory:
Advances and Applications, Vol. 103

Factorization of lower triangular unitary operators with finite Kronecker index into elementary factors

M.A. KAASHOEK and D.R. PIK

It is shown that a block lower triangular unitary operator with Kronecker index $N < \infty$ admits a factorization into a product of N block lower triangular unitary operators with Kronecker index equal to 1. If the block lower triangular operator is exponentially decaying with respect to the main diagonal, then the factors can be chosen exponentially decaying too. A description of the elementary factors is also given. The proofs are based on the fact that the operators involved appear as input output operators of finite dimensional time varying unitary systems, and a cascade decomposition of such systems.

1. Introduction

This article concerns the factorization of lower triangular unitary operators on $\ell^2(\mathbb{C}^m)$, the Hilbert space of all doubly infinite square summable sequences with entries in $\mathbb{C}^m$, into elementary factors. First the operators are assumed to belong to the class $\mathcal{LEK}$, which consists of (see [BenGK]) all operators $T = [t_{j,k}]_{j,k=-\infty}^{\infty}$ on $\ell^2(\mathbb{C}^m)$, satisfying

(i) T is *lower triangular*, i.e., $t_{j,k} = 0$ for all $k > j$,

(ii) T is *exponentially dominated with respect to the main diagonal*, i.e., there exist positive constants $M > 0$ and $\alpha \in (0,1)$ such that $\|t_{j,k}\| \leq M\alpha^{j-k}$ for all $j \geq k$, and

(iii) the number

$$\kappa(T) := \sup_p \operatorname{rank} \begin{bmatrix} t_{p,p-1} & t_{p,p-2} & t_{p,p-3} & \cdots \\ t_{p+1,p-1} & t_{p+1,p-2} & t_{p+1,p-3} & \cdots \\ t_{p+2,p-1} & t_{p+2,p-2} & t_{p+2,p-3} & \cdots \\ \vdots & \vdots & \vdots & \end{bmatrix}$$

is finite.

The number $\kappa(T)$ in (iii) will be called the *Kronecker index* of the operator T. The following theorems will be proved.

Theorem 1.1. *Each unitary operator T in $\mathcal{LEK}$ can be decomposed into a finite product, $T = T_1 T_2 \cdots T_N$, where N is the Kronecker index of T, and each T_i is unitary, belongs to the class $\mathcal{LEK}$, and has Kronecker index equal to 1.*

Theorem 1.2. *An operator $T : \ell^2(\mathbb{C}^m) \to \ell^2(\mathbb{C}^m)$ is unitary, belongs to the class $\mathcal{LEK}$, and has Kronecker index equal to 1, if and only if T is of the following form:*

$$T = T_{(\alpha)(\Gamma)}U \tag{1.1}$$

where $T_{(\alpha)(\Gamma)}$ is given by

$$I - \begin{bmatrix} \ddots & & & & & \\ \cdots & \frac{1}{1+|\alpha_{-1}|}\Gamma_{-1}\Gamma_{-1}^* & & & & \\ \cdots & \Gamma_0\mathrm{sgn}(\alpha_{-1})\Gamma_{-1}^* & \boxed{\frac{1}{1+|\alpha_0|}\Gamma_0\Gamma_0^*} & & & \\ \cdots & \Gamma_1\alpha_0\mathrm{sgn}(\alpha_{-1})\Gamma_{-1}^* & \Gamma_1\mathrm{sgn}(\alpha_0)\Gamma_0^* & \frac{1}{1+|\alpha_1|}\Gamma_1\Gamma_1^* & & \\ \cdots & \Gamma_2\alpha_1\alpha_0\mathrm{sgn}(\alpha_{-1})\Gamma_{-1}^* & \Gamma_2\alpha_1\mathrm{sgn}(\alpha_0)\Gamma_0^* & \Gamma_2\mathrm{sgn}(\alpha_1)\Gamma_1^* & \frac{1}{1+|\alpha_2|}\Gamma_2\Gamma_2^* & \\ & \vdots & \vdots & \vdots & \vdots & \ddots \end{bmatrix}.$$

Here $U : \ell^2(\mathbb{C}^m) \to \ell^2(\mathbb{C}^m)$ is a unitary block diagonal operator and the operators $\Gamma_n : \mathbb{C} \to \mathbb{C}^m$, $\alpha_n : \mathbb{C} \to \mathbb{C}$ satisfy

(j) $\Gamma_n^*\Gamma_n + |\alpha_n|^2 = 1$ *for all* $n \in \mathbb{Z}$,

(jj) $\limsup_{k\to\infty} \sup_{n\in\mathbb{Z}} |\alpha_n\alpha_{n-1}\alpha_{n-2}\cdots\alpha_{n-k+1}|^{1/k} < 1$.

We say that an operator T on $\ell^2(\mathbb{C}^m)$ belongs to the class $\mathcal{LK}$ if conditions (i) and (iii) above are fulfilled. The analogs of Theorems 1.1 and 1.2 for unitary operators in the class $\mathcal{LK}$, but not necessarily in $\mathcal{LEK}$, also hold. In fact, if in the above theorems one replaces $\mathcal{LEK}$ by the larger class $\mathcal{LK}$, then Theorem 1.1 remains true without any further change, and Theorem 1.2 remains true if one replaces condition (jj) by

(jj)' $\prod_{j=0}^{\infty} \alpha_{n-j} = 0$ *and* $\prod_{j=0}^{\infty} \alpha_{n+j} = 0$ *for each* $n \in \mathbb{Z}$.

For the (scalar) Toeplitz case Theorems 1.1 and 1.2 reduce to the statement that a rational function $\phi(z)$ without poles on $\mathbb{D}$ and with unitary values on $\mathbb{T}$ may be written as a product of a finite number of Blaschke factors, a factor z^n and a constant of modulus 1. For the block Toeplitz case Theorems 1.1 and 1.2 follow from the factorization theorems for unitary rational matrix functions given in [AG1] and [AG2].

The method to prove Theorems 1.1 and 1.2 used in this paper is a variation of the state space method employed in [GKLe], which views lower triangular (not necessarily bounded) operators as input output operators of time-varying systems. In fact we show that unitary operators in the class $\mathcal{LEK}$ may be represented as input output operators of exponentially stable time-varying systems, that are unitary and have a finite dimensional state space (see Theorem 4.1). This allows us to use in a time-varying setting classical theorems (see [Br]) about factorization of unitary systems and invariant subspaces of the state space operator. The extension of these results to time-varying unitary systems is given in Section 4. Section 5 contains the proofs of the main theorems for unitary operators in the class $\mathcal{LEK}$. The analogs of Theorems 1.1 and 1.2 for unitary operators in the class $\mathcal{LK}$ are proved in Section 6, by using unitary systems of which the sequence of main operators satisfies a weaker form of forward stability. The first two sections contain preliminaries about unitary time-varying systems.

2. Unitary time varying systems

Let us consider a time-variant system Σ with discrete time n:

$$\Sigma \left\{ \begin{array}{rcl} x_{n+1} & = & A(n)x_n + B(n)u_n, \\ y_n & = & C(n)x_n + D(n)u_n, \end{array} \right. \qquad n \in \mathbb{Z}. \tag{2.1}$$

Here $A(n) : H_n \to H_{n+1}$, $B(n) : K_n \to H_{n+1}$, $C(n) : H_n \to L_n$, and $D(n) : K_n \to L_n$ are bounded linear operators acting between Hilbert spaces. We call the operator $A(n)$ the *main operator*, the operator $B(n)$ the *input operator*, the operator $C(n)$ the *output operator*, and the operator $D(n)$ the *feed-through operator* of the system at time instant n. By $M_\Sigma(n)$ we denote the *system matrix at time* n, i.e.,

$$M_\Sigma(n) = \left[\begin{array}{cc} A(n) & B(n) \\ C(n) & D(n) \end{array} \right] : H_n \oplus K_n \to H_{n+1} \oplus L_n.$$

Instead of (2.1), we will also use the notation $\Sigma = (A(n), B(n), C(n), D(n); H_n, K_n, L_n)$.

A time-varying system $\Sigma = (A(n), B(n), C(n), D(n); H_n, K_n, L_n)$ is called *contractive* (or *dissipative*) if for each sequence of inputs $u_n, u_{n+1}, u_{n+2}, \ldots$ and each initial state $x_n \in H_n$ the following inequality holds for all $j \geq n$:

$$\|u_j\|^2 - \|y_j\|^2 \geq \|x_{j+1}\|^2 - \|x_j\|^2, \tag{2.2}$$

where x_{j+1} and y_j are determined from u_j and x_j via the equations in (2.1). Formula (2.2) can be restated as $\|x_{j+1}\|^2 + \|y_j\|^2 \leq \|x_j\|^2 + \|u_j\|^2$. So Σ is contractive if and only if the system matrix $M_\Sigma(j)$ is a contraction for all $j \in \mathbb{Z}$. A time-varying system is called *unitary* (or *conservative*) if the system matrix is a unitary operator for all j.

Systems of the type (2.1) have a natural input output mapping. To define this transformation we need some additional notations. For an ordered sequence of vector spaces $\mathcal{V} = (V_n)_{n\in\mathbb{Z}}$ we denote by $\ell(V)$ the vector space consisting of all sequences $(v_n)_{n\in\mathbb{Z}}$ with $v_n \in V_n$. If $\mathcal{V}$ happens to be a sequence of Hilbert spaces, then we will denote by $\ell^2(\mathcal{V})$ the Hilbert space direct sum $\bigoplus_{n\in\mathbb{Z}} V_n$. Thus $\ell^2(\mathcal{K})$ and $\ell^2(\mathcal{L})$ will stand for the Hilbert space direct sums $\bigoplus_{n\in\mathbb{Z}} K_n$ and $\bigoplus_{n\in\mathbb{Z}} L_n$, respectively. In the sequel blank spots in operator matrices denote zero entries.

Assume we start operating the system in (2.1) at a time instant n, with initial state x_n, and by feeding in an input sequence $u_n, u_{n+1}, u_{n+2}, \ldots$. Then the output at time $n+k$, $k>0$, is given by

$$\begin{aligned} y_{n+k} = {} & C(n+k)\tau_A(n+k,n)x_n \\ & + \sum_{j=0}^{k-1} C(n+k)\tau_A(n+k,n+j+1)B(n+j)u_{n+j} + D(n+k)u_{n+k}. \end{aligned} \tag{2.3}$$

Here $\tau_A(k,l) : H_l \to H_k$ is defined by

$$\tau_A(k,l) = \begin{cases} A(k-1)A(k-2)\cdots A(l+1)A(l), & k>l, \\ I_{H_l}, & k=l, \\ 0, & k<l. \end{cases} \tag{2.4}$$

The operator $\tau_A(k,l)$ is called the *transition operator* associated with the difference equation $x_{n+1} = A(n)x_n$.

If the system is at rest at time n, i.e., $x_n = 0$, then (2.3) reduces to

$$y_{n+k} = \sum_{j=0}^{k-1} C(n+k)\tau_A(n+k,n+j+1)B(n+j)u_{n+j} + D(n+k)u_{n+k}. \tag{2.5}$$

Let us introduce the linear block lower triangular map

$$\begin{aligned} T_\Sigma &= [t_{i,j}]_{i=-\infty;j=-\infty}^{\infty} \\ &= \begin{bmatrix} \ddots & & & & & \\ \cdots & D(-1) & & & & \\ \cdots & C(0)B(-1) & \boxed{D(0)} & & & \\ \cdots & C(1)\tau_A(1,0)B(-1) & C(1)B(0) & D(1) & & \\ \cdots & C(2)\tau_A(2,0)B(-1) & C(2)\tau_A(2,1)B(0) & C(2)B(1) & D(2) & \\ & \vdots & \vdots & \vdots & & \ddots \end{bmatrix} \\ &: \ell_+(\mathcal{K}) \to \ell_+(\mathcal{L}). \end{aligned} \tag{2.6}$$

Here $\ell_+(\mathcal{K})$ denotes the linear subspace of $\ell(\mathcal{K})$, consisting of all sequences $(u_n)_{n\in\mathbb{Z}}$, $u_n \in K_n$, of the form $(\ldots,0,0,u_n,u_{n+1},u_{n+2},\ldots)$, where n may be any time instant. Then T_Σ is well defined by the usual block matrix multiplication, and we can summarize (2.5) as $y = T_\Sigma u$, with $y = (\ldots,0,0,y_n,y_{n+1},y_{n+2},\ldots)$ and $u = (\ldots,0,0,u_n,u_{n+1},u_{n+2},\ldots)$. The map T_Σ will be called the *input output map* of the time-varying system Σ.

It is well known (see for instance [ACC], page 31, or [AKP]) that for a time-varying contractive system $\Sigma = (A(n),B(n),C(n),D(n);H_n,K_n,L_n)$ the input output map T_Σ induces a bounded linear operator acting from $\ell^2(\mathcal{K})$ into $\ell^2(\mathcal{L})$. This induced operator, which is denoted by T_Σ again, is in fact a contraction. In particular, if Σ is a unitary system, then its input output map acts as a bounded linear operator from $\ell^2(\mathcal{K})$ into $\ell^2(\mathcal{L})$.

A system $\Sigma = (A(n), B(n), C(n), D(n); H_n, K_n, L_n)$ will be called *(forward) exponentially stable* if

$$\limsup_{\nu \to \infty} \left(\sup_j \|\tau_A(j+\nu, j)\| \right)^{1/\nu} < 1 \tag{2.7}$$

An operator $T : \ell^2(\mathcal{K}) \to \ell^2(\mathcal{L})$ has a canonical matrix representation as a double infinite operator matrix, $T = [t_{j,k}]_{j,k=-\infty}^{\infty}$, where $t_{j,k}$ acts from K_k into L_j. We call T *block lower triangular* if for all $j < k$ we have $t_{j,k} = 0$, and in this case T is said to be *exponentially dominated*, if there exist constants $M > 0$ and $\alpha \in (0,1)$ such that $\|t_{j,k}\| \le M\alpha^{j-k}$ for all $j \ge k$. The class of operators satisfying these two properties is denoted by $\mathcal{LE}$.

Theorem 2.1. *Let $\Sigma = (A(n), B(n), C(n), D(n); H_n, K_n, L_n)$ be a system which is unitary and exponentially stable. Then the input output map T_Σ is unitary and $T_\Sigma \in \mathcal{LE}$.*

For the input output operator of a unitary system to be unitary, it is not necessary to require the system to be exponentially stable. In fact, in the next section (see Theorem 3.1) we shall see that a weaker form of stability suffices.

To prove Theorem 2.1 we reduce the results to a time-invariant setting using techniques from [FFGK], Chapter II. For operators $F(n) : K_n \to L_n$, $n \in \mathbb{Z}$, uniformly bounded in the operator norm, we define $\Delta_F : \ell^2(\mathcal{K}) \to \ell^2(\mathcal{L})$ to be the block diagonal operator $\operatorname{diag}_{i \in \mathbb{Z}}(F_i)$. For a uniformly bounded sequence of operators $G(n) : K_n \to L_{n+1}$ we define the *weighted shift* S_G by

$$S_G = \begin{bmatrix} \ddots & & & & \\ & G(-1) & \boxed{0} & & \\ & & G(0) & & \\ & & & G(1) & \\ & & & & \ddots \end{bmatrix} : \ell^2(\mathcal{K}) \to \ell^2(\mathcal{L}). \tag{2.8}$$

Thus with a contractive system $\Sigma = (A(n), B(n), C(n), D(n); H_n, K_n, L_n)$ we may associate the contractions S_A, S_B, Δ_C and Δ_D:

$$\begin{aligned} S_A &: \ell^2(\mathcal{H}) \to \ell^2(\mathcal{H}), & S_B &: \ell^2(\mathcal{K}) \to \ell^2(\mathcal{H}), \\ \Delta_C &: \ell^2(\mathcal{H}) \to \ell^2(\mathcal{L}), & \Delta_D &: \ell^2(\mathcal{K}) \to \ell^2(\mathcal{L}), \end{aligned}$$

which allow us to define the time-invariant system

$$\widehat{\Sigma} \left\{ \begin{aligned} \vec{x}_{n+1} &= S_A \vec{x}_n + S_B \vec{u}_n, \\ \vec{y}_n &= \Delta_C \vec{x}_n + \Delta_D \vec{u}_n. \end{aligned} \right. \tag{2.9}$$

Here $\vec{x}_n \in \ell^2(\mathcal{H})$, $\vec{u}_n \in \ell^2(\mathcal{K})$, and $\vec{y}_n \in \ell^2(\mathcal{L})$. We will call $\widehat{\Sigma}$ *the time-invariant system associated with* Σ.

Define the unitary operators

$$(2.10)\qquad \begin{aligned} &Z_{H,K} : \ell^2(\mathcal{H}) \oplus \ell^2(\mathcal{K}) \to \textstyle\bigoplus_{j\in\mathbb{Z}}(H_j \oplus K_j),\\ &Z_{H,K}((h_j)_{j=-\infty}^{\infty}, (k_j)_{j=-\infty}^{\infty}) = (h_j, k_j)_{j=-\infty}^{\infty},\\ &\widetilde{Z}_{H,L} : \ell^2(\mathcal{H}) \oplus \ell^2(\mathcal{L}) \to \textstyle\bigoplus_{j\in\mathbb{Z}}(H_{j+1} \oplus L_j),\\ &\widetilde{Z}_{H,L}((h_j)_{j=-\infty}^{\infty}, (l_j)_{j=-\infty}^{\infty}) = (h_{j+1}, l_j)_{j=-\infty}^{\infty}. \end{aligned}$$

Then we have for the system matrix $M_{\widehat{\Sigma}}$ of the system $\widehat{\Sigma}$

$$(2.11)\qquad \widetilde{Z}_{H,L} M_{\widehat{\Sigma}} Z_{H,K}^{-1} = \mathrm{diag}_{n\in\mathbb{Z}} M_{\Sigma}(n),$$

where $M_{\Sigma}(n)$ is the system matrix of Σ at time n. So $M_{\widehat{\Sigma}}$ is a contraction, and the system $\widehat{\Sigma}$ is contractive.

If Σ is contractive and exponentially stable, then the spectral radius of S_A is strictly less than one, and

$$(2.12)\qquad T_{\Sigma} = \Delta_D + \Delta_C(I - S_A)^{-1} S_B.$$

Proof (of Theorem 2.1). Part (a). First we show that T_{Σ} is unitary. Let $\widehat{\Sigma}$ be the time-invariant system associated with Σ. From the unitary equivalence in (2.11) it follows that $\widehat{\Sigma}$ is also unitary. Take $\lambda \in \mathbb{D}$ and consider

$$(2.13)\qquad \theta_{\widehat{\Sigma}}(\lambda) = \Delta_D + \lambda\Delta_C(I - \lambda S_A)^{-1} S_B.$$

Notice that $\theta_{\widehat{\Sigma}}$ is the transfer function of $\widehat{\Sigma}$. Since $\widehat{\Sigma}$ is unitary, the following two identities hold (see, e.g., formulas (9a) and (9b) in Section XXVIII.2 of [GGK2]):

$$(2.14)\qquad I - \theta_{\widehat{\Sigma}}(\lambda)^*\theta_{\widehat{\Sigma}}(\lambda) = (1 - |\lambda|^2) S_B^*(I - \bar{\lambda} S_A^*)^{-1}(I - \lambda S_A)^{-1} S_B,$$

$$(2.15)\qquad I - \theta_{\widehat{\Sigma}}(\lambda)\theta_{\widehat{\Sigma}}(\lambda)^* = (1 - |\lambda|^2)\Delta_C(I - \lambda S_A)^{-1}(I - \bar{\lambda} S_A^*)^{-1}\Delta_C^*.$$

Now use the fact that Σ is exponentially stable. Thus the spectral radius of S_A is strictly less than one, and hence in the two identities above we can take the limit for $\lambda \to 1$. This shows that $\theta_{\widehat{\Sigma}}(1)$ is unitary. But, by (2.12), the operator $\theta_{\widehat{\Sigma}}(1)$ is precisely the input output operator of Σ. Thus T_{Σ} is unitary.

Part (b). Now we will show that the operator T_{Σ} is exponentially dominated. Since Σ is exponentially stable, $\limsup_{\nu\to\infty} \left(\sup_j \|\tau_A(j+\nu, j)\|\right)^{1/\nu} = \rho$ for some $\rho \in [0, 1)$. Hence there exists a number $N \geq 1$ such that for all $\nu > N$ we have

$$(2.16)\qquad \sup_{j\in\mathbb{Z}} \|\tau_A(j+\nu, j)\|^{1/\nu} < (1+\rho)/2 = \alpha.$$

Note that $0 \leq \rho < (1+\rho)/2 < 1$. Set $M = \alpha^{-N-1}$. Then $M > 1$. Let us write $T_{\Sigma} = (t_{k,n})_{k,n=-\infty}^{\infty}$. We have to consider three cases, namely $k - n = 0$,

$1 \le k-n \le N$, and $N < k-n$. If $k-n=0$, we have $\|t_{n,n}\| = \|D(n)\| \le 1 < M$. For $1 \le k-n \le N$ we get

$$\|t_{k,n}\| = \|C(k)\tau_A(k,n+1)B(n)\| \le 1 < \alpha^{-1} \le \alpha^{-1-(N-(k-n))} = M\alpha^{k-n},$$

and for $N < k-n$ we find, using (2.16),

$$\begin{aligned} \|t_{k,n}\| &= \|C(k)\tau_A(k,n+1)B(n)\| \\ &\le \|\tau_A(k,n+1)\| \le \alpha^{k-n-1} \le \alpha^{k-n-1-N} \\ &= M\alpha^{k-n}. \end{aligned}$$

The operator T_Σ is exponentially decaying. □

Two time-varying systems Σ_1 and Σ_2,

$$\Sigma_\nu = (A_\nu(n), B_\nu(n), C_\nu(n), D_\nu(n); H_{\nu,n}, K_{\nu,n}, L_{\nu,n}), \qquad \nu = 1,2, \tag{2.17}$$

are said to be *unitarily equivalent* if their feed-through operators coincide, that is $K_{1,n} = K_{2,n}$, $L_{1,n} = L_{2,n}$, and $D_1(n) = D_2(n)$ for all $n \in \mathbb{Z}$, and if there exists a sequence of unitary operators $J_n : H_{1,n} \to H_{2,n}$ such that

$$A_2(n) = J_{n+1}A_1(n)J_n^{-1}, \quad B_2(n) = J_{n+1}B_1(n), \quad C_2(n) = C_1(n)J_n^{-1}. \tag{2.18}$$

From (2.18) it follows that for all $n \in \mathbb{Z}$ and $k \in \mathbb{N}$ we have

$$C_2(n)\tau_{A_2}(n, n-k+1)B_2(n-k) = C_1(n)\tau_{A_1}(n, n-k+1)B_1(n-k).$$

Therefore we have $T_{\Sigma_1} = T_{\Sigma_2}$ for time-varying systems, if Σ_1 and Σ_2 are unitarily equivalent. Under certain additional conditions, the converse statement is also true, but we do not need this fact here.

3. Observability and controllability of unitary time varying systems

For latter purposes (see the proof of Theorem 7.1) we need the notions of observability and controllability. Let $\Sigma = (A(n), B(n), C(n), D(n); H_n, K_n, L_n)$ be a time-varying system. For each integer n, two types of subspaces of the state space H_n will be defined. The *unobservable subspace at time* $n \in \mathbb{Z}$ is defined by

$$\operatorname{Ker}(C|A;n) := \bigcap_{k \ge n} \operatorname{Ker} C(k)\tau_A(k,n). \tag{3.1}$$

Here $\tau_A(\cdot,\cdot)$ is the transition operator defined by (2.4). We use $\operatorname{Im}(A|B;n)$ to denote the linear span of the set

$$\{\tau_A(n,k+1)B(k)u_k \mid u_k \in K_k, k \le n-1\}.$$

The *controllable subspace at time* n is by definition the closure of this set, i.e.:

$$\overline{\operatorname{Im}(A|B;n)} = \bigvee_{k\leq n-1} \operatorname{Im} \tau_A(n,k+1)B(k). \tag{3.2}$$

Note that both $\operatorname{Ker}(C|A;n)$ and $\overline{\operatorname{Im}(A|B;n)}$ are subspaces of H_n.

A time-varying system is called *observable at time* n if $\operatorname{Ker}(C|A;n) = (0)$ and *controllable at time* n if $\overline{\operatorname{Im}(A|B;n)} = H_n$. A system Σ will be called (*completely*) *observable* if Σ is observable at all times n and (*completely*) *controllable* if Σ is controllable at all times n.

Suppose a unitary system Σ has a unitary input output map. Then the observability and controllability of the system Σ are completely determined by the stability properties of the sequence of main operators $(A(n) : H_n \to H_{n+1})_{n\in\mathbb{Z}}$. This relation is the main result of this section. Similar results are well known in the time-invariant setting (see for instance [Sz.-NF], page 257, [Ber], page 115, or [Ar]).

To derive this relation for the time-variant case we first develop a time-varying analog of the $C_{..}$ classification of contractions (see [Sz.-NF], Section II.4). Let $(A(n) : H_n \to H_{n+1})_{n\in\mathbb{Z}}$ be a sequence of contractions, and $\tau_A(n,m) : H_m \to H_n$ its state transition operator. Fix a time instant $n \in \mathbb{Z}$. We call the sequence to be *of class* $C_{0\cdot}$ if for each integer n and for all $h \in H_n$ we have

$$\lim_{M\to\infty} \|\tau_A(M+n,n)h\| = 0. \tag{3.3}$$

Similarly the sequence is said to be *of class* $C_{\cdot 0}$ if for each integer n and for all $h \in H_n$ we have

$$\lim_{M\to\infty} \|\tau_A(n,n-M)^*h\| = 0. \tag{3.4}$$

Finally, a sequence is called *of class* C_{00} if it is both of class $C_{\cdot 0}$ and of class $C_{0\cdot}$.

Let $\Sigma = (A(n), B(n), C(n), D(n); H_n, K_n, L_n)$ be a unitary system. We call Σ (*internally*) *stable* if its sequence of main operators is in the class $C_{0\cdot}$; similarly, Σ is called (*internally*) $*$-*stable* if the sequence of main operators is in the class $C_{\cdot 0}$.

Notice that (3.3) and (3.4) hold for each $n \in \mathbb{Z}$ if condition (2.7) is satisfied. Thus if $\Sigma = (A(n), B(n), C(n), D(n); H_n, K_n, L_n)$ is an exponentially stable unitary system, then Σ is both stable and $*$-stable.

Theorem 3.1. *Let* $\Sigma = (A(n), B(n), C(n), D(n); H_n, K_n, L_n)$ *be a unitary time-varying system, and let* T *be its input output map. Then the following statements are equivalent.*

(j) *The operator* T *is unitary and the system* Σ *is observable and controllable.*

(jj) *The sequence of main operators* $(A(n) : H_n \to H_{n+1})_{n\in\mathbb{Z}}$ *of the system* Σ *belongs to the class* C_{00}.

The proof of this theorem follows from the next two propositions.

Proposition 3.2. *Let $\Sigma = (A(n), B(n), C(n), D(n); H_n, K_n, L_n)$ be a unitary time-varying system. If the main operator sequence $(A(n) : H_n \to H_{n+1})_{n\in\mathbb{Z}} \in C_{0\cdot}$, then the system Σ is observable. If the main operator sequence $(A(n) : H_n \to H_{n+1})_{n\in\mathbb{Z}} \in C_{\cdot 0}$, then the system Σ is controllable. In particular, if $(A(n) : H_n \to H_{n+1})_{n\in\mathbb{Z}} \in C_{00}$, then Σ is observable and controllable.*

Proof. Let us first consider the case when $(A(n) : H_n \to H_{n+1})_{n\in\mathbb{Z}} \in C_{0\cdot}$. Fix a time instant $n \in \mathbb{Z}$. Take arbitrary $x \in \bigcap_{k\geq n} \operatorname{Ker} C(k)\tau_A(k,n)$. Because Σ is unitary, we have

$$x = (I_{H_n} - C(n)^*C(n))x = A(n)^*A(n)x = \tau_A(n+1,n)^*\tau_A(n+1,n)x.$$

For the same reason, for any given integer $m \geq 1$ we have

$$\begin{aligned}
&\tau_A(n+m,n)^*\tau_A(n+m,n)x \\
&= \tau_A(n+m,n)^* \left(I_{H_{n+m}} - C(n+m)^*C(n+m)\right) \tau_A(n+m,n)x \\
&= \tau_A(n+m+1,n)^*\tau_A(n+m+1,n)x.
\end{aligned}$$

Thus for all $m \geq 1$ we get $x = \tau_A(n+m,n)^*\tau_A(n+m,n)x$. Now

$$\|x\| = \|\tau_A(n+m,n)^*\tau_A(n+m,n)x\| \leq \|\tau_A(n+m,n)x\|.$$

The last term tends to 0 if $m \to \infty$ because $(A(n) : H_n \to H_{n+1})_{n\in\mathbb{Z}} \in C_{0\cdot}$. Thus $x = 0$, and we conclude that the system Σ is observable.
Secondly, suppose $(A(n) : H_n \to H_{n+1})_{n\in\mathbb{Z}} \in C_{\cdot 0}$. Fix a time instant $n \in \mathbb{Z}$, and take arbitrary $x \in \operatorname{Im}(A|B;n)^\perp = \bigcap_{k\leq n-1} \operatorname{Ker} B(k)^*\tau_A(n,k+1)^*$. We have

$$x = (I_{H_n} - B(n-1)B(n-1)^*) = A(n-1)A(n-1)^*x = \tau_A(n,n-1)\tau_A(n,n-1)^*x$$

since Σ is unitary. For all $m \geq 1$ we have

$$\begin{aligned}
&\tau_A(n,n-m)\tau_A(n,n-m)^*x \\
&= \tau_A(n,n-m)(I_{H_{n-m}} - B(n-m-1)B(n-m-1)^*)\tau_A(n,n-m)^*x \\
&= \tau_A(n,n-m-1)\tau_A(n,n-m-1)^*x.
\end{aligned}$$

Thus we get $x = \tau_A(n,n-m)\tau_A(n,n-m)^*x$ for all $m \geq 1$. Now

$$\|x\| = \|\tau_A(n,n-m)\tau_A(n,n-m)^*x\| \leq \|\tau_A(n,n-m)^*x\|.$$

Since $(A(n) : H_n \to H_{n+1})_{n\in\mathbb{Z}} \in C_{\cdot 0}$, the last term tends to 0 if $m \to \infty$. Therefore $x = 0$, and we conclude that the system Σ is controllable. □

Proposition 3.3. *Let $\Sigma = (A(n), B(n), C(n), D(n); H_n, K_n, L_n)$ be a unitary time-varying system, and let T be its input output map.*

(j) *Suppose the system Σ is controllable. Then T is an isometry if and only if the main operator sequence $(A(n) : H_n \to H_{n+1})_{n\in\mathbb{Z}} \in C_{0\cdot}$. In this case, Σ is also observable.*

(jj) *Suppose the system Σ is observable. Then T is a co-isometry if and only if the main operator sequence $(A(n) : H_n \to H_{n+1})_{n\in\mathbb{Z}} \in C_{\cdot 0}$. In this case, Σ is also controllable.*

In the proof of Proposition 3.3 we will use the following notation. For $n, m \in \mathbb{Z} \cup \{-\infty, \infty\}$, $n \leq m$, with $m = -\infty$, and $n = \infty$ excluded, we set $\ell^2(\mathcal{K}; n, m) = \bigoplus_{j=n}^{m} K_j$ and $\ell^2(\mathcal{L}; n, m) = \bigoplus_{j=n}^{m} L_j$. Define the canonical embeddings

$$(3.5) \qquad \iota_{\mathcal{K}}(n,m) : \ell^2(\mathcal{K}; n, m) \to \ell^2(\mathcal{K}), \qquad \iota_{\mathcal{L}}(n,m) : \ell^2(\mathcal{L}; n, m) \to \ell^2(\mathcal{L}).$$

By $I_{\mathcal{K}}(n,m)$ and $I_{\mathcal{L}}(n,m)$ we will denote the identity map on $\ell^2(\mathcal{K};\, n, m)$ and $\ell^2(\mathcal{L}; n, m)$, respectively, and we let $T(n,m)$ be the operator defined by

$$(3.6) \qquad T(n,m) = \iota_{\mathcal{L}}(n,m)^*\, T\, \iota_{\mathcal{K}}(n,m).$$

Furthermore, given a system $\Sigma = (A(n), B(n), C(n), D(n); H_n, K_n, L_n)$, we define the operators

$$(3.7) \quad \Lambda_{n,m} = \Lambda_{n,m}(\Sigma) = \mathrm{row}\,(\tau_A(m+1, j+1)B(j))_{j=n}^{m} : \ell^2(\mathcal{K}; n, m) \to H_{m+1},$$

$$(3.8) \quad \Gamma_{n,m} = \Gamma_{n,m}(\Sigma) = \mathrm{col}\,(C(j)\tau_A(j,n))_{j=n}^{m} : H_n \to \ell^2(\mathcal{L}; n, m),$$

where $n \leq m$. For the proof of Proposition 3.3 we need the following lemma, which is taken from [AKP]. For the sake of completeness we give a proof here.

Lemma 3.4. *Let $\Sigma = (A(n), B(n), C(n), D(n); H_n, K_n, L_n)$ be a unitary time-varying system and $T : \ell^2(\mathcal{K}) \to \ell^2(\mathcal{L})$ its input output map. Then we have for all $n, m \in \mathbb{Z}$, with $n \leq m$, the following two identities.*

$$(3.9) \qquad I_{\mathcal{K}}(n,m) - T(n,m)^* T(n,m) = \Lambda_{n,m}^* \Lambda_{n,m},$$

and

$$(3.10) \qquad I_{\mathcal{L}}(n,m) - T(n,m) T(n,m)^* = \Gamma_{n,m} \Gamma_{n,m}^*.$$

Proof. Part (a). Consider the system

$$\left\{ \begin{array}{lcl} x_{j+1} & = & A(j)x_j + B(j)u_j, \\ y_j & = & C(j)x_j + D(j)u_j, \qquad j = n, n+1, \ldots, m-1, m, \\ x_n & = & 0. \end{array} \right.$$

Let $u = (u_n, u_{n+1}, \ldots, u_{m-1}, u_m) \in \ell^2(\mathcal{K}; n, m)$ be an input sequence, and let $y = (y_n, y_{n+1}, \ldots, y_{m-1}, y_m)$ be the corresponding output sequence. Then $y =$

$T(n,m)u$. Since the system matrix $M_\Sigma(j) = \begin{bmatrix} A(j) & B(j) \\ C(j) & D(j) \end{bmatrix}$ is an isometry for each $j \in \mathbb{Z}$, we have:

(3.11) $\|x_{j+1}\|^2 - \|x_j\|^2 = \|u_j\|^2 - \|y_j\|^2, \qquad j = n, n+1, \ldots, m-1, m.$

Hence, using $x_n = 0$, we see that

(3.12) $$\|x_{m+1}\|^2 = \|x_{m+1}\|^2 - \|x_n\|^2 = \sum_{j=n}^{m} \|u_j\|^2 - \sum_{j=n}^{m} \|y_j\|^2,$$

and therefore

(3.13) $$\langle (I - T(n,m)^* T(n,m))u, u \rangle = \sum_{j=n}^{m} \|u_j\|^2 - \sum_{j=n}^{m} \|y_j\|^2 = \|x_{m+1}\|^2.$$

On the other hand, the state vector x_{m+1} at time $m+1$ is given by

(3.14) $$x_{m+1} = \Lambda_{n,m} u,$$

and we may conclude that

$$\langle (I - T(n,m)^* T(n,m))u, u \rangle = \langle \Lambda_{n,m}^* \Lambda_{n,m} u, u \rangle,$$

which yields (3.9).
Part (b). To prove (3.10) we consider the following system given by

$$\begin{cases} x_j &= A(j)^* x_{j+1} + C(j)^* y_j, \\ u_j &= B(j)^* x_{j+1} + D(j)^* y_j, \qquad j = m, m-1, \ldots, n+1, n, \\ x_{m+1} &= 0, \end{cases}$$

which operates in backward time (recall that $n \le m$). Let us feed in the input sequence $y_m, y_{m-1}, \ldots, y_{n+1}, y_n$, and let $u_m, u_{m-1}, \ldots, u_{n+1}, u_n$ be the corresponding output sequence. Put $y = \operatorname{col}(y_j)_{j=n}^m$ and $u = \operatorname{col}(u_j)_{j=n}^m$. Then $u = T(n,m)^* y$.
Using this remark and arguments similar to the ones used in the previous part (with time direction reversed), one proves (3.10). □

P r o o f (of Proposition 3.3). Let $T : \ell^2(\mathcal{K}) \to \ell^2(\mathcal{L})$ be a lower triangular contraction, and let $\Sigma = (A(n), B(n), C(n), D(n); H_n, K_n, L_n)$ be a unitary time-varying system with input output map T.
Part (a). First we prove statement (j). Therefore in this part we suppose that Σ is controllable. Fix $m \in \mathbb{Z}$, and set

(3.15) $$\mathcal{M} = \operatorname{span}\left\{ \sum_{j=n}^{m} \tau_A(m+1, j+1) B(j) k_j \mid k_j \in K_j, n \le m \right\}.$$

Since Σ is controllable, $\mathcal{M}$ is dense in H_{m+1}. Let x be an arbitrary vector in $\mathcal{M}$. So

$$x = \sum_{j=n}^{m} \tau_A(m+1, j+1)B(j)k_j,$$

for some $n \leq m$ and $k_j \in K_j$. By (3.9) we obtain for arbitrary $p \geq 0$:

$$\|\tau_A(m+p+1, m+1)x\|^2 = \|\sum_{j=n}^{m} \tau_A(m+p+1, j+1)B(j)k_j\|^2$$

$$(3.16) \qquad = \langle \Lambda_{n,m+p}^* \Lambda_{n,m+p} \begin{bmatrix} k_n \\ k_{n+1} \\ \vdots \\ k_{m-1} \\ k_m \\ 0_{m+1} \\ \vdots \\ 0_{m+p} \end{bmatrix}, \begin{bmatrix} k_n \\ k_{n+1} \\ \vdots \\ k_{m-1} \\ k_m \\ 0_{m+1} \\ \vdots \\ 0_{m+p} \end{bmatrix} \rangle$$

$$= \langle (I_{\mathcal{K}}(n, m+p) - T(n, m+p)^* T(n, m+p)) \begin{bmatrix} k_n \\ k_{n+1} \\ \vdots \\ k_{m-1} \\ k_m \\ 0_{m+1} \\ \vdots \\ 0_{m+p} \end{bmatrix}, \begin{bmatrix} k_n \\ k_{n+1} \\ \vdots \\ k_{m-1} \\ k_m \\ 0_{m+1} \\ \vdots \\ 0_{m+p} \end{bmatrix} \rangle.$$

Now we are going to show the equivalence in statement (j). So, assume first that T is an isometry. For each $p \geq 0$ decompose the operator $T(n, \infty)$ as follows:

$$(3.17) \qquad \begin{aligned} T(n, \infty) &= \begin{bmatrix} T_{1,1} & 0 & 0 \\ T_{2,1}(p) & T_{2,2}(p) & 0 \\ T_{3,1}(p) & T_{3,2}(p) & T_{3,3}(p) \end{bmatrix} \\ &: \ell^2(\mathcal{K}; n, m) \oplus \ell^2(\mathcal{K}; m+1, m+p) \oplus \ell^2(\mathcal{K}; m+p+1, \infty) \\ &\to \ell^2(\mathcal{L}; n, m) \oplus \ell^2(\mathcal{L}; m+1, m+p) \oplus \ell^2(\mathcal{L}; m+p+1, \infty). \end{aligned}$$

Since T is an isometry, $T(n, \infty)$ is an isometry. Put $k = (k_n, k_{n+1}, \ldots, k_{m-1}, k_m)^T \in \ell^2(\mathcal{K}; n, m)$. The right hand side of (3.17) equals

$$(3.18) \qquad \begin{aligned} &\langle (I - \begin{bmatrix} T_{1,1}^* & T_{2,1}(p)^* \\ 0 & T_{2,2}(p)^* \end{bmatrix} \begin{bmatrix} T_{1,1} & 0 \\ T_{2,1}(p) & T_{2,2}(p) \end{bmatrix}) \begin{bmatrix} k \\ 0 \end{bmatrix}, \begin{bmatrix} k \\ 0 \end{bmatrix} \rangle \\ &= \langle (I - T_{1,1}^* T_{1,1} - T_{2,1}(p)^* T_{2,1}(p))k, k \rangle \\ &= \langle T_{3,1}(p)^* T_{3,1}(p)k, k \rangle = \|T_{3,1}(p)k\|^2. \end{aligned}$$

The last term tends to 0 for $p \to \infty$ because for each p the vector $T_{3,1}(p)k$ is the projection of the vector $T(n,\infty)(k,0,0)^T$ along the first $m+p+1-n$ coordinates onto the remaining ones. Hence, for all $x \in \mathcal{M}$ we have the equality $\lim_{p\to\infty} \|\tau_A(m+p+1,m+1)x\| = 0$.
Take $w \in H_{m+1}$, and let $\varepsilon > 0$. Since the set $\mathcal{M}$ is dense in H_{m+1}, we can find a vector $x \in \mathcal{M}$ such that $\|x - w\| < \varepsilon$. For this x there exists an M such that for all $p > M$ we have $\|\tau_A(m+p,m+1)x\| < \varepsilon$. Then

$$\begin{aligned} \|\tau_A(m+p,m+1)w\| &\leq \|\tau_A(m+p,m+1)(w-x)\| \\ &\quad + \|\tau_A(m+p,m+1)x\| < 2\varepsilon. \end{aligned} \tag{3.19}$$

Since m was arbitrarily chosen, we have $(A(n) : H_n \to H_{n+1})_{n\in\mathbb{Z}} \in C_{0\cdot\cdot}$.
Let us now assume that $(A(n) : H_n \to H_{n+1})_{n\in\mathbb{Z}} \in C_{0\cdot\cdot}$. Take arbitrary $m, n \in \mathbb{Z}$, $m \geq n$, and take $k = (k_n, k_{n+1}, \ldots, k_{m-1}, k_m)^T \in \ell^2(\mathcal{K}; n, m)$. We will use the decomposition of $T(n,\infty)$ given in (3.17). From (3.17) we see:

$$\begin{aligned} 0 &= \lim_{p\to\infty} \langle I - T(n,m+p)^*T(n,m+p) \begin{bmatrix} k \\ 0 \end{bmatrix}, \begin{bmatrix} k \\ 0 \end{bmatrix} \rangle \\ &= \lim_{p\to\infty} \langle (I - T_{1,1}^*T_{1,1} - T_{2,1}(p)^*T_{2,1}(p))k, k \rangle \\ &= \lim_{p\to\infty} \left(\|k\|^2 - \left\| \begin{bmatrix} T_{1,1} & 0 & 0 \\ T_{2,1}(p) & T_{2,2}(p) & 0 \\ T_{3,1}(p) & T_{3,2}(p) & T_{3,3}(p) \end{bmatrix} \begin{bmatrix} k \\ 0 \\ 0 \end{bmatrix} \right\|^2 + \|T_{3,1}(p)k\|^2 \right) \\ &= \|k\|^2 - \left\|T(n,\infty)(k,0,0)^T\right\|^2 + \lim_{p\to\infty} \|T_{3,1}(p)k\|^2 \\ &- \|k\|^2 - \left\|T(n,\infty)(k,0,0)^T\right\|^2. \end{aligned}$$

In the last step we used the fact that $T(n,\infty)$ is a contraction and that for each p the vector $T_{3,1}(p)k$ is the projection of the vector $T(n,\infty)(k,0,0)^T$ along the first $m+p+1-n$ coordinates onto the remaining ones. Since T is lower triangular, we have for all integers n, m, with $n \leq m$,

$$\begin{aligned} &\|T(\ldots,0,k_n,k_{n+1},\ldots,k_{m-1},k_m,0,\ldots)\| \\ &= \|T(n,\infty)(k_n,k_{n+1},\ldots,k_{m-1},k_m,0,\ldots)\| \\ &= \|(k_n,k_{n+1},\ldots,k_{m-1},k_m,0,\ldots)\| \\ &= \|(\ldots,0,k_n,k_{n+1},\ldots,k_{m-1},k_m,0,\ldots)\|. \end{aligned}$$

This implies that T is an isometry. If one of the assertions in the equivalence statements of (j) holds true, then, in particular, the main operator sequence $(A(n) : H_n \to H_{n+1})_{n\in\mathbb{Z}}$ is in the class $C_{0\cdot\cdot}$. It follows from Proposition 3.2 that the system Σ is also observable.
Part (b). Next we consider (jj). Suppose Σ is observable. Fix $n \in \mathbb{Z}$. Let a vector

$x \in H_n$ be given by

$$x = \sum_{j=n}^{m} \tau_A(j,n)^* C(j)^* l_j, \tag{3.20}$$

where $l_j \in L_j$ for each j. Since Σ is observable, vectors of this type are dense in H_n. For arbitrary $p \geq 0$ we get:

$$\|\tau_A(n, n-p)^* x\|^2 = \| \sum_{j=n}^{m} \tau_A(j, n-p)^* C(j)^* l_j \|^2$$

$$= \langle \Gamma_{n-p,m} \Gamma^*_{n-p,m} \begin{bmatrix} 0_{n-p} \\ \vdots \\ 0_{n-1} \\ l_n \\ l_{n+1} \\ \vdots \\ l_{m-1} \\ l_m \end{bmatrix}, \begin{bmatrix} 0_{n-p} \\ \vdots \\ 0_{n-1} \\ l_n \\ l_{n+1} \\ \vdots \\ l_{m-1} \\ l_m \end{bmatrix} \rangle$$

$$= \langle (I_{\mathcal{L}}(n-p,m) - T(n-p,m)T(n-p,m)^*) \begin{bmatrix} 0_{n-p} \\ \vdots \\ 0_{n-1} \\ l_n \\ l_{n+1} \\ \vdots \\ l_{m-1} \\ l_m \end{bmatrix}, \begin{bmatrix} 0_{n-p} \\ \vdots \\ 0_{n-1} \\ l_n \\ l_{n+1} \\ \vdots \\ l_{m-1} \\ l_m \end{bmatrix} \rangle.$$

The equivalence in (jj) and the statement about the controllability can now be obtained in precisely the same way as in the proof of (j). □

Proof (of Theorem 3.1). If the unitary system Σ is observable and controllable, and its input output operator is unitary, then by Proposition 3.3 its main operator sequence $(A(n) : H_n \to H_{n+1})_{n \in \mathbb{Z}}$ is in the class C_{00}. On the other hand, if $(A(n) : H_n \to H_{n+1})_{n \in \mathbb{Z}}$ belongs to the class C_{00}, then the system Σ is observable and controllable by Proposition 3.2. The input output map T of Σ is unitary by Proposition 3.3. □

4. A realization theorem

It can be shown (see [ACC], [P]) that for any block lower contraction $T : \ell^2(\mathcal{K}) \to \ell^2(\mathcal{L})$ there exists a unitary time varying system Σ which has the operator T as input output map. In this case Σ is said to be a *realization in the class of unitary time varying systems* of the operator T. Here we restrict the attention to block lower triangular *unitary* operators. The next theorem provides in a direct way a unitary realization for such an operator.

Theorem 4.1. *Let T be the block lower triangular unitary operator*

$$T = \begin{bmatrix} \ddots & & & & \\ \cdots & t_{-1,-1} & & & \\ \cdots & t_{0,-1} & \boxed{t_{0,0}} & & \\ \cdots & t_{1,-1} & t_{1,0} & t_{1,1} & \\ & \vdots & \vdots & \vdots & \ddots \end{bmatrix} : \ell^2(\mathcal{K}) \to \ell^2(\mathcal{L}).$$

Let the operator $G(n)$ be given by

$$(4.1) \qquad G(n) = \begin{bmatrix} t_{n,n-1} & t_{n,n-2} & t_{n,n-3} & \cdots \\ t_{n+1,n-1} & t_{n+1,n-2} & t_{n+1,n-3} & \cdots \\ t_{n+2,n-1} & t_{n+2,n-2} & t_{n+2,n-3} & \cdots \\ \vdots & \vdots & \vdots & \end{bmatrix}$$
$$: K_{n-1} \oplus K_{n-2} \oplus \cdots \to L_n \oplus L_{n+1} \oplus \cdots .$$

Then for all n, the subspace $\operatorname{Im} G(n)$ is closed. Define the state space sequence $(H_n)_{n\in\mathbb{Z}}$ by

$$(4.2) \qquad H_n = \operatorname{Im} G(n).$$

Let $A(n)$, $B(n)$, $C(n)$, $D(n)$ be the operators

$$(4.3) \quad A(n) : H_n \to H_{n+1}, \quad A(n)\left(G(n) \begin{bmatrix} k_{n-1} \\ k_{n-2} \\ k_{n-3} \\ \vdots \end{bmatrix} \right) = G(n+1) \begin{bmatrix} 0 \\ k_{n-1} \\ k_{n-2} \\ \vdots \end{bmatrix},$$

$$(4.4) \quad B(n) : K_n \to H_{n+1}, \quad B(n)k_n = \begin{bmatrix} t_{n+1,n}k_n \\ t_{n+2,n}k_n \\ \vdots \end{bmatrix},$$

$$(4.5) \quad C(n) : H_n \to L_n, \quad C(n)\left(G(n) \begin{bmatrix} k_{n-1} \\ k_{n-2} \\ k_{n-3} \\ \vdots \end{bmatrix} \right) = \sum_{j=1}^{\infty} t_{n,n-j}k_{n-j},$$

$$(4.6) \quad D(n) : K_n \to L_n, \quad D(n)k_n = t_{n,n}k_n.$$

Then $\Sigma = (A(n), B(n), C(n), D(n); H_n, K_n, L_n)$ is a controllable and observable unitary time varying system such that $T_\Sigma = T$, and its main operator sequence $(A(n) : H_n \to H_{n+1})_{n\in\mathbb{Z}}$ belongs to the class C_{00}. Moreover, if $T \in \mathcal{LE}$, then the system Σ is exponentially stable.

Proof. Part (a). Let us first decompose T for all $n \in \mathbb{Z}$ as follows:

$$T = \begin{bmatrix} * & 0 \\ G(n) & T(n) \end{bmatrix}. \tag{4.7}$$

Since T is unitary, the following equalities hold for all n.

$$T(n)T(n)^* = I - G(n)G(n)^*, \quad T(n)^*G(n) = 0, \quad T(n)^*T(n) = I. \tag{4.8}$$

Thus

$$\operatorname{Im} G(n) \subset \operatorname{Ker} T(n)^* = \operatorname{Ker} T(n)T(n)^* = \operatorname{Ker}(I - G(n)G(n)^*) \subset \operatorname{Im} G(n).$$

Hence $\operatorname{Im} T(n)^{\perp} = \operatorname{Im} G(n)$. This shows that $\operatorname{Im} G(n)$ is a closed subspace of $\bigoplus_{i \geq n} L_i$.
Part (b). Let us now check the fact that Σ is a well defined unitary system. First note that the operators $A(n)$, $B(n)$ and $C(n)$ are well defined. For $(k_{n-1}, k_{n-2}, \ldots) \in \bigoplus_{j=1}^{\infty} K_{n-j}$ we have:

$$G(n) \begin{bmatrix} k_{n-1} \\ k_{n-2} \\ k_{n-3} \\ \vdots \end{bmatrix} = \begin{bmatrix} \sum_{j=1}^{\infty} t_{n,n-j} k_{n-j} \\ G(n+1) \begin{bmatrix} 0 \\ k_{n-1} \\ k_{n-2} \\ \vdots \end{bmatrix} \end{bmatrix}. \tag{4.9}$$

Take $x \in H_n$ and let x be represented as the left side of (4.9). Then

$$\begin{aligned} \|x\|^2 &= \left\| G(n) \begin{bmatrix} k_{n-1} \\ k_{n-2} \\ k_{n-3} \\ \vdots \end{bmatrix} \right\|^2 = \left\| \sum_{j=1}^{\infty} t_{n,n-j} k_{n-j} \right\|^2 + \left\| G(n+1) \begin{bmatrix} 0 \\ k_{n-1} \\ k_{n-2} \\ \vdots \end{bmatrix} \right\|^2 \\ &= \|C(n)x\|^2 + \|A(n)x\|^2. \end{aligned} \tag{4.10}$$

So the operators $A(n)$ and $C(n)$ are bounded. Since T is an isometry, we have

$$I = \begin{bmatrix} t_{n,n}^* & t_{n+1,n}^* & \cdots \end{bmatrix} \begin{bmatrix} t_{n,n} \\ t_{n+1,n} \\ \vdots \end{bmatrix} = D(n)^*D(n) + B(n)^*B(n). \tag{4.11}$$

Thus the operators $B(n)$ and $D(n)$ are bounded. The identity $T(n)^*G(n) = 0$ implies

$$(B(n)^*A(n) + D(n)^*C(n))G(n) \begin{bmatrix} k_{n-1} \\ k_{n-2} \\ k_{n-3} \\ \vdots \end{bmatrix}$$

$$= \begin{bmatrix} t^*_{n+1,n} & t^*_{n+2,n} & \cdots \end{bmatrix} G(n+1) \begin{bmatrix} 0 \\ k_{n-1} \\ k_{n-2} \\ \vdots \end{bmatrix} + t^*_{n,n} \sum_{j=1}^{\infty} t_{n,n-j} k_{n-j}$$

$$(4.12) \qquad = \begin{bmatrix} t^*_{n,n} & t^*_{n+1,n} & t^*_{n+2,n} & \cdots \end{bmatrix} \left(\begin{bmatrix} 0 \\ G(n+1) \begin{bmatrix} 0 \\ k_{n-1} \\ k_{n-2} \\ \vdots \end{bmatrix} \end{bmatrix} + \begin{bmatrix} \sum_{j=1}^{\infty} t_{n,n-j} k_{n-j} \\ 0 \end{bmatrix} \right)$$

$$= \begin{bmatrix} t^*_{n,n} & t^*_{n+1,n} & t^*_{n+2,n} & \cdots \end{bmatrix} G(n) \begin{bmatrix} k_{n-1} \\ k_{n-2} \\ k_{n-3} \\ \vdots \end{bmatrix} = 0.$$

The identities (4.10), (4.11), and (4.13) imply that the system matrix of the system Σ is an isometry.
Part (c). Let us write

$$(4.13) \qquad \widetilde{M}(n) = \begin{bmatrix} D(n) & C(n) \\ B(n) & A(n) \end{bmatrix} : K_n \oplus H_n \to L_n \oplus H_{n+1}.$$

Now

$$(4.14) \qquad \widetilde{M}(n) \begin{bmatrix} k_n \\ G(n) \begin{bmatrix} k_{n-1} \\ k_{n-2} \\ \vdots \end{bmatrix} \end{bmatrix} = \begin{bmatrix} t_{n,n} & t_{n,n-1} & t_{n,n-2} & \cdots \\ t_{n+1,n} & t_{n+1,n-1} & t_{n+1,n-2} & \cdots \\ t_{n+2,n} & t_{n+2,n-1} & t_{n+2,n-2} & \cdots \\ \vdots & \vdots & \vdots & \end{bmatrix} \begin{bmatrix} k_n \\ k_{n-1} \\ k_{n-2} \\ \vdots \end{bmatrix}.$$

The fact that $TT^* = I$ implies

$$(4.15) \qquad \sum_{j=0}^{\infty} t_{n,n-j} t^*_{n,n-j} = I_{L_n} \quad \text{and} \quad \sum_{j=0}^{\infty} t_{n+k,n-j} t^*_{n,n-j} = 0,$$

for all $k > 0$. The second equality of (4.15) is equivalent with the statement that

$$G(n+1)\begin{bmatrix} t^*_{n,n} \\ t^*_{n,n-1} \\ \vdots \end{bmatrix} = 0.$$

Let arbitrary $l_n \in L_n$ and $k_{n-j} \in K_{n-j}$ be given for all $j \geq 0$. Put

$$\widetilde{k}_{n-i} = t^*_{n,n-i}\left(l_n - \sum_{j=0}^{\infty} t_{n,n-j}k_{n-j}\right), \qquad i \geq 0.$$

Then

$$\widetilde{M}(n)\begin{bmatrix} k_n + \widetilde{k}_n \\ G(n)\begin{bmatrix} k_{n-1}+\widetilde{k}_{n-1} \\ k_{n-2}+\widetilde{k}_{n-2} \\ \vdots \end{bmatrix} \end{bmatrix} =$$

$$= \begin{bmatrix} \begin{bmatrix} t_{n,n} & t_{n,n-1} & t_{n,n-2} & \cdots \end{bmatrix} \\ G(n+1) \end{bmatrix} \left(\begin{bmatrix} k_n \\ k_{n-1} \\ \vdots \end{bmatrix} + \begin{bmatrix} \widetilde{k}_n \\ \widetilde{k}_{n-1} \\ \vdots \end{bmatrix} \right)$$

$$= \begin{bmatrix} \sum_{j=0}^{\infty} t_{n,n-j}k_{n-j} \\ G(n+1)\begin{bmatrix} k_n \\ k_{n-1} \\ \vdots \end{bmatrix} \end{bmatrix} + \begin{bmatrix} l_n - \sum_{j=0}^{\infty} t_{n,n-j}k_{n-j} \\ 0 \end{bmatrix}$$

$$= \begin{bmatrix} l_n \\ G(n+1)\begin{bmatrix} k_n \\ k_{n-1} \\ \vdots \end{bmatrix} \end{bmatrix}.$$

This is a typical element of $L_n \oplus H_{n+1}$. Thus the operator $\widetilde{M}(n)$ and hence the system matrix $M(n)$ are surjective. We have shown that the system Σ is unitary.
Part (d). The system Σ is indeed a realization of T. This follows from the equalities

$$\begin{aligned} &C(n+m)\tau_A(n+m,n+1)B(n)k_n \\ &= C(n+m)\tau_A(n+m,n+1)G(n+1)(\delta_{n,j}k_n)_{j=n}^{\infty} \\ &= C(n+m)G(n+m)(\delta_{n,j}k_n)_{j=n-m+1}^{\infty} = t_{n+m,n}k_n. \end{aligned}$$

Part (e). If the system Σ is observable and controllable, then the main operator sequence $(A(n) : H_n \to H_{n+1})_{n\in\mathbb{Z}}$ belongs to the class C_{00} by Theorem 3.1. So, we will show that the system Σ is observable and controllable. Consider the system at an arbitrary time instant $l \in \mathbb{Z}$. Assume $u \in \operatorname{Ker}(C|A;l)$, and let us write u in the form

$$u = G(l)\begin{bmatrix} k_{l-1} \\ k_{l-2} \\ \vdots \end{bmatrix}. \tag{4.16}$$

Then we have for all $n \geq l$:

$$\begin{aligned} 0 &= C(n)\tau_A(n,l)u = C(n)\tau_A(n,l)G(l)(k_{l-1}, k_{l-2}, \dots)^T \\ &= C(n)G(n)(0_{n-1}, \dots, 0_l, k_{l-1}, k_{l-2}, \dots)^T \\ &= \sum_{j=1}^{\infty} t_{n,l-j}k_{l-j}. \end{aligned}$$

According to (4.16), the last term is precisely the n-th entry of u. So $u = 0$. The system Σ is observable.

Next, take $v \in \operatorname{Im}(A|B;l)^\perp$. Then for all $n \leq l-1$ and arbitrary $k_n \in K_n$ we have

$$\begin{aligned} 0 &= \langle v, \tau_A(l, n+1)B(n)k_n\rangle \\ &= \langle v, \tau_A(l, n+1)G(n+1)(k_n, 0, 0, \dots)^T\rangle \\ &= \langle v, G(l)(0, \dots, 0, k_n, 0, 0, \dots)^T\rangle. \end{aligned}$$

Hence $G(l)^* v = 0$. On the other hand, $v \in H_l = \operatorname{Im} G(l)$. Thus $v = 0$. The system Σ is controllable.

Part (f). Suppose now that $T \in \mathcal{LE}$. So there exist an $\alpha \in (0,1)$ and a positive number M such that for all entries $t_{i,j}$ we have $\|t_{i,j}\| \leq M\alpha^{|i-j|}$. We will show that for this α we have

$$\limsup_{p\to\infty} \sup_n \|\tau_A(n+p-1, n)\|^{1/p} \leq \alpha < 1. \tag{4.17}$$

Fix $l \in H_n$ with $\|l\| \leq 1$. We see from (4.2) that l is of the form $l = (l_n, l_{n+1}, l_{n+2}, \dots)$ with $l_j \in L_j$. Take arbitrary $k = (k_n, k_{n+1}, \dots) \in \oplus_{j\geq n}K_j$. Since $H_n \perp \operatorname{Im} T(n)$, we have

$$\begin{aligned} 0 &= \langle T(n)k, l\rangle \\ &= \left\langle \begin{bmatrix} \ddots & & & & \\ \cdots & t_{n-1,n-1} & & & \\ \cdots & t_{n,n-1} & \boxed{t_{n,n}} & & \\ \cdots & t_{n+1,n-1} & t_{n+1,n} & t_{n+1,n+1} & \\ & \vdots & \vdots & \vdots & \ddots \end{bmatrix} \begin{bmatrix} \vdots \\ 0 \\ 0 \\ \boxed{k_n} \\ k_{n+1} \\ \vdots \end{bmatrix}, \begin{bmatrix} \vdots \\ 0 \\ 0 \\ \boxed{l_n} \\ l_{n+1} \\ \vdots \end{bmatrix} \right\rangle \\ &= \langle (\dots, 0, 0, \boxed{k_n}, k_{n+1}, \dots), T^*(\dots, 0, 0, \boxed{l_n}, l_{n+1}, \dots)\rangle. \end{aligned}$$

Therefore we have

$$T^*(\ldots,0,0,\boxed{l_n},l_{n+1},\ldots) = (\ldots,f_{n-2},f_{n-1},\boxed{0},0,\ldots),$$

for some $f = (\ldots,f_{n-2},f_{n-1},\boxed{0},0,\ldots) \in \ell^2(\mathcal{K},n)$. Since T is unitary, we have $l = Tf$, and $\sum_{j=1}^{\infty} \|f_{n-j}\|^2 = \|l\|^2 \leq 1$. The equality $l = Tf$ implies for each $j \geq 0$:

$$\begin{aligned} \|l_{n+j}\| &= \left\| \sum_{k=1}^{\infty} t_{n+j,n-k} f_{n-k} \right\| \leq \sum_{k=1}^{\infty} \|t_{n+j,n-k} f_{n-k}\| \\ &\leq \sum_{k=1}^{\infty} \|t_{n+j,n-k}\| \leq M \sum_{k=1}^{\infty} \alpha^{j+k} = \widetilde{M}\alpha^j, \end{aligned}$$

where $\widetilde{M} = M\alpha(1-\alpha)^{-1}$.
Since the operator $A(n)$ acts as a restricted backward shift on elements of H_n, i.e.,

$$A(n)(l_n,l_{n+1},l_{n+2},\ldots) = (l_{n+1},l_{n+2},l_{n+3}\cdots),$$

we obtain:

$$\begin{aligned} \left\| \tau_A(n+p,n) \begin{bmatrix} l_n \\ l_{n+1} \\ l_{n+2} \\ \vdots \end{bmatrix} \right\|^2 &= \left\| \begin{bmatrix} l_{n+p} \\ l_{n+p+1} \\ \vdots \end{bmatrix} \right\|^2 = \sum_{t=0}^{\infty} \|l_{n+p+t}\|^2 \\ &\leq \sum_{t=0}^{\infty} \widetilde{M}^2 \alpha^{2(p+t)} = \widetilde{M}^2 \frac{\alpha^{2p}}{1-\alpha^2}. \end{aligned}$$

This estimate yields $\sup_n \|\tau_A(n+p-1,n)\| \leq \widetilde{M}\frac{\alpha^p}{\sqrt{1-\alpha^2}}$ and hence

$$\limsup_{p\to\infty} \sup_n \|\tau_A(n+p-1,n)\|^{1/p} \leq \alpha < 1. \tag{4.18}$$

□

5. Cascade connection and factorization

We consider two contractive time varying systems

$$\Sigma_\nu = (A_\nu(n), B_\nu(n), C_\nu(n), D_\nu(n); H_{\nu,n}, K_{\nu,n}, L_{\nu,n}), \qquad \nu = 1,2.$$

We assume that for each n the output space $L_{2,n}$ at time n of Σ_2 is equal to the input space $K_{1,n}$ at time n of Σ_2. Define a new system Σ by

$$\begin{bmatrix} x_{n+1}^{(1)} \\ x_{n+1}^{(2)} \\ y_n^{(1)} \end{bmatrix} = \begin{bmatrix} A_1(n) & B_1(n)C_2(n) & B_1(n)D_2(n) \\ 0 & A_2(n) & B_2(n) \\ C_1(n) & D_1(n)C_2(n) & D_1(n)D_2(n) \end{bmatrix} \begin{bmatrix} x_n^{(1)} \\ x_n^{(2)} \\ u_n^{(2)} \end{bmatrix}. \tag{5.1}$$

The system Σ in (5.1) is called *the cascade connection* or *product* of Σ_1 and Σ_2. It will be denoted by $\Sigma_1\Sigma_2$. In fact we have

$$\begin{aligned}\Sigma_1\Sigma_2 &= \left(\begin{bmatrix} A_1(n) & B_1(n)C_2(n) \\ 0 & A_2(n) \end{bmatrix}, \begin{bmatrix} B_1(n)D_2(n) \\ B_2(n) \end{bmatrix}, \right. \\ &\left. \begin{bmatrix} C_1(n) & D_1(n)C_2(n) \end{bmatrix}, D_1(n)D_2(n); H_{1,n} \oplus H_{2,n}, K_{2,n}, L_{1,n} \right). \end{aligned} \tag{5.2}$$

The equality $T_{\Sigma_1\Sigma_2} = T_{\Sigma_1}T_{\Sigma_2}$ follows immediately from the construction of the cascade connection.

Theorem 5.1. *If Σ_1 and Σ_2 are unitary time varying systems, then the cascade connection is again a unitary time varying system. The product $\widehat{\Sigma_1}\widehat{\Sigma_2}$ of the associated time invariant systems $\widehat{\Sigma_1}$ and $\widehat{\Sigma_2}$ is unitarily equivalent to the time invariant system $\widehat{\Sigma_1\Sigma_2}$ associated with the product $\Sigma_1\Sigma_2$.*

Proof. From the equality

$$\begin{aligned} &\begin{bmatrix} A_1(n) & B_1(n) & 0 \\ C_1(n) & D_1(n) & 0 \\ 0 & 0 & I_{H_{2,n+1}} \end{bmatrix} \begin{bmatrix} I_{H_{1,n}} & 0 & 0 \\ 0 & D_2(n) & C_2(n) \\ 0 & B_2(n) & A_2(n) \end{bmatrix} \\ &= \begin{bmatrix} A_1(n) & B_1(n)D_2(n) & B_1(n)C_2(n) \\ C_1(n) & D_1(n)D_2(n) & D_1(n)C_2(n) \\ 0 & B_2(n) & A_2(n) \end{bmatrix}, \end{aligned} \tag{5.3}$$

we see that $\Sigma_1\Sigma_2$ is a unitary time varying system. Since $\widehat{\Sigma_1}$ and $\widehat{\Sigma_2}$ are unitary systems, the system

$$\begin{aligned}\widehat{\Sigma_1}\widehat{\Sigma_2} &= \left(\begin{bmatrix} S_{A_1} & S_{B_1}\Delta_{C_2} \\ 0 & S_{A_2} \end{bmatrix}, \begin{bmatrix} S_{B_1}\Delta_{D_2} \\ S_{B_2} \end{bmatrix}, \begin{bmatrix} \Delta_{C_1} & \Delta_{D_1}\Delta_{C_2} \end{bmatrix}, \Delta_{D_1}\Delta_{D_2}; \right. \\ &\left. \ell^2(\mathcal{H}_1) \oplus \ell^2(\mathcal{H}_2), \ell^2(\mathcal{K}_2), \ell^2(\mathcal{L}_1) \right) \end{aligned}$$

is again unitary. The system $\widehat{\Sigma_1}\widehat{\Sigma_2}$ and the system

$$\begin{aligned}\widehat{\Sigma_1\Sigma_2} &= \left(S_{\begin{bmatrix} A_1(n) & B_1(n)C_2(n) \\ 0 & A_2(n) \end{bmatrix}}, S_{\begin{bmatrix} B_1(n)D_2(n) \\ B_2(n) \end{bmatrix}}, \right. \\ &\left. \Delta_{\begin{bmatrix} C_1(n) & D_1(n)C_2(n) \end{bmatrix}}, \Delta_{D_1(n)D_2(n)}; \bigoplus_{n\in\mathbb{Z}}(H_{1,n} \oplus H_{2,n}), \ell^2(\mathcal{K}_2), \ell^2(\mathcal{L}_1) \right) \end{aligned}$$

are unitarily equivalent, the equivalence being given by the unitary transformation

$$\mathcal{U} : \ell^2(\mathcal{H}_1) \oplus \ell^2(\mathcal{H}_2) \to \bigoplus_{n\in\mathbb{Z}}(H_{1,n} \oplus H_{2,n}),$$

$$\mathcal{U}\Big((\ldots,x_{-1},\boxed{x_0},x_1,\ldots),(\ldots,y_{-1},\boxed{y_0},y_1,\ldots)\Big)$$
$$=\Big(\ldots,(x_{-1},y_{-1}),\boxed{(x_0,y_0)},(x_1,y_1),\ldots\Big).$$

For a sequence of operators $(A(n) : H_n \to H_{n+1})_{n\in\mathbb{Z}}$ we call the sequence $(K_n)_{n\in\mathbb{Z}}$ of subspaces $K_n \subset H_n$ an *invariant subspace sequence* if for all $n \in \mathbb{Z}$ we have the inclusion

(5.4) $$A(n)K_n \subset K_{n+1}.$$

Let $\Sigma = \Sigma_1\Sigma_2$ be a product of two unitary time varying systems. Then, according to (5.1), the state operator sequence $(A(n) : H_n \to H_{n+1})_{n\in\mathbb{Z}}$ admits the partitioning

(5.5) $$A(n) = \begin{bmatrix} A_1(n) & * \\ 0 & A_2(n) \end{bmatrix} : H_{1,n} \oplus H_{2,n} \to H_{1,n+1} \oplus H_{2,n+1}.$$

It follows that the sequence $(H_{1,n})_{n\in\mathbb{Z}}$ is an invariant subspace sequence of $(A(n) : H_n \to H_{n+1})_{n\in\mathbb{Z}}$. We will show that any invariant subspace sequence of $(A(n) : H_n \to H_{n+1})_{n\in\mathbb{Z}}$ yields a factorization of Σ into a product of two unitary time varying systems. To this end, first we will prove the following embedding lemma.

Lemma 5.2. *Let $A(n) : H_n \to H_{n+1}$ and $C(n) : H_n \to L_n$ be Hilbert space operators for each $n \in \mathbb{Z}$. In order that for $n \in \mathbb{Z}$ the map $A(n)$ is the main operator and $C(n)$ is the output operator, both at time n, of a unitary time-varying system it is necessary and sufficient that*

(5.6) $$A(n)^*A(n) + C(n)^*C(n) = I_{H_n}, \qquad n \in \mathbb{Z}.$$

Assume that the latter conditions are fulfilled, and for $n \in \mathbb{Z}$ put

$$K_{0,n} = \left\{\begin{bmatrix} x \\ y \end{bmatrix} \in H_{n+1} \oplus L_n \mid A(n)^*x + C(n)^*y = 0\right\},$$

(5.7) $$B_0(n) : K_{0,n} \to H_{n+1};\ B_0(n)\begin{bmatrix} x \\ y \end{bmatrix} = x,$$

(5.8) $$D_0(n) : K_{0,n} \to L_n;\ D_0(n)\begin{bmatrix} x \\ y \end{bmatrix} = y.$$

Then any unitary time-varying system Σ with main operator $A(n)$ and output operator $C(n)$, $n \in \mathbb{Z}$, is given by

(5.9) $$\Sigma = (A(n), B_0(n)U_n, C(n), D_0(n)U_n; H_n, K_n, L_n)$$

where $U_n : K_n \to K_{0,n}$, $n \in \mathbb{Z}$, are unitary operators. Moreover, there is a one-one correspondence between the unitary operators $(U_n)_{n\in\mathbb{Z}}$ and the unitary time varying systems with state operators $(A(n))_{n\in\mathbb{Z}}$, and output operators $(C(n))_{n\in\mathbb{Z}}$.

The unitary time varying system

$$\Sigma_0 = (A(n), B_0(n), C(n), D_0(n); H_n, K_{0,n}, L_n) \tag{5.10}$$

described in Lemma 5.2 is called the *standard unitary time varying system associated with* $(A(n), C(n))_{n\in\mathbb{Z}}$.

Proof (of Lemma 5.2). The requirement (5.6) is equivalent with the identity $S_A^*S_A + \Delta_C^*\Delta_C = I_{\ell^2(\mathcal{H})}$. Now, by Lemma XXVIII 7.1 of [GGK2] this is equivalent for S_A and Δ_C to be the main operator and the output operator, respectively, of a unitary time invariant system.
Let us suppose that $S_A^*S_A + \Delta_C^*\Delta_C = I_{\ell^2(\mathcal{H})}$. We will construct the unitary system, following the above mentioned lemma. Let us define the space

$$\mathcal{K}_0 = \left\{ \begin{bmatrix} \vec{x} \\ \vec{y} \end{bmatrix} \in \ell^2(\mathcal{H}) \oplus \ell^2(\mathcal{L}) \mid S_A^*\vec{x} + \Delta_C^*\vec{y} = 0 \right\}, \tag{5.11}$$

and the operators

$$\mathcal{B}_0 : \mathcal{K}_0 \to \ell^2(\mathcal{H}); \mathcal{B}_0 \begin{bmatrix} \vec{x} \\ \vec{y} \end{bmatrix} = \vec{x}, \tag{5.12}$$

and

$$\mathcal{D}_0 : \mathcal{K}_0 \to \ell^2(\mathcal{L}); \mathcal{D}_0 \begin{bmatrix} \vec{x} \\ \vec{y} \end{bmatrix} = \vec{y}. \tag{5.13}$$

Now

$$\mathcal{K}_0 = \left\{ \begin{bmatrix} \vec{x} \\ \vec{y} \end{bmatrix} \mid A(n)^*x_{n+1} + C(n)^*y_n = 0,\ n \in \mathbb{Z} \right\}, \tag{5.14}$$

so the space $\mathcal{K}_0$ can be identified with the Hilbert space direct sum $\bigoplus_{n\in\mathbb{Z}} K_n$, where

$$K_n = \left\{ \begin{bmatrix} x_{n+1} \\ y_n \end{bmatrix} \in H_{n+1} \oplus L_n \mid A(n)^*x_{n+1} + C(n)^*y_n = 0 \right\}.$$

Therefore we can identify the operator $\mathcal{B}_0$ with the weighted shift

$$S' : \bigoplus_{n\in\mathbb{Z}} K_n \to \ell^2(\mathcal{H}), \qquad S' \begin{bmatrix} \vdots \\ (x_0, y_{-1}) \\ \boxed{(x_1, y_0)} \\ (x_2, y_1) \\ \vdots \end{bmatrix} = \begin{bmatrix} \vdots \\ x_{-1} \\ \boxed{x_0} \\ x_1 \\ \vdots \end{bmatrix}, \tag{5.15}$$

and the operator $\mathcal{D}_0$ can be identified with

$$(5.16)\qquad \Delta' : \bigoplus_{n\in\mathbb{Z}} K_n \to \ell^2(\mathcal{L}), \qquad \Delta' \begin{bmatrix} \vdots \\ (x_0, y_{-1}) \\ \boxed{(x_1, y_0)} \\ (x_2, y_1) \\ \vdots \end{bmatrix} = \begin{bmatrix} \vdots \\ y_{-1} \\ \boxed{y_0} \\ y_1 \\ \vdots \end{bmatrix}.$$

Notice that $S' = S_{B_0}$ and $\Delta' = \Delta_{D_0}$ where B_0 and D_0 are given by (5.7) and (5.8).
Since $\widehat{\Sigma_0} = (S_A, S_{B_0}, \Delta_C, \Delta_{D_0})$ is unitary, arguments similar to the ones used in the second paragraph preceding the proof of Theorem 2.1 show that the time varying system in (5.10) is also unitary.
Let now $\Sigma = (A(n), B(n), C(n), D(n))$ be an arbitrary unitary system. By Lemma XXVIII 7.1 in [GGK2] there exists a unique unitary operator U such that

$$(5.17)\qquad \begin{bmatrix} S_A^* & \Delta_C^* \\ S_B^* & \Delta_D^* \end{bmatrix} \begin{bmatrix} S_A & S_{B_0} \\ \Delta_C & \Delta_{D_0} \end{bmatrix} = \begin{bmatrix} I & 0 \\ 0 & U \end{bmatrix}.$$

It remains to show that U is diagonal. But this follows from the identity $U = S_B^* S_{B_0} + \Delta_D^* \Delta_{D_0}$. □

Now let $\Sigma = (A(n), B(n), C(n), D(n); H_n, K_n, L_n)$ be a unitary time varying system and suppose

$$(5.18)\qquad (H_{l,n})_{n\in\mathbb{Z}}$$

is an $(A(n) : H_n \to H_{n+1})_{n\in\mathbb{Z}}$ invariant subspace sequence. Put

$$(5.19)\qquad H_{r,n} = H_n \ominus H_{l,n}, \qquad K_{r,n} = K_n, \qquad L_{l,n} = L_n.$$

Here the subscript l stands for "left", and the subscript r for "right". We partition:

$$A(n) = \begin{bmatrix} A_{ll}(n) & A_{rl}(n) \\ 0 & A_{rr}(n) \end{bmatrix} : H_{l,n} \oplus H_{r,n} \to H_{l,n+1} \oplus H_{r,n+1},$$

$$B(n) = \begin{bmatrix} B_{rl}(n) \\ B_{rr}(n) \end{bmatrix} : K_{r,n} \to H_{l,n+1} \oplus H_{r,n+1},$$

$$C(n) = \begin{bmatrix} C_{ll}(n) & C_{lr}(n) \end{bmatrix} : H_{l,n} \oplus H_{r,n} \to L_{l,n}.$$

Thus

$$(5.20)\qquad \begin{bmatrix} A(n) & B(n) \\ C(n) & D(n) \end{bmatrix} = \begin{bmatrix} A_{ll}(n) & A_{lr}(n) & B_{rl}(n) \\ 0 & A_{rr}(n) & B_{rr}(n) \\ C_{ll}(n) & C_{lr}(n) & D(n) \end{bmatrix}$$
$$: H_{l,n} \oplus H_{r,n} \oplus K_{r,n} \to H_{l,n+1} \oplus H_{r,n+1} \oplus L_{l,n}.$$

We conclude that the right hand side of (5.20) is unitary for each $n \in \mathbb{Z}$. So,

$$A_{ll}(n)^* A_{ll}(n) + C_{ll}(n)^* C_{ll}(n) = I_{H_{l,n}}, \qquad n \in \mathbb{Z}. \tag{5.21}$$

Now we can define

$$\Sigma_l = (A_{ll}(n), B_l(n), C_{ll}(n), D_l(n); H_{l,n}, K_{l,n}, L_{l,n}) \tag{5.22}$$

to be the standard unitary time varying system associated with $(A_{ll}(n), C_{ll}(n))_{n\in\mathbb{Z}}$.

The unitary time varying system Σ_l is called the *left projection* of Σ associated with the invariant subspace sequence $(H_{l,n})_{n\in\mathbb{Z}}$.

Next, consider the product

$$\begin{bmatrix} A_{ll}(n)^* & 0 & C_{ll}(n)^* \\ 0 & I_{H_{r,n+1}} & 0 \\ B_l(n)^* & 0 & D_l(n)^* \end{bmatrix} \left[\begin{array}{cc|c} A_{ll}(n) & A_{lr}(n) & B_{rl}(n) \\ 0 & A_{rr}(n) & B_{rr}(n) \\ \hline C_{ll}(n) & C_{lr}(n) & D(n) \end{array}\right] \tag{5.23}$$
$$: H_{l,n} \oplus H_{r,n} \oplus K_{r,n} \to H_{l,n} \oplus H_{r,n+1} \oplus K_{l,n}.$$

Since the right hand side of (5.20) is a unitary operator, (5.23) is also unitary. Therefore, (5.23) equals

$$\begin{bmatrix} I_{H_{l,n}} & 0 & 0 \\ 0 & A_{rr}(n) & B_{rr}(n) \\ 0 & C_r(n) & D_r(n) \end{bmatrix}, \tag{5.24}$$

for certain operators $D_r(n) : K_{r,n} \to K_{l,n}$ and $C_r(n) : H_{r,n} \to K_{l,n}$. Let us define $L_{r,n} = K_{l,n}$. Thus the system

$$\Sigma_r = (A_{rr}(n), B_{rr}(n), C_r(n), D_r(n); H_{r,n}, K_{r,n}, L_{r,n}) \tag{5.25}$$

is a time varying unitary system. It is called the *right projection* of Σ, associated with the invariant subspace sequence $(H_{l,n})_{n\in\mathbb{Z}}$.

Theorem 5.3. *Let Σ be a unitary time varying system and $(M_n)_{n\in\mathbb{Z}}$ an invariant subspace sequence for the sequence of state operators of Σ. Then the left projection Σ_l and the right projection Σ_r of Σ associated with $(M_n)_{n\in\mathbb{Z}}$ are unitary time varying systems. Moreover, $\Sigma = \Sigma_l \Sigma_r$.*

Proof. Put $H_{l,n} = M_n$. We already showed that Σ_l and Σ_r are unitary time varying systems. So, for all $n \in \mathbb{Z}$, the operators

$$\begin{bmatrix} A_{ll}(n) & C_{ll}(n) & 0 \\ B_l(n) & D_l(n) & 0 \\ 0 & 0 & I_{H_{r,n+1}} \end{bmatrix} \tag{5.26}$$

are unitary. As in Theorem 5.1, the fact that the operator defined by the product in (5.23) is equal to the operator in (5.24) yields $\Sigma = \Sigma_l \Sigma_r$. □

6. Proof of Theorems 1.1 and 1.2

Consider the block lower triangular operator $T = (t_{i,j})_{i,j=-\infty}^{\infty} : \ell^2(\mathcal{K}) \to \ell^2(\mathcal{L})$. Define the *generalized Hankel operator at time* p to be the operator

$$(6.1)\qquad G_T(p) = \begin{bmatrix} t_{p,p-1} & t_{p,p-2} & t_{p,p-3} & \cdots \\ t_{p+1,p-1} & t_{p+1,p-2} & t_{p+1,p-3} & \cdots \\ t_{p+2,p-1} & t_{p+2,p-2} & t_{p+2,p-3} & \cdots \\ \vdots & \vdots & \vdots & \end{bmatrix} : \bigoplus_{j<p} K_j \to \bigoplus_{j\geq p} L_j.$$

The (possibly infinite) number $\kappa(T) = \sup_p \operatorname{rank}(G_T(p))$ will be called the *Kronecker index* of the operator T. The class of operators with finite Kronecker index will be denoted by $\mathcal{K}$.

P r o o f (of Theorem 1.1). We divide the proof into four parts.
Part (a). In this part we show that for the case considered here all generalized Hankel operators $G_T(p)$ have the same rank. Let

$$(6.2)\qquad \Sigma = (A(n), B(n), C(n), D(n); H_n, \mathbb{C}^m, \mathbb{C}^m)$$

be the unitary time varying realization of T constructed in Theorem 4.1. Since T has finite Kronecker index, $\dim H_n = \operatorname{rank} G_n$ is finite for each n. The system matrices

$$(6.3)\qquad M_\Sigma(n) = \begin{bmatrix} A(n) & B(n) \\ C(n) & D(n) \end{bmatrix} : H_n \oplus \mathbb{C}^m \to H_{n+1} \oplus \mathbb{C}^m$$

are unitary, so $\dim H_n = N$ for all n. In particular, $\operatorname{rank} G_n = N$ for each n.
Part (b). In this part we show that the system Σ in (6.2) is unitarily equivalent to a unitary time varying system $\Upsilon = (R(n), \widetilde{B}(n), \widetilde{C}(n), D(n); \mathbb{C}^N, \mathbb{C}^m, \mathbb{C}^m)$ with the additional property that for each $n \in \mathbb{Z}$ the matrix of the operator $R(n)$ with respect to the standard basis of $\mathbb{C}^N$ is upper triangular.
First recall that any complex square matrix A admits a left QR-factorization, $A = QR$, where Q is unitary and R is upper triangular, and also a right RQ-factorization $A = \widetilde{R}\widetilde{Q}$, where $\widetilde{Q}$ is unitary and $\widetilde{R}$ is upper triangular. Consider now the difference equation $x_{n+1} = A(n)x_n$. As in [VDS], we define a unitary matrix $Q(n)$, and an upper triangular matrix $R(n)$ recursively by

$$(6.4)\qquad \begin{cases} A(0) =: Q(1)R(0), & \\ A(n)Q(n) =: Q(n+1)R(n), & n = 1, 2, \ldots, \end{cases}$$
$$\begin{cases} A(-1) =: R(-1)Q(-1)^{-1}, & \\ Q(n+1)^{-1}A(n) =: R(n)Q(n)^{-1}, & n = -2, -3, \ldots, \end{cases}$$

$$Q(0) := I_{\mathbb{C}^N}.$$

Since $Q(n)$ is unitary, and $Q(n+1)^{-1}A(n)Q(n) = R(n)$ for each $n \in \mathbb{Z}$, we can define the system

$$\Upsilon = (R(n), Q(n+1)B(n), C(n)Q(n)^{-1}, D(n); \mathbb{C}^N, \mathbb{C}^m, \mathbb{C}^m). \tag{6.5}$$

which is unitarily equivalent to Σ, and $R(n)$ is upper triangular for each $n \in \mathbb{Z}$.
Part (c). Assume that $R(n)$ is given by

$$R(n) = \begin{bmatrix} \alpha_{11}(n) & * & * & \cdots & * \\ & \alpha_{22}(n) & * & \cdots & * \\ & & \alpha_{33}(n) & \cdots & * \\ & & & \ddots & \vdots \\ & & & & \alpha_{NN}(n) \end{bmatrix}. \tag{6.6}$$

Put $H_{l,n} = \operatorname{span}\{e_1\}$, where e_1 is the first vector in the standard basis of $\mathbb{C}^N$. From (6.6) it follows that $(H_{l,n})_{n\in\mathbb{Z}}$ is an invariant subspace sequence for the operators $(R(n))_{n\in\mathbb{Z}}$. Let

$$\Upsilon_l = (\alpha_{11}(n), B_l(n), C_l(n), D_l(n)) \tag{6.7}$$

be the associated left projection and let

$$\Upsilon_r = (A_r(n), B_r(n), C_r(n), D_r(n)) \tag{6.8}$$

be the associated right projection. Thus

$$A_r(n) = \begin{bmatrix} \alpha_{22}(n) & * & \cdots & * \\ & \alpha_{33}(n) & & * \\ & & \ddots & \vdots \\ & & & \alpha_{NN}(n) \end{bmatrix}. \tag{6.9}$$

Here we identify $H_{l,n} = \operatorname{span}\{e_1\}$ with $\mathbb{C}$ and $H_{r,n} = \mathbb{C}^N \ominus \operatorname{span}\{e_1\}$ with $\mathbb{C}^{N-1}$. Now we will show that the input output operator T_l of Υ_l belongs to the class $\mathcal{LEK}$, is unitary, and that its Kronecker index is less than or equal to 1.
Since T is unitary and an element of the class $\mathcal{LEK}$, the system Σ is exponentially stable by Theorem 4.1. The systems Σ and Υ are unitarily equivalent, and hence

$$|\tau_{\alpha_{11}}(n, n-k+1)|^{1/k} \le \|\tau_R(n, n-k+1)\|^{1/k} = \|\tau_A(n, n-k+1)\|^{1/k}.$$

Hence the system Υ_l is exponentially stable.
Since Υ_l is unitary, we may apply Theorem 2.1 to conclude that its input output map T_l is unitary and in the class $\mathcal{LE}$. The generalized Hankel operator $G_l(n)$ at time n of the input output map T_l decomposes into a product of bounded operators,

(6.10) $$G_l(n) = \Gamma_{n,\infty}(\Upsilon_l)\widetilde{\Lambda}_{n-1,-\infty}(\Upsilon_l),$$

where $\Gamma_{n,\infty}(\Upsilon_l)$ is defined by (3.8), and

(6.11) $$\widetilde{\Lambda}_{n-1,-\infty}(\Upsilon_l) = \operatorname{row}\left(\tau_{\alpha_{11}}(n, j+1)B(j)\right)_{j=n-1}^{-\infty}.$$

Notice that the boundedness of the factors in the right hand side of this identity follows from the fact that Υ_l is unitary and exponentially stable. It follows that $\kappa(T_l) = \operatorname{rank} G_l(n) \leq 1$.
Part (d). In this part we show that the input output operator T_r of the right projection Υ_r is a unitary operator in the class $\mathcal{LEK}$. Furthermore, we prove that $\kappa(T_r) = N-1$ and $\kappa(T_l) = 1$.
Since $T = T_lT_r$, where the operators T and T_l are unitary, we can conclude that T_r is also unitary. According to (6.6) we may partition $R(n)$ in the following way:

(6.12) $$R(n) = \begin{bmatrix} \alpha_{11}(n) & * \\ 0 & A_r(n) \end{bmatrix},$$

where $A_r(n)$ is given by (6.9). This holds for each $n \in \mathbb{Z}$. It follows that

$$R(n)R(n-1)\cdots R(n-k+1) = \begin{bmatrix} \tau_{\alpha_{11}}(n, n-k+1) & * \\ 0 & \tau_{A_r}(n, n-k+1) \end{bmatrix}.$$

Hence

$$\|\tau_{A_r}(n, n-k+1)\|^{1/k} \leq \|\tau_R(n, n-k+1)\|^{1/k} = \|\tau_A(n, n-k+1)\|^{1/k}.$$

Thus the system Υ_r is exponentially stable. From (6.8) and (6.9) we see that the Kronecker index of the map T_r is less than or equal to $N-1$. By Theorem 2.1, the input output operator T_r is a unitary operator in the class $\mathcal{LEK}$.
It remains to show that $\kappa(T_r) = N-1$ and $\kappa(T_l) = 1$. For each $p \in \mathbb{Z}$ let the operators $X(p)$ and $Y(p)$ be defined by the decompositions

(6.13) $$T_l = \begin{bmatrix} * & 0 \\ G_l(p) & Y(p) \end{bmatrix}, \qquad T_r = \begin{bmatrix} X(p) & 0 \\ G_r(p) & * \end{bmatrix},$$

where $G_l(p)$ and $G_r(p)$ are the generalized Hankel operators of the input output maps T_l and T_r, respectively. Since $T = T_lT_r$, we have $G_T(p) = G_{T_l}(p)X(p) + Y(p)G_{T_r}(p)$, $p \in \mathbb{Z}$. It follows that

$$N = \kappa(T) \leq \kappa(T_l) + \kappa(T_r) \leq N.$$

Hence $\kappa(T_l) = 1$ and $\kappa(T_r) = N-1$. We may finish the proof by induction. □

P r o o f (of Theorem 1.2). Theorem 4.1 provides in a unitary realization Σ of T with one dimensional state space,

(6.14) $$\Sigma = (\alpha_n, B(n), \Gamma_n, D(n); \mathbb{C}, \mathbb{C}^m, \mathbb{C}^m),$$

which is exponentially stable. Let us introduce the system

$$\widetilde{\Sigma} = (\alpha_n, \widetilde{B}(n), \Gamma_n, \widetilde{D}(n); \mathbb{C}, \mathbb{C}^m, \mathbb{C}^m), \tag{6.15}$$

where $\widetilde{B}(n) = -\mathrm{sgn}(\alpha_n)\Gamma_n^*$, $\widetilde{D}(n) = I_{\mathbb{C}^m} - (1+|\alpha_n|)^{-1}\Gamma_n\Gamma_n^*$ and

$$\mathrm{sgn}(x) = \begin{cases} 1, & x = 0, \\ x/|x|, & x \neq 0. \end{cases} \tag{6.16}$$

First note that $|\alpha_n|^2 + \Gamma_n^*\Gamma_n = 1$. Since

$$(I - (1+|\alpha_n|)^{-1}\Gamma_n\Gamma_n^*)\Gamma_n = |\alpha_n|\Gamma_n, \quad (I - (1+|\alpha_n|)^{-1}\Gamma_n\Gamma_n^*)^2 = I - \Gamma_n\Gamma_n^*,$$

it is straightforward to check that $\widetilde{\Sigma}$ is unitary.
Note that the input output map of $\widetilde{\Sigma}$ is precisely the operator given by the matrix in the right hand side of (1.1). Since Σ is exponentially stable, the same holds for $\widetilde{\Sigma}$, and hence condition (jj) has been fulfilled. Both systems Σ and $\widetilde{\Sigma}$ are unitary with main operator α_n at time n and output operator Γ_n at time n. Therefore, by Lemma 5.2, there are unitary operators $U_n : \mathbb{C}^m \to \mathbb{C}^m$ such that $B(n) = \widetilde{B}(n)U_n$ and $D(n) = \widetilde{D}(n)U_n$.
On the other hand, if we start with an operator $T = T_{(\alpha)(\Gamma)}U$ as given in (1.1), then the system $\widetilde{\Sigma}$, given by (6.15), is a unitary realization of of $T_{(\alpha)(\Gamma)}$ by (j). Let us denote the unitary block diagonal operator U by $\mathrm{diag}(U_j)_{j=-\infty}^{\infty} : \ell^2(\mathbb{C}^m) \to \ell^2(\mathbb{C}^m)$. It is clear that $T = T_{(\alpha)(\Gamma)}U$ is a lower triangular unitary map and that the system

$$\Sigma = (\alpha_n, \widetilde{B}(n)U_n, \Gamma_n, \widetilde{D}(n)U(n); \mathbb{C}, \mathbb{C}^m, \mathbb{C}^m), \tag{6.17}$$

is a unitary realization of T, which is exponentially stable by (jj). By Theorem 2.1, the operator T is a unitary operator in the class $\mathcal{LE}$. It has Kronecker index less than or equal to 1, since it admits a one dimensional realization. Its Kronecker index cannot be equal to 0, because this would imply for all integers n that $(1 + |\alpha_n|)^{-1}\Gamma_n\Gamma_n^* = 0$, $\Gamma_n = 0$ and hence, by condition (j), $|\alpha_n| = 1$. This contradicts condition (jj). □

7. Main theorems for operators in the class $\mathcal{LK}$

A unitary block Laurent operator $T : \ell^2(\mathbb{C}^m) \to \ell^2(\mathbb{C}^m)$ in the class $\mathcal{LK}$ is exponentially decaying off the main diagonal. This follows from the fact that in this case the symbol is an inner rational matrix function, and hence analytic on an open annulus containing the unit circle. In the time varying setting, there exist lower triangular unitary maps with finite Kronecker index, which are not exponentially decaying off the main diagonal. To see this we present the following example.

Define the one dimensional system Σ by

$$\Sigma = (\alpha_n, -\sqrt{1-|\alpha_n|^2}, \sqrt{1-|\alpha_n|^2}, \alpha_n; \mathbb{C}, \mathbb{C}, \mathbb{C}), \tag{7.1}$$

where the main operator sequence is given by

$$\alpha_n = \begin{cases} e^{-\frac{1}{|n|}}, & n \neq 0, \\ 0, & n = 0. \end{cases} \tag{7.2}$$

Let $T = (t_{i,j})_{i,j=-\infty}^{\infty}$ be the input output map of the system Σ. Then

$$t_{i,j} = \begin{cases} 0, & i < j, \\ \alpha_i, & i = j, \\ -\sqrt{1-|\alpha_i|^2}\, \alpha_{i-1}\alpha_{i-2}\dots\alpha_{j+1}\, \sqrt{1-|\alpha_j|^2}, & i > j. \end{cases}$$

Clearly, the map T is in the class $\mathcal{LK}$. The system Σ is unitary, and using the identities

$$\prod_{k=1}^{\infty} \alpha_k = \prod_{k=1}^{\infty} \alpha_{-k} = \prod_{k=1}^{\infty} e^{-\frac{1}{k}} = 0,$$

it follows from Theorem 3.1 that the input output map T of Σ is unitary.

Suppose T is exponentially decaying. That is, there exist numbers $M > 0$ and $a \in (0,1)$ such that for all integers j, k with $j \geq k$ we have

$$|t_{j,k}| \leq M a^{j-k}. \tag{7.3}$$

Let us define for all $n \in \mathbb{N}$ the numbers $d_{2n-1} := t_{3n-1,n}$. So, d_{2n-1} is a value on the $(2n-1)$-th diagonal of T. Since

$$\lim_{n\to\infty} \prod_{j=n+1}^{3n-2} \alpha_j = \lim_{n\to\infty} \exp\left(-\sum_{j=n+1}^{3n-2} \frac{1}{j}\right) = \frac{1}{3},$$

we have

$$\begin{aligned} \lim_{n\to\infty} (2n-1) d_{2n-1} &= -\lim_{n\to\infty} (2n-1)\sqrt{(1-e^{-\frac{2}{3n-1}})(1-e^{-\frac{2}{n}})} \prod_{j=n+1}^{3n-2} \alpha_j \\ &= -\frac{4}{9}\sqrt{3}. \end{aligned} \tag{7.4}$$

The inequality (7.3) implies $|(2n-1)d_{2n-1}| \leq (2n-1)Ma^{2n-1}$. From (7.4) it follows that for all n large enough $\frac{1}{2} \leq |(2n-1)d_{2n-1}|$, so $\frac{1}{2} \leq (2n-1)Ma^{2n-1}$. This is impossible, because the right hand term of the last inequality tends to zero. Thus the operator T is in the class $\mathcal{LK}$, but not in the class $\mathcal{LEK}$, and hence the class $\mathcal{LK}$ is strictly larger than the class $\mathcal{LEK}$.

We state now our main theorems for unitary operators in the class $\mathcal{LK}$.

Theorem 7.1. *Each unitary operator T in the class $\mathcal{LK}$ can be decomposed into a finite product $T = T_1T_2\cdots T_N$, where N is the Kronecker index of T and each T_i is unitary, belongs to the class $\mathcal{LK}$ and has Kronecker index equal to* 1.

Theorem 7.2. *An operator $T : \ell^2(\mathbb{C}^m) \to \ell^2(\mathbb{C}^m)$ is unitary, belongs to the class $\mathcal{LK}$ and has Kronecker index equal to* 1 *if and only if T is of the form*

$$T = T_{(\alpha)(\Gamma)}U \tag{7.5}$$

with $T_{(\alpha)(\Gamma)}$ given by

$$I - \begin{bmatrix} \ddots & & & & & \\ \cdots & \frac{1}{1+|\alpha_{-1}|}\Gamma_{-1}\Gamma_{-1}^* & & & & \\ \cdots & \Gamma_0\mathrm{sgn}(\alpha_{-1})\Gamma_{-1}^* & \boxed{\frac{1}{1+|\alpha_0|}\Gamma_0\Gamma_0^*} & & & \\ \cdots & \Gamma_1\alpha_0\mathrm{sgn}(\alpha_{-1})\Gamma_{-1}^* & \Gamma_1\mathrm{sgn}(\alpha_0)\Gamma_0^* & \frac{1}{1+|\alpha_1|}\Gamma_1\Gamma_1^* & & \\ \cdots & \Gamma_2\alpha_1\alpha_0\mathrm{sgn}(\alpha_{-1})\Gamma_{-1}^* & \Gamma_2\alpha_1\mathrm{sgn}(\alpha_0)\Gamma_0^* & \Gamma_2\mathrm{sgn}(\alpha_1)\Gamma_1^* & \frac{1}{1+|\alpha_2|}\Gamma_2\Gamma_2^* & \\ & \vdots & \vdots & \vdots & \vdots & \ddots \end{bmatrix}.$$

Here $U : \ell^2(\mathbb{C}^m) \to \ell^2(\mathbb{C}^m)$ is a unitary block diagonal operator and the operators $\Gamma_n : \mathbb{C} \to \mathbb{C}^m$ and $\alpha_n : \mathbb{C} \to \mathbb{C}$ satisfy the following conditions:

(i) $\Gamma_n^*\Gamma_n + |\alpha_n|^2 = 1$ *for each $n \in \mathbb{Z}$,*

(ii) $\prod_{j=0}^{\infty} \alpha_{n-j} = 0$ *and* $\prod_{j=0}^{\infty} \alpha_{n+j} = 0$, *for each $n \in \mathbb{Z}$.*

The proofs of both theorems are essentially the same as those of Theorems 1.1 and 1.2.

Proof (of Theorem 7.1). Let $\Sigma = (A(n), B(n), C(n), D(n); H_n, \mathbb{C}^m, \mathbb{C}^m)$ be the unitary time varying realization of T constructed in Theorem 4.1. In particular, the main operator sequence $(A(n) : H_n \to H_{n+1})_{n\in\mathbb{Z}}$ belongs to the class C_{00}. Using the same arguments as in Parts (a) and (b) of the proof of Theorem 1.1, we see that we can find a unitary system

$$\Upsilon = (R(n), \widetilde{B}(n), \widetilde{C}(n), D(n); \mathbb{C}^N, \mathbb{C}^m, \mathbb{C}^m))$$

which is unitarily equivalent to Σ, such that for each $n \in \mathbb{Z}$ the operator $R(n)$ is given by

$$R(n) = \begin{bmatrix} \alpha_{11}(n) & * & * & \cdots & * \\ & \alpha_{22}(n) & * & \cdots & * \\ & & \alpha_{33}(n) & \cdots & * \\ & & & \ddots & \vdots \\ & & & & \alpha_{NN}(n) \end{bmatrix}. \tag{7.6}$$

By unitary equivalence the sequence $(R(n) : \mathbb{C}^N \to \mathbb{C}^N)_{n\in\mathbb{Z}}$ belongs to the class C_{00}.
Put $H_{l,n} = \text{span}\{e_1\}$, where e_1 is the first vector of the standard basis of $\mathbb{C}^N$. It follows from (7.6) that $(H_{l,n})_{n\in\mathbb{Z}}$ is an invariant subspace sequence for the operators $(R(n) : \mathbb{C}^N \to \mathbb{C}^N)_{n\in\mathbb{Z}}$. Let

$$\Upsilon_l = (\alpha_{11}(n), B_l(n), C_l(n), D_l(n)) \tag{7.7}$$

be the associated left projection, and let

$$\Upsilon_r = (A_r(n), B_r(n), C_r(n), D_r(n)) \tag{7.8}$$

be the associated right projection. Thus

$$A_r(n) = \begin{bmatrix} \alpha_{22}(n) & * & \cdots & * \\ & \alpha_{33}(n) & & * \\ & & \ddots & \vdots \\ & & & \alpha_{NN}(n) \end{bmatrix}. \tag{7.9}$$

Again we identify $H_{l,n} = \text{span}\{e_1\}$ with $\mathbb{C}$ and $H_{r,n} = \mathbb{C}^N \ominus \text{span}\{e_1\}$ with $\mathbb{C}^{N-1}$. We will show that the input output operator T_l of Υ_l belongs to the class $\mathcal{LK}$, is unitary, and that its Kronecker index is less than or equal to 1.
The main operator sequence $(A(n) : H_n \to H_{n+1})_{n\in\mathbb{Z}}$ of the unitary realization Σ of the unitary map T belongs to the class C_{00}, by Theorem 4.1. We will show that the operator sequence $(\alpha_{11}(n) : \mathbb{C} \to \mathbb{C})_{n\in\mathbb{Z}}$ also belongs to the class C_{00}. Fix a time instant $M \in \mathbb{Z}$. Take $x \in \mathbb{C}$ and put $\widetilde{x} = (x, 0, \ldots, 0) \in \mathbb{C}^N$. Then we have for all $n \in \mathbb{N}$:

$$|\tau_{\alpha_{11}}(M+n, M)x| = \|\tau_R(M+n, M)\widetilde{x}\|. \tag{7.10}$$

The last term tends to 0 for $n \to \infty$. Because M was arbitrarily chosen, we have $(\alpha_{11}(n) : \mathbb{C} \to \mathbb{C})_{n\in\mathbb{Z}} \in C_{0\cdot}$. Fix again a time instant $M \in \mathbb{Z}$. Take $x \in \mathbb{C}$ and write $\widetilde{x} = (x, 0, \ldots, 0) \in \mathbb{C}^N$. From the inequality

$$|\tau_{\alpha_{11}}(M, M-n)^* x| \le \|\tau_R(M, M-n)^* \widetilde{x}\|$$

follows that $(\alpha_{11}(n) : \mathbb{C} \to \mathbb{C})_{n\in\mathbb{Z}} \in C_{\cdot 0}$. We conclude that $(\alpha_{11}(n) : \mathbb{C} \to \mathbb{C})_{n\in\mathbb{Z}} \in C_{00}$.
From Theorem 3.1 it follows that the input output map T_l of Υ_l is unitary. The generalized Hankel operator $G_l(n)$ at time n of the input output map T_l decomposes into a product of the operators

$$G_l(n) = \Gamma_{n,\infty}(\Upsilon_l)\widetilde{\Lambda}_{n-1,-\infty}(\Upsilon_l), \tag{7.11}$$

where $\Gamma_{n,\infty}(\Upsilon_l)$ is defined by (3.8), and $\widetilde{\Lambda}_{n-1,-\infty}(\Upsilon_l)$ is defined by (6.11). The boundedness of the first factor on the right side follows from

$$\begin{aligned}\|\Gamma_{n,\infty}(\Upsilon_l)\| &= \lim_{m\to\infty}\sum_{j=0}^{m} C_l(n+j)^*|\tau_{\alpha_{11}}(n+j,n)|^2 C_l(n+j)\\ &= \lim_{m\to\infty}\left((1-|\alpha_n|^2)+\sum_{j=1}^{m}(1-|\alpha(n+j)|^2)\prod_{i=0}^{j-1}|\alpha_{11}(n+i)|^2\right)\\ &= \lim_{m\to\infty}\left(1-\prod_{i=0}^{m}|\alpha_{11}(n+i)|^2\right)=1.\end{aligned}$$

Here we used the fact that Υ_l is unitary and that $(\alpha_{11}(n):\mathbb{C}\to\mathbb{C})_{n\in\mathbb{Z}}\in C_{\cdot 0}$. We conclude that the rank of $G_l(n)$ is less than or equal to 1. Hence $\kappa(T_l)\leq 1$.
The operators T_l and T are unitary, and the input output operator T_r of the right projection Σ_r satisfies $T=T_lT_r$. Therefore T_r is unitary. Because $R(n)$ is upper triangular for all n, we have

$$\tau_R(n,m)=\begin{bmatrix}\tau_{\alpha_{11}}(n,m) & * \\ 0 & \tau_{A_r}(n,m)\end{bmatrix},$$

for all $n>m$. Fix a time instant $M\in\mathbb{Z}$. Take $y\in\mathbb{C}^{N-1}$ and put $\widetilde{y}=(0,y)\in\mathbb{C}^N$. Then

$$|\tau_{A_r}(M+n,M)y| = \|\tau_R(M+n,M)\widetilde{y}\|, \tag{7.12}$$

and

$$|\tau_{A_r}(M,M-n)^*y| \leq \|\tau_R(M,M-n)^*\widetilde{y}\|. \tag{7.13}$$

Therefore the sequence $(A_r(n):\mathbb{C}^{N-1}\to\mathbb{C}^{N-1})_{n\in\mathbb{Z}}$ belongs to C_{00}.
From (7.9) and (7.8) and a decomposition similar to (7.11) it follows that the Kronecker index of the map T_r is less than or equal to $N-1$. The input output operator T_r is a unitary operator in the class $\mathcal{LK}$. The system Υ_r is a realization of T_r with main operator sequence in the class C_{00}. Finally, by the same reasoning as in the proof of Theorem 1.1, the Kronecker index of T_l equals 1 and that of T_r equals $N-1$. We can proceed by induction to finish the proof. □

P r o o f (of Theorem 7.2). Suppose the operator $T:\ell^2(\mathbb{C}^m)\to\ell^2(\mathbb{C}^m)$ is unitary in the class $\mathcal{LK}$ and has Kronecker index equal to 1. Then Theorem 4.1 provides us with a unitary realization $\Sigma=(\alpha_n,B(n),\Gamma_n,D(n);\mathbb{C},\mathbb{C}^m,\mathbb{C}^m)$ of T, with one dimensional state space and main operator sequence $(\alpha_n:\mathbb{C}\to\mathbb{C})_{n\in\mathbb{Z}}$ in the class C_{00}.
As in the proof of Theorem 1.2, we can introduce the system

$$\widetilde{\Sigma}=(\alpha_n,\widetilde{B}(n),\Gamma_n,\widetilde{D}(n);\mathbb{C},\mathbb{C}^m,\mathbb{C}^m), \tag{7.14}$$

where $\widetilde{B}(n) = -\mathrm{sgn}(\alpha_n)\Gamma_n^*$, $\widetilde{D}(n) = I_{\mathbb{C}^m} - (1+|\alpha_n|)^{-1}\Gamma_n\Gamma_n^*$. The system $\widetilde{\Sigma}$ is a unitary system with input output map equal to $T_{(\alpha)(\Gamma)}$.
By Lemma 5.2, there are unitary operators $U_n : \mathbb{C}^m \to \mathbb{C}^m$ such that $B(n) = \widetilde{B}(n)U_n$ and $D(n) = \widetilde{D}(n)U_n$. Thus we can factor $T = T_{(\alpha)(\Gamma)}U$, with the required properties (i) and (ii) satisfied.
On the other hand, let us start with an operator $T = T_{(\alpha)(\Gamma)}U$ as given in (7.5). Then by condition (i) of the theorem, the system (7.14) is a unitary realization of $T_{(\alpha)(\Gamma)}$. Let us write $U = \mathrm{diag}_{j\in\mathbb{Z}}(U_j) : \ell^2(\mathbb{C}^m) \to \ell^2(\mathbb{C}^m)$. It is clear that the operator $T = T_{(\alpha)(\Gamma)}U$ is a lower triangular unitary map and that the system

$$\Sigma = (\alpha_n, \widetilde{B}(n)U_n, \Gamma_n, \widetilde{D}(n)U(n); \mathbb{C}, \mathbb{C}^m, \mathbb{C}^m), \tag{7.15}$$

is a unitary realization of T. By condition (ii) we have $(\alpha_n : \mathbb{C} \to \mathbb{C})_{n\in\mathbb{Z}} \in C_{00}$. Obviously the Kronecker index of T is less than or equal to 1. By a similar reasoning as in the proof of Theorem 1.2 the equality $\kappa(T) = 0$ would imply that α_n is of modulus 1 for all n, which contradicts the fact that $(\alpha_n, \mathbb{C} \to \mathbb{C})_{n\in\mathbb{Z}} \in C_{00}$. □

After this paper was completed, A. L. Sakhnovich proved in a more algebraic way a somewhat different version of Theorem 1.1, which does not require the lower triangular operator to be bounded, but assumes that the generalized Hankel operators have uniformly the same rank [Sa].

References

[AG1] Alpay, D., Gohberg, I.C.: Unitary rational matrix functions; in: Topics in Interpolation Theory of Rational Matrix-valued Functions, Operator Theory: Adv. Appl. 33, Birkhäuser Verlag, Basel 1988, 175–222.

[AG2] Alpay, D., Gohberg, I.C.: On orthogonal matrix polynomials; in: Orthogonal Matrix-valued polynomials and Applications, Operator Theory: Adv. Appl. 34, Birkhäuser Verlag, Basel 1988, 25–46.

[Ar] Arov, D.Z.: Passive linear stationary dynamic systems; Sibirsk. Mat. Zh. 20:2 (1979), 211–228 (in Russian).

[AKP] Arov, D.Z., Kaashoek, M.A., Pik, D.R.: Minimal systems and optimal systems; Minimal and optimal linear discrete time-varying dissipative scattering systems; Rapportnr. WS-473, Department of Mathematics and Computer Science, Vrije Universiteit, Amsterdam 1997.

[ACC] Arsene, Gr., Ceausescu, Z., Constantinescu, T.: Schur analysis of some completion problems; Linear Algebra Appl. 109 (1988), 1–35.

[BenGK] Ben-Artzi, A., Gohberg, I.C., Kaashoek, M.A.: Exponentially dominated infinite block matrices of finite Kronecker rank; Integral Equations Operator Theory 18 (1994), 30–77.

[Ber] Bercovici, H.: Operator theory and arithmetic in H^∞; Math. Surveys Monographs 26, American Mathematical Society, Providence, Rhode Island 1988.

[Br] BRODSKII, M.S.: Unitary operator colligations and their characteristic functions, Uspekhi Mat. Nauk 33:4 (1978), 141–168.

[FFGK] FOIAS, C., FRAZHO, A.E., GOHBERG, I.C., KAASHOEK, M.A.: Discrete timevariant interpolation as classical interpolation with an operator argument; Integral Equations Operator Theory 26 (1996), 371–403.

[GGK2] GOHBERG, I.C., GOLDBERG, S., KAASHOEK, M.A.: Classes of linear operators, Volume 2, Operator Theory: Adv. Appl. 63, Birkhäuser Verlag, Basel 1993.

[GKLe] GOHBERG, I.C., KAASHOEK, M.A., LERER, L.: Factorization of banded lower triangular infinite matrices; Linear Algebra Appl. 247 (1996), 371–403.

[P] PIK, D.R.: Time varying dissipative systems; Doctoraal scriptie (Master thesis), Vrije Universiteit Amsterdam, Amsterdam 1994.

[Sa] SAKHNOVICH, A.L.: private communication, 1995.

[Sz.-NF] SZÖKEFALVI-NAGY, B., FOIAS, C.: Analyse harmonique des operateurs de l'espace de Hilbert; Akademiai Kiado, Budapest 1967.

[VDS] VAN DOOREN, P., SREEDHAR, J.: When is a periodic discrete time system equivalent to a time invariant one?; Linear Algebra Appl. 212/213 (1994) 131–151.

Faculteit Wiskunde en Informatica, Vrije Universiteit
De Boelelaan 1081a
1081 HV Amsterdam
The Netherlands
kaash@cs.vu.nl, drpik@cs.vu.nl

1991 Mathematics Subject Classification: Primary 47A48, 47A68; Secondary 47B15, 93B28

Submitted: May 2, 1996

Operator Theory:
Advances and Applications, Vol. 103

Fredholm theory of interpolation morphisms

M. KRAUSE

Let (V_0, V_1) and (W_0, W_1) be compatible couples of Banach spaces and $T : V_i \to W_i$ be a bounded linear operator, $i = 0, 1$. The abstract Fredholm theory of T as an interpolation morphism is studied and the results are applied to the real interpolation method. In this application it is shown that, whenever T_{Θ_0,p_0} is a Fredholm operator (for some $0 < \Theta_0 < 1$ and $1 \leq p_0 < \infty$) then $T_{\Theta,p}$ is a Fredholm operator for all Θ in a neighborhood of Θ_0 and all $1 \leq p < \infty$ and the null space and the deficiency of these operators are constant.

1. Introduction

A disadvantage of Banach algebras is that in this setting one cannot study the behavior of operators between two different spaces. To circumvent this restriction, we will develop Fredholm theory in the more general framework of paraalgebras which were first introduced in [Rol68].

A paraalgebra (with unit) $\mathcal{M}$ is a quadruple of (complex) Banach spaces

$$\mathcal{M} = \begin{pmatrix} & M_{10} & \\ M_{00} & & M_{11} \\ & M_{01} & \end{pmatrix}$$

with associative multiplications $\cdot : M_{ij} \times M_{jk} \to M_{ik}$ for all $i, j, k = 0, 1$, which are distributive with respect to the addition, and elements $e_0 \in M_{00}, e_1 \in M_{11}$ such that

$$ae_j = a = e_i a \ , \ \|e_0\| = 1 = \|e_1\|,$$

$$\lambda(ab) = (\lambda a)b = a(\lambda b)$$

and

$$\|a \cdot b\| \leq \|a\| \, \|b\|$$

for all $\lambda \in \mathbb{C}$ and all $a \in M_{ij}$, $b \in M_{jk}$, $i, j, k = 0, 1$.

A typical example for the previous definition is the paraalgebra of all bounded linear operators between the normed spaces X and Y:

$$\mathcal{M} = \begin{pmatrix} & \mathrm{BL}(X,Y) & \\ \mathrm{BL}(X) & & \mathrm{BL}(Y) \\ & \mathrm{BL}(Y,X) & \end{pmatrix}.$$

Before we state some well-known definitions from the theory of Banach algebras in the framework of paraalgebras, we must clarify a few notational difficulties: if we make the identification

$$m \longleftrightarrow (m,i,j) \tag{1.1}$$

for all $m \in M_{ij}$, $i,j = 0,1$, then we can look at the paraalgebra $\mathcal{M}$ as a set. The four zeros and two units of the paraalgebra are denoted by $\mathbf{0}$ and $\mathbf{1}$, respectively. Moreover, we understand by "ab" that the multiplication between a and b is defined and that the result of the multiplication is ab. The same convention is made for the addition.

An ideal of a paraalgebra $\mathcal{M}$ is a subset $I \subset \mathcal{M}$ such that $\mathbf{0} \subset I$, $I + I = I$, $I\mathcal{M} \subset I$ and $\mathcal{M}I \subset I$. Other notations like minimal, maximal, left or right ideal transfer similarly and the radical of a paraalgebra is defined by

$$\mathrm{Rad}(\mathcal{M}) = \{m \in \mathcal{M} \mid \mathbf{1} - m\mathcal{M} \subset \mathrm{Inv}(\mathcal{M})\},$$

where $\mathrm{Inv}(\mathcal{M})$ denotes the invertible elements in $\mathcal{M}$.

The elements of $\mathcal{M}$ which are invertible modulo an ideal I are called the I–Fredholm elements of $\mathcal{M}$ and are denoted by $\Phi_I(\mathcal{M})$. For a general discussion of Fredholm theory in Banach algebras see [BMSW82].

If the paraalgebra is semi-simple, the span of all minimal left ideals and the span of all minimal right ideals coincide and are called the socle $\mathrm{Soc}(\mathcal{M})$ of $\mathcal{M}$. The socle is an ideal and it turns out that the $\mathrm{Soc}(\mathcal{M})$–Fredholm elements of $\mathcal{M}$, which are simply called the Fredholm elements of $\mathcal{M}$, have many "nice and natural" properties.

The aim is to characterize the Fredholm elements of a special paraalgebra of interpolation morphisms and to use the results to show some stability properties of usual Fredholm operators under real interpolation.

If X and Y are Banach spaces, $\mathrm{L}(X,Y)$ denotes the set of all linear operators and $\mathrm{BL}(X,Y)$ the set of all bounded linear operators between X and Y. If $T \in \mathrm{BL}(X,Y)$ is a Fredholm operator, $n(T)$, $d(T)$ and $\mathrm{ind}(T)$ denote the nullity, the deficiency, and the index of T, respectively. $N(T)$ is the kernel and $R(T)$ is the range of T. $\mathbb{R}^+$ are the positive real numbers.

2. Fredholm theory in a paraalgebra of interpolation morphisms

Firstly let us recall some basic terms of interpolation theory (for a general introduction see [BL76]). A couple $\mathbf{V} = (V_0, V_1)$ of (complex) Banach spaces is called compatible if there is a Hausdorff topological vector space U such that V_0 and V_1 are continuously embedded in U. Then the intersection and the sum are defined by

$$\begin{aligned} V_\Delta &= V_0 \cap V_1, \\ \|x\|_{V_\Delta} &= \max\left(\|x\|_{V_0}, \|x\|_{V_1}\right) \end{aligned}$$

and

$$\begin{aligned} V_\Sigma &= V_0 + V_1, \\ \|x\|_{V_\Sigma} &= \inf_{x=x_0+x_1} (\|x_0\|_{V_0} + \|x_1\|_{V_1}), \end{aligned}$$

respectively, both of which are Banach spaces. An operator $T \in \mathrm{L}(V_\Sigma, W_\Sigma)$ is called interpolation morphism between $\mathbf{V}$ and $\mathbf{W}$ if

$$T|_{V_i} \in \mathrm{BL}(V_i, W_i), \qquad i = 0, 1.$$

The set of all interpolation morphisms between $\mathbf{V}$ and $\mathbf{W}$ is denoted by $\mathrm{Mor}(\mathbf{V},\mathbf{W})$, or briefly $\mathrm{Mor}(\mathbf{V})$ for $\mathrm{Mor}(\mathbf{V}, \mathbf{V})$, and is a Banach space with the norm $\|T\| = \max(\|T|_{V_0}\|, \|T|_{V_1}\|)$. For notational convenience, we set

$$T_i : V_i \to W_i, \qquad x \mapsto Tx, \qquad i = 0, 1, \Delta, \Sigma,$$

for all $T \in \mathrm{Mor}(\mathbf{V}, \mathbf{W})$. Finally, note that T_Δ and T_Σ are bounded linear operators as well.

We use the following notation: if $\mathbf{V}$ is a compatible couple of Banach spaces $\overline{V_0}, \overline{V_1}$ and $\overline{V_\Delta}$ denote the closures of V_0, V_1 and V_Δ in V_Σ, respectively.

The result of the next lemma, which provides us with one technical tool for our following investigations, is a bit surprising since the intersection of closures is in general not the closure of the intersection.

Lemma 2.1. *Suppose (V_0, V_1) is a compatible couple of Banach spaces then*

$$\overline{V_\Delta} = \overline{V_0} \cap \overline{V_1}.$$

Proof. It is evident that $\overline{V_\Delta} \subset \overline{V_0} \cap \overline{V_1}$. To prove the reverse inclusion we take an arbitrary $x \in \overline{V_0} \cap \overline{V_1}$ and choose sequences $(v_i^n)_{n\in\mathbb{N}} \subset V_i$, $i = 0, 1$, such that

$$\sum_{n=0}^{\infty} \|v_i^n\|_{V_\Sigma} < \infty, \qquad i = 0, 1,$$

and

$$x = \sum_{n=0}^{\infty} v_i^n, \qquad i = 0, 1.$$

Here, and in the rest of the proof, all vector-valued limits are taken in V_Σ. Taking decompositions $v_i^n = x_i^n + y_i^n$, $x_i^n \in V_0, y_i^n \in V_1$, with

$$\|x_i^n\|_{V_0} + \|y_i^n\|_{V_1} \le \|v_i^n\|_{V_\Sigma} + \frac{1}{2^n}, \qquad i = 0, 1, \ n \in \mathbb{N},$$

we get $y_0^n, x_1^n \in V_\Delta$ and thus

$$\begin{aligned} x_0 &:= \sum_{n=0}^{\infty} x_0^n \in V_0, & y_0 &:= \sum_{n=0}^{\infty} y_0^n \in V_1 \cap \overline{V_\Delta}, \\ x_1 &:= \sum_{n=0}^{\infty} x_1^n \in V_0 \cap \overline{V_\Delta}, & y_1 &:= \sum_{n=0}^{\infty} y_1^n \in V_1. \end{aligned}$$

Furthermore, we have

$$x_0 - x_1 = y_1 - y_0 \in V_\Delta \subset \overline{V_\Delta},$$

which implies $x_0 \in \overline{V_\Delta}$ and finally $x = x_0 + y_0 \in \overline{V_\Delta}$. □

The following paraalgebra, which as we see later is semi-simple under certain circumstances, is the central object in our investigations.

Definition 2.2. Let $\mathbf{V}, \mathbf{W}$ be compatible couples of Banach spaces. Then

$$\mathcal{M}(\mathbf{V},\mathbf{W}) = \left(\text{Mor}(\mathbf{V},\mathbf{V}) \quad \begin{matrix} \text{Mor}(\mathbf{V},\mathbf{W}) \\ \\ \text{Mor}(\mathbf{W},\mathbf{V}) \end{matrix} \quad \text{Mor}(\mathbf{W},\mathbf{W}) \right)$$

endowed with the composition as multiplication will be called a paraalgebra of interpolation morphisms.

Lemma 2.3. *Suppose $\mathcal{M}(\mathbf{V},\mathbf{W})$ is a paraalgebra of interpolation morphisms. A morphism $T \in \text{Mor}(\mathbf{V},\mathbf{W})$ is then in the radical of $\mathcal{M}(\mathbf{V},\mathbf{W})$ if and only if*

$$TV_\Sigma \subset \overline{W_\Delta} \tag{2.1}$$

and

$$TV_\Delta \subset \{0\}. \tag{2.2}$$

Proof. "⇐": Let $S \in \text{Mor}(\mathbf{W},\mathbf{V})$. To show that $I - ST$ is invertible in $\mathcal{M}(\mathbf{V},\mathbf{W})$, we observe that (2.1), (2.2) and $SW_\Delta \subset V_\Delta$ yields $(S_\Sigma T_\Sigma)^2 = 0$. This implies $(I - ST)(I + ST) = I$.
"⇒": Assuming that (2.1) is not satisfied there then exists $x \in V_0$ or $x \in V_1$ with $Tx \notin \overline{W_\Delta} = \overline{W_0} \cap \overline{W_1}$ (see Lemma 2.1). W.l.o.g. we let $x \in V_0$ and hence $Tx \notin \overline{W_1}$. Now choose $x' \in W'_\Sigma$, satisfying $x'(Tx) = 1$ and $x'|_{\overline{W_1}} = 0$, and set $S = x \otimes x' \in \text{Mor}(\mathbf{W},\mathbf{V})$. Then $I - ST$ is not injective and hence not invertible in $\mathcal{M}(\mathbf{V},\mathbf{W})$, which is a contradiction to the fact that T is in the radical. To prove the second condition, assume that there exists $x \in V_\Delta$ with $Tx \neq 0$ and then choose $x' \in V'_\Sigma$, satisfying $x'(Tx) = 1$, and set $S = x \otimes x' \in \text{Mor}(\mathbf{W},\mathbf{V})$. Thus, we get a contradiction as above. □

Suppose $\mathcal{M} = \mathcal{M}(\mathbf{V},\mathbf{W})$ is a paraalgebra of interpolation morphisms auch that V_Δ is dense in V_i and W_Δ is dense in W_i, $i = 0, 1$. From the previous lemma we then easily see that $\mathcal{M}$ is semi-simple and thus that the socle of $\mathcal{M}$ is well-defined. For this reason we make the following assumption:

Assumption 2.4. If $\mathcal{M}(\mathbf{V},\mathbf{W})$ is a paraalgebra of interpolation morphisms then V_Δ is dense in V_i and W_Δ is dense in W_i, $i = 0, 1$.

In interpolation theory it is normally the case that the intersection of a compatible couple of Banach spaces is dense in both elements of the couple. So the above assumption is no major restriction.

Provided that Assumption 2.4 holds, an elementary calculation yields that the minimal left and minimal right ideals in $\mathcal{M}(\mathbf{V},\mathbf{W})$ are all generated by one-dimensional idempotent interpolation morphisms. From Assumption 2.4 we also get $R(T) \subset W_\Delta$ for all finite dimensional $T \in \mathrm{Mor}(\mathbf{V},\mathbf{W})$ and consequently

$$\mathrm{Soc}(\mathcal{M}(\mathbf{V},\mathbf{W})) = \{T \in \mathcal{M}(\mathbf{V},\mathbf{W}) \mid T \text{ finite dimensional}\}.$$

Let us write $\Phi(\mathbf{V},\mathbf{W})$ for the Fredholm elements of $\mathcal{M}(\mathbf{V},\mathbf{W})$ in $\mathrm{Mor}(\mathbf{V},\mathbf{W})$.

In order to prove a characterization of $\Phi(\mathbf{V},\mathbf{W})$ we require two lemmas. The first is a slight generalization of [Sdl87] Lemma 2.10.

Lemma 2.5. *Let $X, \widetilde{X}, Y$ and $\widetilde{Y}$ be Banach spaces, such that $\widetilde{X}$ is continuously embedded in X and $\widetilde{Y}$ is continuously embedded and dense in Y. Let $T : X \to Y$ and $\widetilde{T} : \widetilde{X} \to \widetilde{Y}$ be bounded Fredholm operators with $Tx = \widetilde{T}x$ for all $x \in \widetilde{X}$ and D a dense subspace of $\widetilde{Y}$. Then there exist subspaces $U \subset \widetilde{U} \subset D \subset \widetilde{Y}$ such that*

$$Y = R(T) \oplus U \tag{2.3}$$

and

$$\widetilde{Y} = R(\widetilde{T}) \oplus \widetilde{U}. \tag{2.4}$$

Furthermore, it holds that

$$d(T) \le d(\widetilde{T}), \qquad n(\widetilde{T}) \le n(T), \qquad \mathrm{ind}(\widetilde{T}) \le \mathrm{ind}(T). \tag{2.5}$$

If in addition $\mathrm{ind}(\widetilde{T}) = \mathrm{ind}(T)$, then

$$\widetilde{U} = U, \qquad N(\widetilde{T}) = N(T). \tag{2.6}$$

Proof. From the facts that D is dense in Y and $R(T)$ is closed, it follows from that there exists a finite dimensional subspace $U \subset D$ such that (2.3) holds. Hence $R(\widetilde{T}) \cap U = \{0\}$ and, since D is dense in $\widetilde{Y}$ and $R(\widetilde{T}) \oplus U$ is closed in $\widetilde{Y}$, we can extend U to a subspace $\widetilde{U} \subset D$ such that (2.4) is fulfilled. With this (2.5) is now easily established. If in addition $\mathrm{ind}(\widetilde{T}) = \mathrm{ind}(T)$, then we get $d(T) = d(\widetilde{T})$ and $n(\widetilde{T}) = n(T)$ from (2.5) and out of this (2.6) is implied. □

Lemma 2.6. *Suppose $\mathbf{V}$ is a compatible couple of Banach spaces and $P : V_\Sigma \to V_\Sigma$ is a continuous projection with finite dimensional kernel. If $P(V_i) \subset V_i$ for some $i = 0, 1$, then it holds that $P|_{V_i} \in BL(V_i)$.*

Proof. Since V_i is continuously embedded in V_Σ and $R((I_{V_\Sigma} - P)|_{V_i})$ is a finite dimensional subspace of V_i, we obtain $(I_{V_\Sigma} - P)|_{V_i} \in BL(V_i)$ and hence $P|_{V_i} \in BL(V_i)$. □

We now state our first main theorem. For a related result concerning some special Lebesgue spaces see [Bar90], Theorem 4.2.

Theorem 2.7. *Assume $\mathcal{M}(\mathbf{V},\mathbf{W})$ is a paraalgebra of interpolation morphisms fulfilling Assumption 2.4. The following assertions are then equivalent:*

1. $T \in \Phi(\mathbf{V},\mathbf{W})$,
2. T_0 *and* T_1 *are Fredholm operators with* $\operatorname{ind}(T_0) = \operatorname{ind}(T_1)$, $N(T_0) = N(T_1)$ *and* $R(T_\Sigma) \cap W_\Delta = R(T_\Delta)$,
3. T_Δ *and* T_Σ *are Fredholm operators with* $\operatorname{ind}(T_\Delta) = \operatorname{ind}(T_\Sigma)$.

If one of these conditions holds and $D \subset W_\Delta$ is dense, then there exists a finite dimensional subspace $H \subset D \subset W_\Delta$ such that

$$N(T_\Delta) = N(T_0) = N(T_1) = N(T_\Sigma)$$

and

$$R(T_i) \oplus H = W_i, \qquad i = 0, 1, \Delta, \Sigma.$$

In particular, $\operatorname{ind}(T_0)$, $\operatorname{ind}(T_1)$, $\operatorname{ind}(T_\Delta)$ *and* $\operatorname{ind}(T_\Sigma)$ *all have the same value, which we will call the index* $\operatorname{ind}_{\mathcal{M}}(T)$ *of T in $\mathcal{M}$.*

Proof. "$1 \Rightarrow 2$": Choosing $S \in \operatorname{Mor}(\mathbf{W},\mathbf{V})$, $E \in \operatorname{Fin}(\mathbf{V})$ and $F \in \operatorname{Fin}(\mathbf{W})$ such that $TS = I + F$ and $ST = I + E$, we get

$$T_i S_i = I_i + F_i, \qquad S_i T_i = I_i + E_i, \qquad i = 0, 1, \Delta, \Sigma.$$

Hence T_i and S_i, $i = 0, 1, \Delta, \Sigma$, are Fredholm operators. From Lemma 2.5 it follows that $\operatorname{ind}(T_\Delta) \le \operatorname{ind}(T_i) \le \operatorname{ind}(T_\Sigma)$ and $\operatorname{ind}(S_\Delta) \le \operatorname{ind}(S_i) \le \operatorname{ind}(S_\Sigma)$, $i = 0, 1$. Thus we obtain

$$\operatorname{ind}(T_\Delta) \le \operatorname{ind}(T_i) = -\operatorname{ind}(S_i) \le -\operatorname{ind}(S_\Delta) = \operatorname{ind}(T_\Delta), \qquad i = 0, 1, \Sigma,$$

and another application of Lemma 2.5 yields

$$N(T_\Delta) = N(T_0) = N(T_1) = N(T_\Sigma),$$

$$d(T_\Delta) = d(T_0) = d(T_1) = d(T_\Sigma)$$

and the existence of a subspace $H \subset W_\Delta$ with

$$R(T_\Sigma) \oplus H = W_\Sigma, \qquad R(T_\Delta) \oplus H = W_\Delta.$$

Now, let $y \in R(T_\Sigma) \cap W_\Delta$ be arbitrary, $x_\Sigma \in V_\Sigma$ such that $y = T_\Sigma(x_\Sigma)$ and $x_\Delta \in V_\Delta$, $h \in H$ such that $y = T_\Delta x_\Delta + h$. This leads to $R(T_\Sigma) \ni T_\Sigma(x_\Sigma) - T_\Delta x_\Delta = h \in H$ and hence to $h = 0$ and $y \in R(T_\Delta)$. Since $R(T_\Delta) \subset R(T_\Sigma) \cap W_\Delta$ trivially holds, we have proven 2.

"$2 \Rightarrow 3$": Let $x_0 \in V_0$ and $x_1 \in V_1$ such that $T_\Sigma(x_0 + x_1) = 0$. Since $W_0 \supset R(T_0) \ni T_0 x_0 = -T_1 x_1 \in R(T_1) \subset W_1$ and $R(T_\Sigma) \cap W_\Delta = R(T_\Delta)$, there is an $x_\Delta \in V_\Delta$ with $T_\Delta x_\Delta = T_0 x_0 = T_1(-x_1)$. Hence $x_0 - x_\Delta \in N(T_0) = N(T_1) \ni x_1 + x_\Delta$ and therefore $x_0 + x_1 = (x_0 - x_\Delta) + (x_1 + x_\Delta) \in N(T_i)$ for $i = 0, 1$. Consequently we obtain

$$N(T_\Delta) = N(T_0) \cap N(T_1) = N(T_0) = N(T_1) \subset N(T_\Sigma) \subset N(T_1).$$

Now, choose $U_i \subset W_\Delta$ finite dimensional such that $R(T_i) \oplus U_i = W_i$, $i = 0, 1$, which is possible since W_Δ is dense both in W_0 and W_1. It follows $W_\Sigma \subset R(T_\Sigma) + U_0 + U_1$ and hence that T_Σ is a Fredholm operator.
Moreover, we have

$$\begin{aligned} R(T_\Delta) + U_0 + U_1 &= (R(T_\Sigma) \cap W_\Delta) + (U_0 \cap W_\Delta) + (U_1 \cap W_\Delta) \\ &= (R(T_\Sigma) + U_0 + U_1) \cap W_\Delta = W_\Sigma \cap W_\Delta = W_\Delta, \end{aligned}$$

which implies that T_Δ is a Fredholm operator.
Taking $U_\Delta \subset W_\Delta$ with $R(T_\Delta) \oplus U_\Delta = W_\Delta$ results in

$$R(T_\Sigma) \cap U_\Delta = R(T_\Sigma) \cap W_\Delta \cap U_\Delta = R(T_\Delta) \cap U_\Delta = \{0\}$$

which proves $d(T_\Sigma) \geq d(T_\Delta)$. Since we already know $d(T_\Sigma) \leq d(T_\Delta)$ from Lemma 2.5, we can conclude $\operatorname{ind}(T_\Sigma) = \operatorname{ind}(T_\Delta)$.
"3 $\Rightarrow$ 1": Let $H \subset W_\Delta$ (according to Lemma 2.5) and $M \subset V_\Sigma$ such that

$$V_\Sigma = M \oplus N(T_\Sigma), \qquad W_\Sigma = R(T_\Sigma) \oplus H, \qquad W_\Delta = R(T_\Delta) \oplus H.$$

Since $N(T_\Sigma)$ and H are finite dimensional, there are projections $P \in BL(V_\Sigma)$ and $Q \in BL(W_\Sigma)$ with the following properties:

$$\begin{aligned} N(P) &= N(T_\Sigma), \qquad & N(Q) &= H, \\ R(P) &= M, & R(Q) &= R(T_\Sigma). \end{aligned}$$

Define $A : M \to R(T_\Sigma)$, $x \mapsto Tx$, which is bijective, and set

$$S : W_\Sigma \to V_\Sigma, \qquad x \mapsto A^{-1}Qx.$$

We now prove that S is a morphism.
To show $R(Q|_{W_0}) = R(T_0)$, let $x \in W_0$ be arbitrary and $x_0 \in V_0$, $x_1 \in V_1$ such that $Qx = T_\Sigma(x_0 + x_1)$. Then

$$W_1 \ni T_1 x_1 = Qx - T_0 x_0 = x - (I_{W_\Sigma} - Q)x - T_0 x_0 \in W_0$$

and hence $T_1 x_1 \in W_\Delta$. Thus there exist $x_\Delta \in V_\Delta$ and $h \in H$ with $T_1 x_1 = T_\Delta x_\Delta + h$. Consequently, from $H \ni h = T_1 x_1 - T_\Delta x_\Delta \in R(T_\Sigma)$ it follows $h = 0$ and therefore $Qx = T_\Sigma(x_0 + x_\Delta) \in R(T_0)$ which proves $R(Q|_{W_0}) = R(T_0)$. Noting that $Q|_{W_0}$ is a projection of W_0 onto $R(T_0)$, we see from Lemma 2.6 that $Q|_{W_0} \in BL(W_0)$ and hence that $R(T_0)$ is closed in W_0.
To show $R(P|_{V_0}) = M \cap V_0$ observe $N(T_0) \subset N(T_\Sigma) = N(T_\Delta) \subset N(T_0)$ and therefore

$$V_0 = (N(T_\Sigma) \oplus M) \cap V_0 = N(T_0) \oplus (M \cap V_0).$$

Again from Lemma 2.6 we now obtain $P|_{V_0} \in BL(V_0)$ and hence that $M \cap V_0$ is closed in V_0.
Since $A_0 : (M \cap V_0, \|\cdot\|_{V_0}) \to (R(T_0), \|\cdot\|_{W_0})$, $x \mapsto Tx$ is a continuous and bijective mapping between Banach spaces and since $Sw = A^{-1}Qw = A_0^{-1}Q|_{W_0}w$ for all $w \in W_0$, we can conclude $R(S|_{W_0}) \subset V_0$ and $S|_{W_0} \in BL(W_0, V_0)$.
For this reason we see that S, $P = I - ST$ and $Q = I - TS$ are morphisms and finally that $T \in \Phi(\mathbf{V}, \mathbf{W})$.
If now one of the conditions 1, 2 or 3 holds, then the first part of the proof yields $N(T_\Delta) = N(T_0) = N(T_1) = N(T_\Sigma)$ and $d(T_\Delta) = d(T_0) = d(T_1) = d(T_\Sigma)$. From Lemma 2.5 follows the existence of a finite dimensional subspace $H \subset D$ such that $W_\Sigma = R(T_\Sigma) \oplus H$. Since $R(T_i) \cap H \subset R(T_\Sigma) \cap H = \{0\}$, the assertion follows. □

Corollary 2.8. *Under the same conditions as in Theorem 2.7 the following assertions are equivalent:*

1. $T \in \mathrm{Inv}(\mathbf{V}, \mathbf{W})$,
2. T_0 *and* T_1 *are invertible and* $R(T_\Sigma) \cap W_\Delta = R(T_\Delta)$,
3. T_Δ *and* T_Σ *are invertible.*

Proof. See the proof of Theorem 2.7. □

3. Interpolation of Fredholm elements

In this section we show that the elements of $\Phi(\mathbf{V}, \mathbf{W})$ behave very predictably under interpolation, which allows us to subsequently show some stability results.

The next result is similar to [Her89], Main Theorem, and [AS89], Lemma 2, but in both articles only the special case where $V_0 \subset V_1$ and $W_0 \subset W_1$ has been considered.

Definition 3.1. Let $\mathbf{V}$ and $\mathbf{W}$ be compatible couples of Banach spaces. Two spaces V and W are called compatible interpolation spaces (with respect to $\mathbf{V}$ and $\mathbf{W}$) if the following three conditions are fulfilled:

1. $V_\Delta \subset V \subset V_\Sigma$ and $W_\Delta \subset W \subset W_\Sigma$ with continuous inclusions,
2. $T|_V \in \mathrm{BL}(V, W)$ for all $T \in \mathrm{Mor}(\mathbf{V}, \mathbf{W})$ and
3. $S|_W \in \mathrm{BL}(W, V)$ for all $S \in \mathrm{Mor}(\mathbf{W}, \mathbf{V})$.

Lemma 3.2. *Suppose $\mathcal{M}(\mathbf{V}, \mathbf{W})$ is a paraalgebra of interpolation morphisms fulfilling Assumption 2.4. If D is a dense subset of W_Δ and $T \in \Phi(\mathbf{V}, \mathbf{W})$, then there exist finite dimensional subspaces $N \subset V_\Delta$ and $H \subset D \subset W_\Delta$ such that for all compatible interpolation spaces V and W (with respect to $\mathbf{V}$ and $\mathbf{W}$)*

$$N(T_V) = N$$

and

$$W = R(T_V) \oplus H,$$

where $T_V : V \to W, x \mapsto Tx$. *In particular,* T_V *is a Fredholm operator and* $\mathrm{ind}(T_V) = \mathrm{ind}(T)$.

Proof. Choose $N \subset V_\Delta$ and $H \subset D$ according to Corollary 2.7 satisfying

$$N = N(T_i), \qquad W_i = R(T_i) \oplus H, \qquad i = 0, 1, \Delta, \Sigma,$$

and take $S \in \Phi(\mathbf{W}, \mathbf{V})$ such that $I - ST$ and $I - TS$ are finite dimensional interpolation morphisms. With this setting the operators $I_V - S|_W T|_V \in \mathrm{BL}(V)$ and $I_W - T|_V S|_W \in \mathrm{BL}(W)$ are both finite dimensional, which shows that T_V is a Fredholm operator. Additionally it holds that $N = N(T_\Delta) \subset N(T|_V) \subset N(T_\Sigma) = N$ and, due to $R(T_\Sigma) \cap H = \{0\}$, we have $R(T|_V) \cap H = \{0\}$. From this it follows that $\mathrm{ind}(T|_V) \leq \mathrm{ind}(T_\Sigma)$ and similarly we get $\mathrm{ind}(S|_W) \leq \mathrm{ind}(S_\Sigma)$, which yields

$$\mathrm{ind}(T|_V) \leq \mathrm{ind}(T_\Sigma) = -\mathrm{ind}(S_\Sigma) \leq -\mathrm{ind}(S|_W) = \mathrm{ind}(T|_V),$$

and therefore finally gives us $\mathrm{ind}(T|_V) = \mathrm{ind}(T_\Sigma)$ and $R(T|_V) \oplus H = W$. □

4. Perturbation results for the real interpolation methods

We now apply the results of the previous section to the real interpolation method, but first of all we must recall some basic definitions.

Definition 4.1. If $\mathbf{V} = (V_0, V_1)$ is a compatible couple of Banach spaces, $1 \leq p < \infty$ and $0 < \Theta < 1$, one can define

$$K(t, x, V_0, V_1) = \inf_{x = x_0 + x_1} (\|x_0\|_{V_0} + t\|x_1\|_{V_1}).$$

Then

$$V_{\Theta,p} = \left\{ x \in V_\Sigma : \int_0^\infty \left(\frac{K(t, x, V_0, V_1)}{t^\Theta} \right)^p \frac{dt}{t} < \infty \right\}$$

with the norm

$$\|x\|_{V_{\Theta,p}} = \left(\int_0^\infty \left(\frac{K(t, x, V_0, V_1)}{t^\Theta} \right)^p \frac{dt}{t} \right)^{\frac{1}{p}},$$

is called the real interpolation space of $\mathbf{V}$ with parameters Θ and p. If $\mathbf{W}$ is another compatible couple and T is a morphism between $\mathbf{V}$ and $\mathbf{W}$, we define

$$T_{\Theta,p} : V_{\Theta,p} \to W_{\Theta,p}, \qquad x \mapsto Tx.$$

For further properties of real interpolation spaces see [BL76].

To prove our results we need a technical lemma which employs a useful relation between the real interpolation method and the functor of intersection and sum functor. This lemma can be found in [Mal86], Corollary 1.

Lemma 4.2. *Suppose* $\mathbf{V}$ *is a compatible couple of Banach spaces,* $1 \le p < \infty$ *and* $0 < \Theta \le \frac{1}{2}$. *Then*

$$\left((V_0, V_1)_{\Theta,p}, (V_0, V_1)_{1-\Theta,p}\right)_\Delta = (V_\Delta, V_\Sigma)_{\Theta,p} \tag{4.1}$$

and

$$\left((V_0, V_1)_{\Theta,p}, (V_0, V_1)_{1-\Theta,p}\right)_\Sigma = (V_\Delta, V_\Sigma)_{1-\Theta,p} \tag{4.2}$$

with equivalent norms.

Lemma 4.3. *Let* $\mathbf{V}$ *and* $\mathbf{W}$ *be compatible couples of Banach spaces,* T *be an interpolation morphism between* $\mathbf{V}$ *and* $\mathbf{W}$, $0 < \Theta_0 < 1$ *and* $1 \le p < \infty$, *such that* $T_{\Theta_0,p}$ *is a Fredholm operator. There then exists* $\varepsilon > 0$ *(with* $0 < \Theta_0 - \varepsilon < \Theta_0 + \varepsilon < 1$*) so that* $T_{\Theta,p}$ *is a Fredholm operator with* $\operatorname{ind}(T_{\Theta,p}) = \operatorname{ind}(T_{\Theta_0,p})$ *for all* $\Theta \in (\Theta_0 - \varepsilon, \Theta_0 + \varepsilon)$.

Proof. The assertion follows from [Alb84], Corollary 4.4, by an application of [BL76], Theorem 4.7.2, and the reiteration theorem. □

The following theorem is the main result in this section. Its usefulness is shown in Corollary 4.6 and Lemma 4.8.

Theorem 4.4. *Suppose* $\mathcal{M}(\mathbf{V}, \mathbf{W})$ *is a paraalgebra of interpolation morphisms fulfilling Assumption* 2.4. *If* $T \in \operatorname{Mor}(\mathbf{V}, \mathbf{W})$, $0 < \Theta_0 < 1$ *and* $1 \le p_0 < \infty$ *such that* T_{Θ_0,p_0} *is a Fredholm operator, then there exists* $\varepsilon > 0$ *(with* $0 < \Theta_0 - \varepsilon < \Theta + \varepsilon < 1$*) so that*

$$T^{AB} : \mathbf{A} \to \mathbf{B}, \qquad x \mapsto Tx$$

is a Fredholm element in $\mathcal{M}(\mathbf{A}, \mathbf{B})$, *where*

$$\mathbf{A} = (V_{\Theta_0-\varepsilon,p_0}, V_{\Theta_0+\varepsilon,p_0})$$

and

$$\mathbf{B} = (W_{\Theta_0-\varepsilon,p_0}, W_{\Theta_0+\varepsilon,p_0}).$$

Proof. Due to the reiteration theorem we can assume that $\Theta_0 = \frac{1}{2}$ without loss of generality. Setting $\widehat{X} = (V_\Delta, V_\Sigma)$ and $\widehat{Y} = (W_\Delta, W_\Sigma)$ and then defining $\widehat{T} \in \operatorname{Mor}(\widehat{X}, \widehat{Y})$ by $\widehat{T}x := Tx$, we obtain from [Mal86], Corollary 1, that $\widehat{T}_{\frac{1}{2},p_0} = T_{\frac{1}{2},p_0}$ is a Fredholm operator. Hence according to Lemma 4.3 there exists $\varepsilon \in (0, \frac{1}{2})$ such that $\widehat{T}_{\frac{1}{2}-\varepsilon,p_0}$ and $\widehat{T}_{\frac{1}{2}+\varepsilon,p_0}$ are Fredholm operators with

$$\operatorname{ind}(\widehat{T}_{\frac{1}{2}-\varepsilon,p_0}) = \operatorname{ind}(\widehat{T}_{\frac{1}{2}+\varepsilon,p_0}). \tag{4.3}$$

From Lemma 4.2 we can thus see

$$X_{\frac{1}{2}-\varepsilon,p_0} = (V_\Delta, V_\Sigma)_{\frac{1}{2}-\varepsilon,p_0} = \left((V_0, V_1)_{\frac{1}{2}-\varepsilon,p_0}, (V_0, V_1)_{\frac{1}{2}+\varepsilon,p_0}\right)_\Delta$$

and

$$X_{\frac{1}{2}+\varepsilon,p_0} = (V_\Delta, V_\Sigma)_{\frac{1}{2}+\varepsilon,p_0} = \Big((V_0,V_1)_{\frac{1}{2}-\varepsilon,p_0}, (V_0,V_1)_{\frac{1}{2}+\varepsilon,p_0}\Big)_\Sigma$$

which with (4.3) implies that T_Δ^{AB} and T_Σ^{AB} are Fredholm operators with the same index. Applying Theorem 2.7 completes the proof. □

Proposition 4.5. *Suppose* $\mathcal{M}(\mathbf{V},\mathbf{W})$ *is as in Theorem* 4.4, $0<\Theta_0<1$, $1\le p_0<\infty$ *and* $T\in \mathrm{Mor}(\mathbf{V},\mathbf{W})$ *such that* T_{Θ_0,p_0} *is invertible. There then exists* $\varepsilon>0$ *(with* $0<\Theta_0-\varepsilon<\Theta+\varepsilon<1$*) so that*

$$T^{AB}:\mathbf{A}\to\mathbf{B}, \qquad x\mapsto Tx,$$

is invertible in $\mathcal{M}(\mathbf{A},\mathbf{B})$*, where* $\mathbf{A}$ *and* $\mathbf{B}$ *are as in Theorem* 4.4.

Proof. The proof runs along similar lines as that of Theorem 4.4, but additionally a special version of Lemma 4.3 for invertible operators, which is also a consequence of [Alb84], Corollary 4.4, is used. □

The next corollary is a generalization of [AS89], Theorem 3, and was the starting point of the work in this section.

Corollary 4.6. *Taking the assumptions from Theorem* 4.4 *there exist* $\varepsilon>0$ *(with* $0<\Theta_0-\varepsilon<\Theta+\varepsilon<1$*) and finite dimensional subspaces* $N\subset V_\Sigma$ *and* $H\subset W_\Delta$*, such that for all* $\Theta\in(\Theta_0-\varepsilon,\Theta_0+\varepsilon)$ *and all* $1\le p<\infty$ *it holds that*

$$N = N(T_{\Theta,p})$$

and

$$W_{\Theta,p} = R(T_{\Theta,p})\oplus H.$$

In particular, if T_{Θ_0,p_0} *is invertible then* $T_{\Theta,p}$ *will still be invertible.*

Proof. Choose $\varepsilon>0$ according to Theorem 4.4. We see from the reiteration theorem that for all $\Theta\in(\Theta_0-\varepsilon,\Theta+\varepsilon)$ and all $p\in[1,\infty)$ the spaces $V_{\Theta,p}$ and $W_{\Theta,p}$ are interpolation spaces with respect to $(V_{\Theta_0-\varepsilon,p_0},V_{\Theta_0+\varepsilon,p_0})$ and $(W_{\Theta_0-\varepsilon,p_0},W_{\Theta_0+\varepsilon,p_0})$. Thus, since V_Δ is dense in $V_{\Theta,p}$ and W_Δ is dense in $W_{\Theta,p}$ for all $0<\Theta<1$ and $p\ge1$, Lemma 3.2 implies the assertion. □

In [Ran86], Definition 2.(ii), Ransford defined the local uniqueness-of-resolvent condition for the complex interpolation method and remarked that this condition may be void. The following is the corresponding definition for the real interpolation method.

Definition 4.7. An interpolation morphism $T\in\mathrm{Mor}(\mathbf{V})$ satisfies the local real uniqueness-of-resolvent (U.R.) condition if and only if the following condition is met: whenever $0<\Theta_0<1$, $1\le p<\infty$, $\lambda\in\mathbb{C}$ and $(\lambda I-T)_{\Theta,p}$ is invertible in some neighborhood of Θ_0, then the inverses of $\lambda I-T_{\Theta,p}$ agree on V_Δ for all Θ in some (possibly smaller) neighborhood of Θ_0.

Proposition 4.8. *Suppose $\mathcal{M}(\mathbf{V},\mathbf{W})$ is a paraalgebra of interpolation morphisms fulfilling Assumption 2.4. If T is an interpolation morphism between $\mathbf{V}$ and $\mathbf{W}$ then T satisfies the local real uniqueness-of-resolvent condition.*

Proof. Assume that $(\lambda I - T)_{\Theta_0,p_0}$ is invertible for some $0 < \Theta_0 < 1,\ 1 \leq p_0 < \infty$ and $\lambda \in \mathbb{C}$. According to, and with the notation of, Theorem 4.5 there exists $\varepsilon > 0$ such that

$$S := (\lambda I - T^{AB}) \in \mathrm{Inv}(\mathcal{M}(\mathbf{A},\mathbf{B}))$$

and hence, by using the reiteration theorem, we obtain for all $\Theta \in (\Theta_0 - \varepsilon, \Theta_0 + \varepsilon)$ that

$$((\lambda I - T)_{\Theta,p})^{-1}|_{V_\Delta} = (S_{\frac{\Theta-(\Theta_0-\varepsilon)}{2\varepsilon}})^{-1}|_{V_\Delta} = S^{-1}|_{V_\Delta}. \qquad \square$$

Acknowledgements

I am grateful to Ernst Albrecht and Karl-Heinz Förster for helpful discussions and suggestions.

References

[AV83] ALBRECHT, E., VASILESCU, F.-H.: Semi-Fredholm complexes; Operator Theory: Adv. Appl. 11 (1983), 15–39.

[Alb84] ALBRECHT, E.: Spectral interpolation; Operator Theory: Adv. Appl. 14 (1984), 13–37.

[AS89] ALBRECHT, E., SCHINDLER, K.: Spectra of Operators on Real Interpolation Spaces; preprint (1989).

[BMSW82] BARNES, B., MURPHY, G., SMYTH, M., WEST, T.: Riesz and Fredholm theory in Banach algebras; Res. Notes Math. 67 (1982).

[Bar90] BARNES, B.: Interpolation of Spectrum of Bounded Operators on Lebesgue Spaces; Rocky Mountain J. Math. 20:2 (1990), 359–378.

[BL76] BERGH, J., LÖFSTRÖM, J.: Interpolation Spaces; Springer Verlag, Berlin Heidelberg New York 1976.

[Her89] HERRERO, D.: One-sided interpolation of Fredholm operators; Proc. Roy. Irish Acad. Sect. A 89: 1 (1989), 79–89.

[Mal86] MALIGRANDA, L.: Interpolation between Sum and Intersection of Banach spaces; J. Approximation Theory 47 (1986), 42–53.

[Ran86] RANSFORD, T.J.: The spectrum of an interpolated operator and analytic multivalued functions; Pacific J. Math. 121:2 (1986), 445–465.

[Rol68] ROLEWICZ, D. AND ST.: Equations in linear spaces; Monografie Mathematyczne, Polish Scientific Publishers, Warsaw 1968.

[Sdl87] SCHINDLER, K.: Spektrale Eigenschaften linearer Operatoren auf Interpolationsräumen; Dissertation, Universität des Saarlandes, Saarbrücken 1987.

[Zaf80] ZAFRAN, M.: Spectral Theory and Interpolation of Operators; J. Funct. Anal. 36 (1980), 185–204.

Technische Universität Berlin
Fachbereich 3 Mathematik
Sekr. MA 6-4
Straße des 17. Juni 136
10623 Berlin
Germany
krause@math.tu-berlin.de

1991 Mathematics Subject Classification: Primary 47A53; Secondary 47A55, 47A57

Submitted: March 21, 1996

Operator Theory:
Advances and Applications, Vol. 103

Resolvents of symmetric operators and the degenerated Nevanlinna-Pick problem

H. LANGER and H. WORACEK

Let $\mathcal{H}$ be an n-dimensional space, which is equipped with a positive semidefinite inner product with a one-dimensional isotropic part. Consider a symmetric operator S in $\mathcal{H}$ with defect index $(1,1)$. We give a parametrization of the Štraus extensions of S acting in a Pontryagin space $\mathcal{P}_c \supseteq \mathcal{H}$ of dimension $n+1$ and of negative index 1, and a parametrization of the corresponding set of u-resolvents of S. These results are applied to a classical Nevanlinna-Pick interpolation problem for which the Pick matrix is positive semidefinite but singular: We obtain explicit formulas for the solutions of this interpolation problem belonging to the class $\mathcal{N}_1$ (see Section 4).

1. Introduction

In various extension, interpolation or moment problems the following question arises: Given a symmetric operator S in an inner product space $\mathcal{H}$ and an element $u \in \mathcal{H}$. Describe the so-called u-resolvents of S, i.e., find a formula which parametrizes the family of the complex functions $[(A-z)^{-1}u, u]$, where A runs through the self-adjoint extensions of S in Pontryagin spaces $\mathcal{P}$ extending $\mathcal{H}$.

If S is a densely defined operator with defect $(1,1)$ and $\mathcal{H}$ is a Pontryagin space, such a description was given e.g. in [6]. There the so-called resolvent matrix was introduced and studied. This matrix defines a fractional linear transformation which parametrizes the family of u-resolvents.

In [7] these results were applied to an extension problem for positive definite functions with an accelerant. There a description of all the extensions is given under the assumption that the endpoint of the original interval is not singular. If the endpoint is singular, the results of [6] cannot be applied, as the inner product on the original space $\mathcal{H}$ is degenerated. It is this situation which is considered here for the particular case that the space $\mathcal{H}$ is finite dimensional. That is, in this note we consider the case that $\mathcal{H}$ is a positive semidefinite, finite dimensional inner product space and we assume that $\dim \mathcal{H}^\circ = 1$, $\mathcal{H}^\circ = \langle h^\circ \rangle$, where $\mathcal{H}^\circ$ denotes the isotropic part of $\mathcal{H}$: $\mathcal{H}^\circ = \mathcal{H} \cap \mathcal{H}^\perp$. Let S be a symmetric operator defined on a hyperplane $\mathcal{D}(S)$ of $\mathcal{H}$, $\dim \mathcal{D}(S) = \dim \mathcal{H} - 1$, and assume that h° is not an eigenvector of S. Let $u \in \mathcal{H}$ and let A run through the self-adjoint extensions of S acting in a Pontryagin space $\mathcal{P}$ with negative index 1 (a π_1-space) extending $\mathcal{H}$. As in the classical case we obtain a description of the family of u-resolvents of S by means of a fractional linear transformation involving a parameter function.

However, the family of parameters occurring here is more complicated than in the classical case.

The present paper is divided into two parts. In the first part, consisting of Sections 2 and 3, a solution of the problem described above is given. The second part, Sections 4 and 5, deals with an application of the derived formulas to an interpolation problem of Nevanlinna-Pick type.

Here is a short summary of the contents of the sections. In Section 2 the Štraus extension associated with a self-adjoint extension of S is considered. In this way we reduce the problem of the description of the u-resolvents to a finite dimensional problem. In Section 3 we give a formula for the family of all u-resolvents. There it is convenient to distinguish between the cases if $\mathcal{D}(S)$ is degenerated or not. In Section 4 we formulate a Nevanlinna-Pick type interpolation problem, recall some results of [11] and discuss the unique positive definite (in some sense minimal) solution. Finally, in Section 5, explicit interpolation formulas are derived.

The notation used in this article is similar to that of [3] and [4].

2. Štraus extensions

Let $(\mathcal{H}, [\cdot,\cdot])$ be a finite dimensional inner product space. Assume that $[\cdot,\cdot]$ is positive semidefinite and that $\dim \mathcal{H}^\circ = 1$, say $\mathcal{H}^\circ = \langle h^\circ\rangle$; here $\langle\cdot\rangle$ denotes the linear span of the elements between the brackets. If $\mathcal{H}_0$ is a nondegenerated hyperplane of $\mathcal{H}$, then $\mathcal{H} = \mathcal{H}_0[\dot{+}]\langle h^\circ\rangle$; here $\dot{+}$ denotes the direct sum and $[\dot{+}]$ denotes the $[\cdot,\cdot]$-orthogonal and direct sum. Further, let h^1 be a formal element, $h^1 \notin \mathcal{H}$.

Definition 2.1. Let $\mathcal{H}_0$ be a nondegenerated hyperplane of $\mathcal{H}$. The vector space

$$\mathcal{P}_c = \mathcal{H}_0[\dot{+}](\langle h^\circ\rangle \dot{+} \langle h^1\rangle) \tag{2.1}$$

endowed with the inner product

$$[x,y] = [x_0,y_0] + \xi_0\overline{\eta_1} + \xi_1\overline{\eta_0}, \tag{2.2}$$

where $x = x_0 + \xi_0 h^\circ + \xi_1 h^1$, $y = y_0 + \eta_0 h^\circ + \eta_1 h^1$, $x_0, y_0 \in \mathcal{H}_0$, $\xi_0,\xi_1,\eta_0,\eta_1 \in \mathbb{C}$, is called a canonical extension of $\mathcal{H}$.

Note that, as a vector space, $\mathcal{P}_c$ is isomorphic to $\mathcal{H}\dot{+}\mathbb{C}$. Any nondegenerated hyperplane $\mathcal{H}_0$ is isometrically isomorphic to $\mathcal{H}/\mathcal{H}^\circ$, thus $\mathcal{P}_c$ is unique up to isometric isomorphisms.

Proposition 2.2. *Let $\mathcal{P}$ be a π_1-space extending $\mathcal{H}$ and let $\mathcal{P}_c$ be a canonical extension of $\mathcal{H}$. Then there is an isometric embedding of $\mathcal{P}_c$ into $\mathcal{P}$.*

Proof. As $\mathcal{H}_0$ is a nondegenerated closed subspace of $\mathcal{P}$, it has an orthogonal complement in $\mathcal{P}$: $\mathcal{H}_0[\dot{+}]\mathcal{H}_0^\perp = \mathcal{P}$. The space $\mathcal{H}_0^\perp$ is itself a Pontryagin space and contains h°. Thus we can find an element $h \in \mathcal{H}_0^\perp$ which is skewly linked with h°. Due to the definition (2.2) of the inner product of $\mathcal{P}_c$, the mapping which acts as the identity on $\mathcal{H}$ and maps h^1 onto h is an isometric embedding of $\mathcal{P}_c$ into $\mathcal{P}$. □

Fixing such an embedding in the following, we can assume that $\mathcal{P}_c \subseteq \mathcal{P}$. Evidently, the orthogonal complement of $\mathcal{P}_c$ in $\mathcal{P}$ is a Hilbert space.

Let S be a symmetric operator in $\mathcal{H}$, defined on a hyperplane $\mathcal{D}(S) \subseteq \mathcal{H}$, $\dim \mathcal{D}(S) = \dim \mathcal{H} - 1$. If A is a self-adjoint relation extending S with $\rho(A) \neq \emptyset$ acting in a π_1-space $\mathcal{P} \supseteq \mathcal{H}$ and $\mathcal{P}_c$ is a canonical extension of $\mathcal{H}$, $\mathcal{P}_c \subseteq \mathcal{P}$, then A is also an extension of S considered in $\mathcal{P}_c$. Evidently, $\dim \mathcal{P}_c/\mathcal{D}(S) = 2$.

Denote the orthogonal projection of $\mathcal{P}$ onto $\mathcal{P}_c$ by P_c. The operator function

$$R(z) = P_c(A-z)^{-1}|_{\mathcal{P}_c}, \qquad z \in \rho(A),$$

is called a generalized resolvent of S, the function

$$T(z) = R(z)^{-1} + z, \qquad z \in \rho(A), \tag{2.3}$$

is called the Štraus extension of S associated with A. Observe that the values of the Štraus extension are in general linear relations in $\mathcal{P}_c$.

We have to adapt some results of [2] to the situation considered here. First let us recall some notation. If $\mathcal{P}_1$ and $\mathcal{P}_2$ are Pontryagin spaces we denote by $\mathcal{S}_0(\mathcal{P}_1, \mathcal{P}_2)$ the set of all meromorphic functions

$$\vartheta : \mathbb{D} = \{z \in \mathbb{C} : \ |z| < 1\} \to \mathcal{L}(\mathcal{P}_1, \mathcal{P}_2),$$

such that the matrix kernel

$$S_\vartheta(z,w) = \begin{pmatrix} \frac{I-\vartheta(w)^+\vartheta(z)}{1-z\overline{w}} & \frac{\vartheta(\overline{z})^+-\vartheta(w)^+}{z-\overline{w}} \\ \frac{\vartheta(z)-\vartheta(\overline{w})}{z-\overline{w}} & \frac{I-\vartheta(\overline{w})\vartheta(\overline{z})^+}{1-z\overline{w}} \end{pmatrix}, \qquad |z|, |w| < 1,$$

is positive semidefinite. For a Pontryagin space $\mathcal{P}$ we call a linear subspace of $\mathcal{P}^2$ a linear relation. Let $\mathcal{N}_0^\tau(\mathcal{P})$ denote the set of all functions T in $\mathbb{C}^+$ with values in the set of linear relations in $\mathcal{P}$ and such that:

(i) for $z_i \in \mathbb{C}^+$ and $(f_i; g_i) \in T(z_i)$, $i = 1, \dots, n$, the matrix

$$\left(\frac{[g_i, f_j] - [f_i, g_j]}{z_i - \overline{z_j}}\right)_{i=1}^n$$

is positive semidefinite,

(ii) there exists a $z_0 \in \mathbb{C}^+$ and a neighbourhood $U(z_0) \subseteq \mathbb{C}^+$ of z_0, such that for $z \in U(z_0)$ the Cayley transform

$$C_{z_0}(T(z)) = \{(g - z_0 f; g - \overline{z_0} f) : \ (f; g) \in T(z)\}$$

is a bounded operator in $\mathcal{P}$ and depends holomorphically on z.

If, in particular, T is a holomorphic $(n \times n)$–matrix function on $\mathbb{C}^+$ such that the Nevanlinna kernel

$$N_T(z,w) = \frac{T(z) - T(w)^*}{z - \overline{w}}$$

is positive semidefinite, we write $T \in \mathcal{N}_0(\mathbb{C}^n)$.

If $\mathcal{P}_1$ and $\mathcal{P}_2$ have the same negative index, which is the case in the following considerations, it is proved in [3] (compare also [2]) that a function $\vartheta \in \mathcal{S}_0(\mathcal{P}_1, \mathcal{P}_2)$ is uniquely determined by its values on any nonempty open subset of $\mathbb{D}$, and if it is extended to $\mathbb{C} \setminus \partial\mathbb{D}$ by

$$\vartheta(\frac{1}{\overline{z}}) = \vartheta(z)^+,$$

the kernel S_ϑ remains positive semidefinite. Furthermore, ϑ is in fact holomorphic. Similarly, a function $T \in \mathcal{N}_0^{\tau}(\mathcal{P})$ is uniquely determined by its values on any open subset of $\mathbb{C}^+$, and if it is extended to $\mathbb{C} \setminus \mathbb{R}$ by

$$T(\overline{z}) = T(z)^+,$$

the kernel N_T remains positive semidefinite.

Let $\mathcal{P}_1 \subseteq \mathcal{P}_2$ be Pontryagin spaces and assume that the extending space $\mathcal{P}_1^\perp$ of P_1 in P_2 is positive. Denote by P_1 the orthogonal projection of $\mathcal{P}_2$ onto $\mathcal{P}_1$. If A is a self-adjoint relation in $\mathcal{P}_2$ with $\rho(A) \neq \emptyset$, we denote by $R(z)$ the compressed resolvent of A:

$$R(z) = P_1(A - z)^{-1}|_{P_1}, \qquad z \in \rho(A),$$

and by $T(z)$, defined as in (2.3), the Štraus extension associated with A. Recall that A is called $\mathcal{P}_1$-minimal if

$$\mathcal{P}_2 = \overline{\langle x, (A - z)^{-1}x : \; x \in \mathcal{P}_1, z \in \rho(A)\rangle}.$$

Lemma 2.3. *Let $\mathcal{P}_1$, $\mathcal{P}_2$, A and $T(z)$ be as above. Then*

$$T(z)(0) := \{x \in \mathcal{P}_1 : \; (0; x) \in T(z)\} = A(0) \cap \mathcal{P}_1.$$

Furthermore,

$$\overline{\mathcal{D}(T(z))} = (A(0) \cap \mathcal{P}_1)^\perp,$$

where the orthogonal companion on the right hand side is considered in $\mathcal{P}_1$.

If $\mathcal{P}_1$ is a Hilbert space the proof of this lemma can be found in [3]. It can be generalized immediately to the situation considered here.

Lemma 2.3 applies to the situation described in the beginning of this section with $\mathcal{P}_1 = \mathcal{P}_c$, the π_1-space $\mathcal{P}_2 = \mathcal{P}$ and a self-adjoint extension A of S.

For the convenience of the reader we formulate a result which is a particular case of [2], Theorem 2.1.

Proposition 2.4. *Let $\mathcal{P}_c$, $\mathcal{P}$, S and A be as before, and let $z_0 \in \rho(A)\setminus\mathbb{R}$ be such that $\mathcal{R}(S - z_0)$ and $\mathcal{R}(S - \overline{z_0})$ are nondegenerated. Then there exists a neighbourhood*

$U(z_0)$ of z_0 with $\Im(z)\Im(z_0) > 0$ for $z \in U(z_0)$, such that the Štraus extension $T(z)$ associated with A is of the form

$$T(z) = S\dot{+}\{((I - \vartheta(\zeta(z)))f, (\overline{z_0} - z_0\vartheta(\zeta(z)))f) : \; f \in \mathcal{R}(S - z_0)^{\perp}\}, \tag{2.4}$$

where $\zeta(z) = \frac{z-z_0}{z-\overline{z_0}}$ and $\vartheta \in \mathcal{S}_0(\mathcal{R}(S - z_0)^{\perp}, \mathcal{R}(S - \overline{z_0})^{\perp})$.
Conversely, if $\vartheta \in \mathcal{S}_0(\mathcal{R}(S - z_0)^{\perp}, \mathcal{R}(S - \overline{z_0})^{\perp})$, then there exists a self-adjoint extension A of S in a π_1-space $\mathcal{P}$ with $z_0 \in \rho(A)$ such that its corresponding Štraus extension is given by (2.4).

In the situation of Proposition 2.4 the Cayley transform

$$F(\zeta) := C_{z_0}(T(z(\zeta))) = (T(z(\zeta)) - z_0)(T(z(\zeta)) - \overline{z_0})^{-1},$$

where $z(\zeta) = \frac{\overline{z_0}\zeta - z_0}{\zeta - 1}$, is given by

$$F(\zeta) = \begin{pmatrix} C_{z_0}(S) & 0 \\ 0 & \vartheta(\zeta) \end{pmatrix} : \begin{pmatrix} \mathcal{R}(S - z_0) \\ [\dot{+}] \\ \mathcal{R}(S - z_0)^{\perp} \end{pmatrix} \longrightarrow \begin{pmatrix} \mathcal{R}(S - \overline{z_0}) \\ [\dot{+}] \\ \mathcal{R}(S - \overline{z_0})^{\perp} \end{pmatrix}. \tag{2.5}$$

As $\mathcal{P}_c$ is finite dimensional, $\mathcal{D}(T(z)) = \overline{\mathcal{D}(T(z))}$ and therefore, according to Lemma 2.3, $T(z)(0)$ as well as $\mathcal{D}(T(z))$ are independent of z. We denote $\mathcal{D} := \mathcal{D}(T(z))$ and $T_\infty := T(z)(0) = \ker(F(\zeta(z)) - I)$. The main result of this section is the following representation of the Štraus extension.

Theorem 2.5. *Let $\mathcal{H}$, S be as above, let $\mathcal{P}$ be a π_1-space , $\mathcal{P}_c \subseteq \mathcal{P}$ a canonical extension of $\mathcal{H}$ and let A be a self-adjoint extension of S with $\rho(A) \neq \emptyset$. Denote by $T(z)$, $z \in \rho(A)$, the Štraus extension given by* (2.3). *Then there exists an operator valued meromorphic function*

$$T_0 : \mathbb{C} \setminus \mathbb{R} \to \mathcal{L}(\mathcal{D}, \mathcal{P}_c),$$

such that

(i) *there exists a $z_0 \in \mathbb{C}^+$ with $\mathcal{R}(T(z_0) - z_0) + \mathcal{D}^{\perp} = \mathcal{P}_c$,*
(ii) *$-T_0 \subset \mathcal{N}_0^{\mathcal{T}}(\mathcal{P}_c)$,*

and for $z \in \rho(A)$ the following relations hold:

(iii) *$S \subseteq T_0(z)$,*
(iv) *$T_0(z) \subseteq T_0(\overline{z})^+$,*
(v) *$T(z) = T_0(z)\dot{+}(0 \times T_\infty)$.*

Conversely, if $\mathcal{D} \supseteq \mathcal{D}(S)$ is a subspace of $\mathcal{P}_c$ and $T_0 : \mathbb{C} \setminus \mathbb{R} \to \mathcal{L}(\mathcal{D}, \mathcal{P}_c)$ is meromorphic and satisfies (i), (ii) *and* (iii)–(v) *for all z in the domain of holomorphy of*

T_0, then there exists a self-adjoint extension A of S in a π_1-space $\mathcal{P}$ with $\rho(A)\neq\emptyset$ such that its corresponding Štraus extension is given by

$$T(z)=T_0(z)\dot{+}(0\times\mathcal{D}^{\perp}).$$

Proof. Let $z_0\in\rho(A)\cap\mathbb{C}^+$ be such that $\mathcal{R}(S-z_0)$ and $\mathcal{R}(S-\overline{z_0})$ are nondegenerated. Then Proposition 2.4 shows that there exists a neighbourhood $U(z_0)$ of z_0 such that $T(z)$ and its Cayley transform $F(\zeta)$ are given by (2.4) and (2.5), respectively. We have

$$\mathcal{D}(C_{z_0}(S))\cap\ker(F(\zeta)-I)=\{0\}.$$

Indeed, if $x\in\mathcal{D}(C_{z_0}(S))$ we have $F(\zeta)x=C_{z_0}(S)x$ and thus $x\in\ker(F(\zeta)-I)$ shows that $x\in\ker(C_{z_0}(S)-I)$. As S is an operator ($S(0)=\{0\}$), we conclude that $x=0$.
Recall that $\ker(F(\zeta)-I)=T_\infty$. We decompose $\mathcal{P}_c$ as

$$\mathcal{P}_c=\mathcal{D}_1\dot{+}T_\infty$$

where $\mathcal{D}_1$ is chosen such that $\mathcal{D}(C_{z_0}(S))\subseteq\mathcal{D}_1$. Observing that $\mathcal{D}=\mathcal{D}(T(z))=\mathcal{R}(F(\zeta)-I)$, the operator $F(\zeta)-I$ induces a bijection

$$G(\zeta)=(F(\zeta)-I)|_{\mathcal{D}_1}$$

from $\mathcal{D}_1$ onto $\mathcal{D}$. As $F(\zeta)$ depends holomorphically on ζ, we find that $G(\zeta)^{-1}$ and thus also

$$T_0(z)=(z_0F(\zeta)-\overline{z_0})G(\zeta)^{-1}$$

depend holomorphically on ζ and on $z=\frac{\overline{z_0}\zeta-z_0}{\zeta-1}$, respectively.
The inclusions $\mathcal{D}(C_{z_0}(S))\subseteq\mathcal{D}_1$ and $G(\zeta)\subseteq F(\zeta)-I$ imply $S\subseteq T_0(z)\subseteq T(z)$. Furthermore, $T(z)=T(\overline{z})^+$, which shows that $T_0(z)\subseteq T_0(\overline{z})^+$. From the relation

$$F(\zeta)-I=G(\zeta)\dot{+}(T_\infty\times\{0\}),$$

where $\dot{+}$ has to be understood as the sum of subspaces of $\mathcal{P}_c^2$, we find

$$T(z)=(z_0F(\zeta)-\overline{z_0})(F(\zeta)-I)^{-1}=T_0(z)\dot{+}(\{0\}\times T_\infty).$$

It is shown in [3] that the kernels $S_{F(\zeta)}$ and N_{-T} have the same number of negative squares. As $\vartheta(\zeta)\in\mathcal{S}_0(\mathcal{R}(S-z_0)^{\perp},\mathcal{R}(S-\overline{z_0})^{\perp})$, obviously $C_{z_0}(T_0(z))$ depends holomorphically on z and thus $T_0(z)\in\mathcal{N}_0^{\mathcal{T}}(\mathcal{P}_c)$. Clearly,

$$\mathcal{R}(T_0(z)-z)+\mathcal{D}^{\perp}=\mathcal{D}_1+T_\infty=\mathcal{P}_c.$$

Conversely, suppose that $T_0(z)$ is given. Choose z_0 as in (i). Then the relation

$$\dim\mathcal{P}_c=\dim\mathcal{R}(T_0(z_0)-z_0)+\dim(\mathcal{D}^{\perp}/\mathcal{R}(T_0(z_0)-z_0)\cap\mathcal{D}^{\perp})$$

$$\leq \dim \mathcal{P}_c - \dim \mathcal{D}^{\perp} + \dim (\mathcal{D}^{\perp} / \mathcal{R}\,(T_0(z_0) - z_0) \cap \mathcal{D}^{\perp})$$

implies that $\mathcal{R}\,(T_0(z_0) - z_0) \cap \mathcal{D}^{\perp} = \{0\}$. Thus, in a sufficiently small neighbourhood $U(z_0)$ of z_0, $C_{z_0}(T_0(z))$ can be extended to $\mathcal{P}_c$ by

$$F(\zeta)x = \begin{cases} C_{z_0}(T_0(z))x & \text{if } x \in \mathcal{R}\,(T_0(z) - z)\,, \\ x & \text{if } x \in \mathcal{D}^{\perp}. \end{cases}$$

Clearly, $F(\zeta) = C_{z_0}(T_0(z)\dot{+}(\{0\} \times \mathcal{D}^{\perp}))$ and $F(\zeta) \in \mathcal{S}_0(\mathcal{P}_c, \mathcal{P}_c)$. It follows as in [3] for the Hilbert space case that $T(z) = T_0(z)\dot{+}(\{0\} \times \mathcal{D}^{\perp})$ is the Štraus extension associated with some self-adjoint extension A of S with $z_0 \in \rho(A)$. □

Remark 2.6. If T_∞ is nondegenerated, Theorem 2.5 can be deduced immediately from the fact that $\mathcal{D}[\dot{+}]T_\infty = \mathcal{P}_c$. In the situation considered here it may actually occur that T_∞ is degenerated.

3. The u-resolvents of S

Let $\mathcal{H}$ and S be as in Section 2 and fix an element $u \in \mathcal{H}$. Let A be any self-adjoint extension of S in a π_1-space $\mathcal{P}$, containing $\mathcal{H}$, and thus containing a canonical extension $\mathcal{P}_c$. In this section we consider the corresponding u-resolvent of S:

(3.1) $$[(A - z)^{-1}u, u], \qquad z \in \rho(A),$$

and find a description of all such u-resolvents. Observe that with the Štraus extension $T(z)$ corresponding to A the u-resolvent (3.1) can be written as

$$[(A - z)^{-1}u, u] = [(T(z) - z)^{-1}u, u].$$

As $T(z) = T_0(z)\dot{+}(0 \times T_\infty)$, in order to find $(T(z) - z)^{-1}u$, it suffices to solve the equation

(3.2) $$(T_0(z) - z)x = u + b$$

for $x \in \mathcal{D}$ and $b \in T_\infty$.

We consider the cases where $\mathcal{D}(S)$ is degenerated or nondegenerated separately.

I. If $\mathcal{D}(S)$ is nondegenerated, we use the decomposition

(3.3) $$\mathcal{P}_c = \mathcal{D}(S)\,[\dot{+}]\,(\langle h^\circ \rangle \dot{+} \langle h^1 \rangle)$$

with skewly linked neutral elements h°, h^1 as in (2.1), $[h^\circ, h^1] = 1$. For $x \in \mathcal{D}(S)$ we write

(3.4) $$Sx = S_d x + [x, a]h^\circ,$$

where $a \in \mathcal{D}(S)$, and decompose u as

(3.5) $$u = u_d + \eta_0 h^\circ.$$

Now let A be a self-adjoint extension of S as above and let $T(z)$ be the corresponding Štraus extension:

$$T(z) = T_0(z)\dot{+}(\{0\} \times T_\infty).$$

Note that $T_\infty \subseteq \langle h^\circ, h^1 \rangle$. We shall treat the cases

(i) $T_\infty = \{0\}$,

(ii) $\dim T_\infty = 1, \quad h^\circ \notin T_\infty$,

(iii) $\dim T_\infty = 1, \quad h^\circ \in T_\infty$,

(iv) $\dim T_\infty = 2$

separately. As the proofs in the cases (i)–(iv) are quite similar we will go into detail only in case (i).

Case (i): Write $T_0(z)$ as a block matrix with respect to the decomposition (3.3):

$$(3.6) \quad T_0(z) = \left(\begin{array}{c|cc} S_d & 0 & a \\ \hline [\cdot, a] & t_{11}(z) & t_{12}(z) \\ 0 & t_{21}(z) & t_{22}(z) \end{array} \right) : \begin{pmatrix} \mathcal{D}(S) \\ + \\ \langle h^\circ \rangle \\ + \\ \langle h^1 \rangle \end{pmatrix} \longrightarrow \begin{pmatrix} \mathcal{D}(S) \\ + \\ \langle h^\circ \rangle \\ + \\ \langle h^1 \rangle \end{pmatrix}.$$

Here the form of the first row and column of the matrix (3.6) follows from the properties (iii) and (iv) of Theorem 2.5. Furthermore, (ii) implies that

$$\Phi(z) = -\begin{pmatrix} t_{12}(z) & t_{11}(z) \\ t_{22}(z) & t_{21}(z) \end{pmatrix} \in \mathcal{N}_0(\mathbb{C}^2).$$

Lemma 3.1. *If $T_\infty = \{0\}$ and $T_0(z)$ and u are given by (3.6) and (3.5), respectively, we have*

$$(3.7) \quad [(A-z)^{-1}u, u] = [(S_d - z)^{-1}u_d, u_d] - \frac{(\eta_0 - [(S_d - z)^{-1}u_d, a])(\overline{\eta_0} - [(S_d - z)^{-1}a, u_d])}{\tau(z) + [(S_d - z)^{-1}a, a]},$$

where

$$(3.8) \quad \tau(z) = \frac{(t_{11}(z) - z)(t_{22}(z) - z)}{t_{21}(z)} - t_{12}(z).$$

If $t_{21} = 0$, then $\tau(z) = \infty$, i.e.,

$$[(A-z)^{-1}u, u] = [(S_d - z)^{-1}u_d, u_d].$$

Conversely, given a function $\Phi(z) \in \mathcal{N}_0(\mathbb{C}^2)$, the right hand side of (3.7) defines a u-resolvent of S if $\tau \neq -[(S_d - z)^{-1}a, a]$.

Proof. In order to compute $[(A-z)^{-1}u,u]$ we solve the equation $(A-z)x=u$, which, for $x = x_d + \xi_0 h^\circ + \xi_1 h^1$, can be written in vector form as

$$\begin{pmatrix} S_d - z & 0 & a \\ [\cdot, a] & t_{11}(z) - z & t_{12}(z) \\ 0 & t_{21}(z) & t_{22}(z) - z \end{pmatrix} \begin{pmatrix} x_d \\ \xi_0 \\ \xi_1 \end{pmatrix} = \begin{pmatrix} u_d \\ \eta_0 \\ 0 \end{pmatrix}. \tag{3.9}$$

Consider first the case that $t_{21} \neq 0$. The third row of (3.9) implies

$$t_{21}(z)\xi_0 + (t_{22}(z) - z)\xi_1 = 0,$$

i.e.,

$$\xi_0 = -\frac{t_{22}(z) - z}{t_{21}(z)}\xi_1. \tag{3.10}$$

The first row gives $(S_d - z)x_d + \xi_1 a = u_d$, or

$$x_d = (S_d - z)^{-1}u_d - \xi_1 (S_d - z)^{-1} a.$$

This implies

$$[x_d, a] = [(S_d - z)^{-1}u_d, a] - \xi_1 [(S_d - z)^{-1}a, a]. \tag{3.11}$$

From the second row of (3.9) we obtain $[x_d, a] + \xi_0(t_{21}(z) - z) + \xi_1 t_{12}(z) = \eta_0$, and by substituting the expressions (3.10) and (3.11) into this relation we find

$$\xi_1 = \frac{[(S_d - z)^{-1}u_d, a] - \eta_0}{\tau(z) + [(S_d - z)^{-1}a, a]}$$

and thus

$$x_d = (S_d - z)^{-1}u_d + \frac{\eta_0 - [(S_d - z)^{-1}u_d, a]}{\tau(z) + [(S_d - z)^{-1}a, a]}(S_d - z)^{-1}a.$$

Using these expressions we compute

$$\begin{aligned} [(A-z)^{-1}u, u] &= \left[\begin{pmatrix} x_d \\ \xi_0 \\ \xi_1 \end{pmatrix}, \begin{pmatrix} u_d \\ \eta_0 \\ 0 \end{pmatrix} \right] = [x_d, u_d] + \xi_1 \overline{\eta_0} \\ &= [(S_d - z)^{-1}u_d, u_d] \\ &\quad - \frac{(\eta_0 - [(S_d - z)^{-1}u_d, a])(\overline{\eta_0} - [(S_d - z)^{-1}a, u_d])}{\tau(z) + [(S_d - z)^{-1}a, a]}. \end{aligned}$$

Suppose now that $t_{21}(z) \equiv 0$. Then, as $\Phi(z) \in \mathcal{N}_0(\mathbb{C}^2)$, $t_{11}(z) - z \not\equiv 0$ and $t_{22}(z) - z \not\equiv 0$. Thus it follows from (3.9) that $\xi_1 = 0$, $x_d = (S_d - z)^{-1}u_d$ and ζ_0 can be computed from the second row of (3.9). We obtain

$$[(A-z)^{-1}u, u] = [(S_d - z)^{\ 1}u_d, u_d].$$

Consider the converse part of the lemma. It is obvious from the definition (3.6) of $T_0(z)$, that conditions (ii)–(iv) of Theorem 2.5 are satisfied. Note that condition (i) holds for some z_0 if and only if $T_0(z_0) - z_0$ is injective. Assume that for each z we have $\ker(T_0(z) - z) \neq \{0\}$, say

$$0 \neq \begin{pmatrix} x_d \\ \xi_0 \\ \xi_1 \end{pmatrix} \in \ker(T_0(z) - z).$$

If $t_{21} \neq 0$ we do the same calculations as above (with $u_d = 0$, $\eta_0 = 0$) up to (3.11), then the second row of (3.9) yields

$$\xi_1(\tau(z) + [(S_d - z)^{-1}a, a]) = 0.$$

In case $\xi_1 = 0$ we obtain $x_d = 0$ and $\xi_0 = 0$, a contradiction. Consequently, $\tau + [(S_d - z)^{-1}a, a] = 0$. If $t_{21}(z) \equiv 0$, again $t_{11}(z) - z \not\equiv 0$ and $t_{22}(z) - z \not\equiv 0$, and therefore (3.9) ($u_d = 0$, $\eta_0 = 0$) shows that $\xi_1 = 0$, $x_d = 0$ and $\xi_0 = 0$. □

Remark 3.2. Note that as a parameter in (3.7) the entries of the matrix $\Phi(z)$ do not appear separately, but only in form of the scalar function $\tau(z)$. The same is true for the representation formulas below.

Case (ii): Let $T_\infty = \langle b \rangle$ with $b = b_0 h^\circ + h^1 \in \mathcal{D}(S)^\perp$. Then

$$T_0(z) = \left(\begin{array}{c|c} S_d & a \\ \hline [\cdot, a] & t_1(z) \\ 0 & t_2(z) \end{array} \right) : \begin{pmatrix} \mathcal{D}(S) \\ + \\ \langle d \rangle \end{pmatrix} \longrightarrow \begin{pmatrix} \mathcal{D}(S) \\ + \\ \langle h^\circ \rangle \\ + \\ \langle h^1 \rangle \end{pmatrix}, \tag{3.12}$$

where $d = -\overline{b_0} h^\circ + h^1$. If $f = f_d + \lambda_f d, h = h_d + \lambda_h d \in \mathcal{D}(T_0(z))$, we find

$$\begin{aligned} \left[\begin{pmatrix} 0 & 0 \\ 0 & N_{t_1}(z,w) \\ 0 & N_{t_2}(z,w) \end{pmatrix} f, h \right] &= \left[\begin{pmatrix} 0 \\ N_{t_1}(z,w)\lambda_f \\ N_{t_2}(z,w)\lambda_f \end{pmatrix}, \lambda_h \begin{pmatrix} 0 \\ -\overline{b_0} \\ 1 \end{pmatrix} \right] \\ &= N_{t_1 - b_0 t_2}(z,w) \lambda_f \overline{\lambda_h}, \end{aligned}$$

which shows with (ii) of Theorem 2.5 that

$$\varphi(z) = -(t_1(z) - b_0 t_2(z)) \in \mathcal{N}_0(\mathbb{C}).$$

Lemma 3.3. *If* $\dim T_\infty = 1$ *and* $h^\circ \notin T_\infty$, *we have*

$$\begin{aligned} [(A - z)^{-1}u, u] &= [(S_d - z)^{-1}u_d, u_d] \\ &\quad - \frac{(\eta_0 - [(S_d - z)^{-1}u_d, a])(\overline{\eta_0} - [(S_d - z)^{-1}a, u_d])}{\tau(z) + [(S_d - z)^{-1}a, a]}, \end{aligned} \tag{3.13}$$

where

$$\tau(z) = -2\Re(b_0)z + \varphi(z), \tag{3.14}$$

with $\varphi(z) = -(t_1(z) - b_0 t_2(z)) \in \mathcal{N}_0(\mathbb{C})$. *Conversely, given a function* $\varphi(z) \in \mathcal{N}_0(\mathbb{C})$ *and* $b_0 \in \mathbb{C}$, *the right hand side of* (3.13) *defines a u-resolvent of* S *if* $\tau(z) \not\equiv -[(S_d - z)^{-1}a, a]$.

Proof. We have to solve the equation

$$\begin{pmatrix} S_d - z & a \\ [\cdot, a] & t_1 + \overline{b_0} z \\ 0 & t_2 - z \end{pmatrix} \begin{pmatrix} x_d \\ \xi_2 \end{pmatrix} = \begin{pmatrix} u_d \\ \eta_0 \\ 0 \end{pmatrix} + \lambda \begin{pmatrix} 0 \\ b_0 \\ 1 \end{pmatrix}$$

for $x = x_d + \xi_2 d$ and λ. This is done by similar computations as in the proof of Lemma 3.1 and yields (3.13). The converse statement is proved analogously as well. □

Case (iii): Let $T_\infty = \langle h^\circ \rangle$. Then

$$T_0(z) = \begin{pmatrix} S_d & 0 \\ [\cdot, a] & t_1(z) \\ 0 & t_2(z) \end{pmatrix} : \begin{pmatrix} \mathcal{D}(S) \\ + \\ \langle h^\circ \rangle \end{pmatrix} \longrightarrow \begin{pmatrix} \mathcal{D}(S) \\ + \\ \langle h^\circ \rangle \\ + \\ \langle h^1 \rangle \end{pmatrix}.$$

Lemma 3.4. *If* $\dim T_\infty = 1$ *and* $h^\circ \in T_\infty$, *we have*

$$[(A - z)^{-1}u, u] = [(S_d - z)^{-1}u_d, u_d]. \tag{3.15}$$

Conversely, the right hand side of (3.15) *defines a u-resolvent of* S.

In order to prove Lemma 3.4, we have to consider the equation

$$\begin{pmatrix} S_d - z & 0 \\ [\cdot, a] & t_1 - z \\ 0 & t_2 \end{pmatrix} \begin{pmatrix} x_d \\ \xi_2 \end{pmatrix} = \begin{pmatrix} u_d \\ \eta_0 \\ 0 \end{pmatrix} + \lambda \begin{pmatrix} 0 \\ 1 \\ 0 \end{pmatrix}.$$

Case (iv): Let $\dim T_\infty = 2$. Then $T_0(z) = S$.

Lemma 3.5. *If* $\dim T_\infty = 2$, *the formula* (3.15) *holds.*

In order to prove Lemma 3.5, consider the equation

$$\begin{pmatrix} S_d - z \\ [\cdot, a] \\ 0 \end{pmatrix} x_d = \begin{pmatrix} u_d \\ \eta_0 \\ 0 \end{pmatrix} + \lambda \begin{pmatrix} 0 \\ 1 \\ 0 \end{pmatrix} + \mu \begin{pmatrix} 0 \\ 0 \\ 1 \end{pmatrix}.$$

Definition 3.6. Let $\mathcal{T}_n$ denote the set of all functions of the form

$$\tau(z) = \frac{(t_{11}(z) - z)(t_{22}(z) - z)}{t_{21}(z)} - t_{12}(z),$$

where

$$\Phi(z) = -\begin{pmatrix} t_{12}(z) & t_{11}(z) \\ t_{22}(z) & t_{21}(z) \end{pmatrix} \in \mathcal{N}_0(\mathbb{C}^2),$$

or $\tau(z) = -b_1 z + \varphi(z)$ where $\varphi(z) \in \mathcal{N}_0(\mathbb{C})$ and $b_1 \in \mathbb{R}$, or $\tau(z) \equiv \infty$.

Note that all functions of $\mathcal{T}_n$ can be obtained as limits of functions of the first kind, if certain entries of the matrix $\Phi(z)$ tend to infinity.

From the Lemmas 3.1 to 3.5 we obtain the following result.

Theorem 3.7. *Let $\mathcal{D}(S)$ be nondegenerated. Then the set of u-resolvents of S is parametrized by the formula*

$$[(A-z)^{-1}u, u] = [(S_d - z)^{-1}u_d, u_d] - \frac{(\eta_0 - [(S_d - z)^{-1}u_d, a])(\overline{\eta_0} - [(S_d - z)^{-1}a, u_d])}{\tau(z) + [(S_d - z)^{-1}a, a]},$$

where $\tau \in \mathcal{T}_n$.

Remark 3.8. The correspondence between u-resolvents and parameters is bijective, whereas different extensions may yield the same u-resolvent. Consider for example Lemma 3.1 with $t_{21} = 0$.

II. If $\mathcal{D}(S)$ is degenerated, we have $h^\circ \in \mathcal{D}(S)$. Assume for the moment that $\dim \mathcal{H} \geq 3$. We choose a decomposition $\mathcal{D}(S) = \mathcal{D}_1[\dot{+}]\langle h^\circ\rangle$ of $\mathcal{D}(S)$ where $\mathcal{D}_1$ is nondegenerated. Let $g \in \mathcal{H}$ be such that $g \perp \mathcal{D}(S)$ and $[g, g] = 1$. Then we can choose $\mathcal{H}_0 = \langle g\rangle[\dot{+}]\mathcal{D}_1$ for the construction of a canonical extension $\mathcal{P}_c$:

$$\mathcal{P}_c = \langle g\rangle[\dot{+}]\underbrace{\mathcal{D}_1[\dot{+}](\langle h^\circ\rangle}_{\mathcal{D}(S)}\dot{+}\langle h^1\rangle) \tag{3.16}$$

with skewly linked elements h°, h^1. For $x = x_d + \xi_0 h^\circ \in \mathcal{D}(S)$ let, according to the decomposition (3.16),

$$Sx = ([x_d, a] + c_1\xi_0)g + S_d x_d + ([x_d, b] + c_2)h^\circ, \tag{3.17}$$

and

$$u = \eta_2 g + u_d + \eta_0 h^\circ. \tag{3.18}$$

Recall our assumption that h° is not an eigenvector of S, thus $c_1 \neq 0$. Similar as in the preceding discussion we distinguish the cases

(i) $T_\infty = \{0\}$,

(ii) $\dim T_\infty = 1, \quad h^\circ \notin T_\infty$,

(iii) $\dim T_\infty = 1, \quad h^\circ \in T_\infty$,

(iv) $\dim T_\infty = 2$.

Again the details will be carried out only in the first case.

Case (i): $T_0(z)$ can be written as a block matrix with respect to the decomposition (3.16) of the space $\mathcal{P}_c$:

$$(3.19)\qquad T_0(z) = \left(\begin{array}{c|cc|c} t_{11}(z) & [\,\cdot\,,a] & c_1 & t_{12}(z) \\ \hline a & S_d & 0 & b \\ t_{21}(z) & [\,\cdot\,,b] & c_2 & t_{22}(z) \\ \hline \overline{c_1} & 0 & 0 & \overline{c_2} \end{array}\right) : \begin{pmatrix} \langle g\rangle \\ + \\ \mathcal{D}_1 \\ + \\ \langle h^\circ\rangle \\ + \\ \langle h^1\rangle \end{pmatrix} \longrightarrow \begin{pmatrix} \langle g\rangle \\ + \\ \mathcal{D}_1 \\ + \\ \langle h^\circ\rangle \\ + \\ \langle h^1\rangle \end{pmatrix}.$$

The form of the second and third column and the second and fourth row of the block matrix (3.19) follows from (iii) and (iv) of Theorem 2.5, (ii) implies that

$$\Phi(z) = -\begin{pmatrix} t_{11}(z) & t_{12}(z) \\ t_{21}(z) & t_{22}(z) \end{pmatrix} \in \mathcal{N}_0(\mathbb{C}^2).$$

Remark 3.9. Note that in this case the operator S_d acts in a space of dimension $\dim \mathcal{H} - 2$, whereas in the case $h^\circ \notin \mathcal{D}(S)$ the operator S_d acts in a space of dimension $\dim \mathcal{H} - 1$.

Lemma 3.10. *If $T_\infty = \{0\}$ and $T_0(z)$ and u are given by (3.19) and (3.18), respectively, we have*

$$(3.20)\qquad [(A-z)^{-1}u,u] = [(S_d-z)^{-1}u_d,u_d]$$
$$-\frac{((c_2-z)\eta_2 - c_1\eta_0 + [(S_d-z)^{-1}u_d,c_{\overline{z}}])((\overline{c_2}-z)\overline{\eta_2} - \overline{c_1\eta_0} - [(S_d-z)^{-1}c_z,u_d])}{\tau(z) + z(c_2-z)(\overline{c_2}-z) + [(S_d-z)^{-1}c_z,c_{\overline{z}}]}$$

where $c_z = \overline{c_1}b - (\overline{c_2}-z)a$ and

$$(3.21)\qquad \begin{aligned}\tau(z) \;=\; & -t_{11}(z)z^2 - 2(c_1t_{21}(z) + \overline{c_1}\overline{t_{12}(\overline{z})} - (c_2+\overline{c_2})t_{11}(z))z \\ & -(|c_2|^2t_{11}(z) + |c_1|^2t_{22}(z) - 2(c_1\overline{c_2}t_{21}(z) + \overline{c_1}c_2\overline{t_{12}(\overline{z})})).\end{aligned}$$

Conversely, given a function $\Phi(z) \in \mathcal{N}_0(\mathbb{C}^2)$, the right hand side of (3.20) defines a u-resolvent of S if

$$(3.22)\qquad \tau(z) \not\equiv -z(c_2-z)(\overline{c_2}-z) + [(S_d-z)^{-1}c_z,c_{\overline{z}}].$$

Proof. We have to solve the equation

$$\tag{3.23} \left(\begin{array}{c|cc|c} t_{11}(z)-z & [\cdot,a] & c_1 & t_{12}(z) \\ \hline a & S_d - z & 0 & b \\ t_{21}(z) & [\cdot,b] & c_2 - z & t_{22}(z) \\ \hline \overline{c_1} & 0 & 0 & \overline{c_2} - z \end{array}\right) \begin{pmatrix} \xi_2 \\ x_d \\ \xi_0 \\ \xi_1 \end{pmatrix} = \begin{pmatrix} \eta_2 \\ u_d \\ \eta_0 \\ 0 \end{pmatrix}.$$

The last row of (3.23) implies that

$$\tag{3.24} \overline{c_1}\xi_2 + (\overline{c_2} - z)\xi_1 = 0.$$

The second row gives $\xi_2 a + (S_d - z)x_d + \xi_1 b = u_d$, i.e.,

$$\tag{3.25} x_d = (S_d - z)^{-1}u_d - \xi_2(S_d - z)^{-1}c_z'$$

where $c_z' = \frac{\overline{c_1}}{\overline{c_2}-z}b - a$. We find

$$[x_d, a] = [(S_d - z)^{-1}u_d, a] + \xi_2[(S_d - z)^{-1}c_z', a],$$
$$[x_d, b] = [(S_d - z)^{-1}u_d, b] + \xi_2[(S_d - z)^{-1}c_z', b].$$

Substituting these expressions into the relation obtained from the first row of (3.23) we get

$$\tag{3.26} \xi_2\left(t_{11}(z) - z - \frac{\overline{c_1}}{\overline{c_2} - z}t_{12}(z) + [(S_d - z)^{-1}c_z', a]\right) + \xi_0 c_1$$
$$= \eta_2 - [(S_d - z)^{-1}u_d, a],$$

and, substituted into the relation obtained from the third row,

$$\tag{3.27} \xi_2\left(t_{21}(z) - \frac{\overline{c_1}}{\overline{c_2} - z}t_{22}(z) + [(S_d - z)^{-1}c_z', b]\right) + \xi_0(c_2 - z)$$
$$= \eta_0 - [(S_d - z)^{-1}u_d, b].$$

From the linear system of equations (3.26) and (3.27) we obtain, using Cramer's rule and (3.24),

$$\xi_0 = \frac{\Delta_0}{\Delta}, \qquad \xi_2 = z\frac{\Delta_2}{\Delta},$$

where Δ, Δ_0 and Δ_2 are given by

$$\Delta = \begin{vmatrix} t_{11}(z) - z - \frac{\overline{c_1}}{\overline{c_2}-z}t_{12}(z) + [(S_d - z)^{-1}c_z, a] & c_1 \\ t_{21}(z) - \frac{\overline{c_1}}{\overline{c_2}-z}t_{22}(z) + [(S_d - z)^{-1}c_z, b] & c_2 - z \end{vmatrix},$$

$$\Delta_0 = \begin{vmatrix} t_{11}(z) - z - \frac{\overline{c_1}}{\overline{c_2}-z}t_{12}(z) + [(S_d - z)^{-1}c_z, a] & \eta_2 - [(S_d - z)^{-1}u_d, a] \\ t_{21}(z) - \frac{\overline{c_1}}{\overline{c_2}-z}t_{22}(z) + [(S_d - z)^{-1}c_z, b] & \eta_0 - [(S_d - z)^{-1}u_d, b] \end{vmatrix},$$

$$\Delta_2 = \begin{vmatrix} \eta_2 - [(S_d - z)^{-1}u_d, a] & c_1 \\ \eta_0 - [(S_d - z)^{-1}u_d, b] & c_2 - z \end{vmatrix}.$$

Finally, we find

$$\xi_1 = -\frac{\overline{c_1}}{\overline{c_2} - z}\frac{\Delta_2}{\Delta},$$

and by (3.25)

$$x_d = (S_d - z)^{-1}u_d + \frac{\Delta_2}{\Delta}(S_d - z)^{-1}c_z'.$$

Using these expressions we compute

$$\begin{aligned}
&[(A-z)^{-1}u, u] \\
&= \left[\begin{pmatrix} \xi_2 \\ x_d \\ \xi_0 \\ \xi_1 \end{pmatrix}, \begin{pmatrix} \eta_2 \\ u_d \\ \eta_0 \\ 0 \end{pmatrix}\right] = \xi_2\overline{\eta_2} + [x_d, u_d] + \xi_1\overline{\eta_0} \\
&= [(S_d - z)^{-1}u_d, u_d] + \frac{\Delta_2}{\Delta}\left(\overline{\eta_2} + [(S_d - z)^{-1}c_z, u_d] - \frac{\overline{c_1}}{\overline{c_2} - z}\right) \\
&= [(S_d - z)^{-1}u_d, u_d] \\
&\quad - \frac{((c_2-z)\eta_2 - c_1\eta_0 + [(S_d - z)^{-1}u_d, c_{\overline{z}}])((\overline{c_2}-z)\overline{\eta_2} - \overline{c_1}\overline{\eta_0} - [(S_d - z)^{-1}c_z, u_d])}{\tau(z) + z(c_2 - z)(\overline{c_2} - z) + [(S_d - z)^{-1}c_z, c_{\overline{z}}]}.
\end{aligned}$$

Consider the converse part of the lemma. As $\Delta \neq 0$ if and only if (3.22) is satisfied, we find, by similar computations as above, that $T_0(z)$ is injective and thus surjective, i.e., condition (i) of Theorem 2.5 is satisfied. All the other conditions of Theorem 2.5 immediately follow from the construction (3.19) of $T_0(z)$. □

Remark 3.11. Again not all the entries of the matrix $\Phi(z)$ occur separately in (3.20), but only the scalar function τ.

Case (ii): We can write $T_\infty = \langle b \rangle$ with $b = g + b_0 h^\circ$, and

$$T_0(z) = \begin{pmatrix} [\,\cdot\,, a] & c_1 & t_1(z) \\ S_d & 0 & b - \overline{b_0}a \\ [\,\cdot\,, b] & c_2 & t_2(z) \\ 0 & 0 & -\overline{b_0} \end{pmatrix} : \begin{pmatrix} \mathcal{D}_1 \\ + \\ \langle h^\circ \rangle \\ + \\ \langle d \rangle \end{pmatrix} \longrightarrow \begin{pmatrix} \langle g \rangle \\ + \\ \mathcal{D}(S) \\ + \\ \langle h^\circ \rangle \\ + \\ \langle h^1 \rangle \end{pmatrix}.$$

Lemma 3.12. *If* $\dim T_\infty = 1$ *and* $h^\circ \notin T_\infty$, *we have*

$$[(A - z)^{-1}u, u] = [(S_d - z)^{-1}u_d, u_d]. \tag{3.28}$$

Conversely, the right hand side of (3.28) *defines a u-resolvent of* S.

In order to prove Lemma 3.12, consider the equation

$$\begin{pmatrix} [\cdot\,,a] & c_1 & t_1(z)+z\overline{b_0} \\ S_d - z & 0 & b-\overline{b_0}a \\ [\cdot\,,b] & c_2 - z & t_2(z) \\ 0 & 0 & -\overline{b_0}-z \end{pmatrix} \begin{pmatrix} x_d \\ \xi_0 \\ \xi_3 \end{pmatrix} = \begin{pmatrix} \eta_2 \\ u_d \\ \eta_0 \\ 0 \end{pmatrix} + \lambda \begin{pmatrix} 1 \\ 0 \\ b_0 \\ 0 \end{pmatrix}.$$

Case (iii): We have

$$T_0(z) = \begin{pmatrix} t_1(z) & [\cdot\,,a] & c_1 \\ a & S_d & 0 \\ t_2(z) & [\cdot\,,b] & c_2 \\ \overline{c_1} & 0 & 0 \end{pmatrix} : \begin{pmatrix} \langle g\rangle \\ + \\ \mathcal{D}_1 \\ + \\ \langle h^\circ\rangle \end{pmatrix} \longrightarrow \begin{pmatrix} \langle g\rangle \\ + \\ \mathcal{D}_1 \\ + \\ \langle h^\circ\rangle \\ + \\ \langle h^1\rangle \end{pmatrix}.$$

Lemma 3.13. *If* $\dim T_\infty = 1$ *and* $h^\circ \in T_\infty$, *the formula* (3.28) *holds.*

In order to prove Lemma 3.13, consider the equation

$$\begin{pmatrix} t_1(z)-z & [\cdot\,,a] & c_1 \\ a & S_d - z & 0 \\ t_2(z) & [\cdot\,,b] & c_2 - z \\ \overline{c_1} & 0 & 0 \end{pmatrix} \begin{pmatrix} \xi_2 \\ x_d \\ \xi_0 \end{pmatrix} = \begin{pmatrix} \eta_2 \\ u_d \\ \eta_0 \\ \eta_1 \end{pmatrix} + \lambda \begin{pmatrix} 0 \\ 0 \\ 1 \\ 0 \end{pmatrix} \tag{3.29}$$

with $\eta_1 = 0$. To prove the converse part note that (3.29) remains solvable if $\eta_1 \in \mathbb{C}$ is arbitrary, and therefore condition (i) of Theorem 2.5 is satisfied.

Case (iv): In this case $T_0(z) = S$ and $\mathcal{D}^\perp = \langle g, h^\circ\rangle$. Thus

$$\mathcal{R}\,(T_0(z)-z) + \mathcal{D}^\perp \subseteq \mathcal{H} \neq \mathcal{P}_c$$

and (i) of Theorem 2.5 cannot be satisfied.

Remark 3.14. Similarly to the case $h^\circ \notin \mathcal{D}(S)$, different extensions may have the same u-resolvent.

For the sake of completeness we consider the case $\dim \mathcal{H} = 2$. Then the component $\mathcal{D}_1$ in the decomposition (3.16) of $\mathcal{P}_c$ does not appear. With similar calculations as in the preceding lemmata we obtain that the formulas developed above remain valid if only the terms involving S_d are deleted.

Definition 3.15. Let $\mathcal{T}_d$ denote the set of all functions τ of the form

$$\begin{aligned} \tau(z) \;&=\; -t_{11}(z)z^2 - 2(c_1 t_{21}(z) + \overline{c_1}\overline{t_{12}(\overline{z})} - (c_2+\overline{c_2})t_{11}(z))z \\ &\quad -(|c_2|^2 t_{11}(z) + |c_1|^2 t_{22}(z) - 2(c_1\overline{c_2}t_{21}(z) + \overline{c_1}c_2\overline{t_{12}(\overline{z})})) \end{aligned}$$

with c_1, c_2 given by (3.17), where

$$\Phi(z) = -\begin{pmatrix} t_{11}(z) & t_{12}(z) \\ t_{21}(z) & t_{22}(z) \end{pmatrix} \in \mathcal{N}_0(\mathbb{C}^2),$$

or $\tau(z) = \infty$.

From the above lemmata we obtain the following result.

Theorem 3.16. *Let $\mathcal{D}(S)$ be degenerated. Then the set of all u-resolvents of S is parametrized by the formula*

$$[(A-z)^{-1}u, u] = [(S_d - z)^{-1}u_d, u_d]$$

$$-\frac{((c_2 - z)\eta_2 - c_1\eta_0 + [(S_d - z)^{-1}u_d, c_{\overline{z}}])((\overline{c_2} - z)\overline{\eta_2} - \overline{c_1\eta_0} + [(S_d - z)^{-1}c_z, u_d])}{\tau(z) + z(c_2 - z)(\overline{c_2} - z) + [(S_d - z)^{-1}c_z, c_{\overline{z}}]},$$

where $\tau \in \mathcal{T}_d$.

4. The degenerated Nevanlinna-Pick problem

By $\mathcal{N}_\kappa$ we denote the class of all complex functions f which are meromorphic in $\mathbb{C}^+$, and such that the kernel

$$N_f(z, w) = \frac{f(z) - \overline{f(w)}}{z - \overline{w}}, \qquad z, w \in \rho(f),$$

has exactly κ negative squares. Here $\rho(f)$ denotes the domain of holomorphy of the function f.

Recall that the classical Nevanlinna-Pick interpolation problem for the upper half plane can be formulated as follows:

Given $n \in \mathbb{N}$, $z_1, \ldots, z_n \in \mathbb{C}^+$ and $w_1, \ldots, w_n \in \mathbb{C}^+$. Find conditions such that there exist functions $f \in \mathcal{N}_0$ which satisfy

$$f(z_i) = w_i, \qquad i = 1, \ldots, n. \tag{4.1}$$

It was shown in [9] that such functions exist if and only if the corresponding so-called Pick matrix

$$\mathbb{P} = \left(\frac{w_j - \overline{w_i}}{z_j - \overline{z_i}}\right)_{i,j=1}^n$$

is positive semidefinite. In fact the problem has a unique solution $f \in \mathcal{N}_0$ if and only if the Pick matrix $\mathbb{P}$ is degenerated, otherwise it has infinitely many solutions, which can be described e.g. by a fractional linear transformation with a parameter running through the set $\mathcal{N}_0 \cup \{\infty\}$.

Here we consider the following problem: Given $n \in \mathbb{N}$, $z_1, \dots, z_n \in \mathbb{C}^+$ and $w_1, \dots, w_n \in \mathbb{C}^+$, and suppose that the corresponding Pick matrix $\mathbb{P}$ is positive semidefinite and degenerated, having zero as a simple eigenvalue. Hence the Nevanlinna-Pick interpolation problem has a unique solution $f \in \mathcal{N}_0$. Describe all functions $f \in \mathcal{N}_1$ which are holomorphic in $z_1, \dots, z_n$ and which are such that (4.1) holds. It was shown in [11] that there exist infinitely many solutions f of this problem. They correspond to the self-adjoint extensions A of a symmetric operator S, where S acts in a finite dimensional space $\mathcal{H}$ with a positive semidefinite inner product, having a one dimensional isotropic subspace, and where A is a self-adjoint extension of S in a π_1-space $\mathcal{P} \supseteq \mathcal{H}$ such that $z_i \in \rho(A)$ for $i = 1, \dots, n$.

Recall (see [11]) that $\mathcal{H}$ and S can be constructed as follows: $\mathcal{H}$ is the linear space of all formal sums

$$\sum_{i=1}^{n} \xi_i e_i, \qquad \xi_i \in \mathbb{C},$$

equipped with the inner product given by

$$[e_i, e_j] = \frac{w_i - \overline{w_j}}{z_i - \overline{z_j}}, \qquad i, j = 1, \dots, n,$$

and S is the operator in $\mathcal{H}$ with domain

$$\mathcal{D}(S) = \left\{ \sum_{i=1}^{n} \xi_i e_i : \sum_{i=1}^{n} \xi_i = 0 \right\}$$

which acts as

$$S\left(\sum_{i=1}^{n} \xi_i e_i \right) = \sum_{i=1}^{n} z_i \xi_i e_i \quad \text{for} \quad \sum_{i=1}^{n} \xi_i e_i \in \mathcal{D}(S).$$

By a straightforward calculation it can be shown that S is symmetric and has no eigenvalues.

The following result is a particular case of Theorem 1 of [11]:

Proposition 4.1. *The family of solutions $f \in \mathcal{N}_1$ of the interpolation problem* (4.1) *corresponds to the family of self-adjoint (relation) extensions A of S in Pontryagin spaces $\mathcal{P} \supseteq \mathcal{H}$ with negative index* 1 *which contain the points $z_1, \dots, z_n$ in their resolvent set and are e_1-minimal. This correspondence is established by the formula $f = Q_A$ with*

$$Q_A(z) = \overline{w_1} + \frac{\Im(w_1)}{\Im(z_1)}(z - \overline{z_1}) + (z - \overline{z_1})(z - z_1)[(A - z)^{-1} e_1, e_1] \tag{4.2}$$

and becomes bijective if we do not distinguish between e_1-unitarily equivalent extensions of S.

There an extension A of S in a Pontryagin space $\mathcal{P}$ with $z_1 \in \rho(A)$ is called e_1-minimal if

$$\overline{\langle e_1, (A-z)^{-1}e_1 : \ z \in \rho(A)\rangle} = \mathcal{P}.$$

Two extensions A_1 and A_2 acting in Pontryagin spaces $\mathcal{P}_1$ and $\mathcal{P}_2$, respectively, are called e_1-unitarily equivalent if there exists a unitary operator $U : \mathcal{P}_1 \to \mathcal{P}_2$ with $Ue_1 = e_1$, such that $UA_1 = A_2U$.

Observe that in the relation (4.2) an e_1-resolvent appears. Moreover, the space $\mathcal{H}$ and the operator S satisfy the assumptions of Sections 2 and 3. Hence, in the results obtained there, we can take $u = e_1$ and decompose this element according to (3.3) or (3.16):

$$e_1 = u_d + \eta_0 h^\circ \quad \text{or} \quad e_1 = \eta_2 g + u_d + \eta_0 h^\circ,$$

respectively, where again $\langle h^\circ \rangle = \mathcal{H}^\circ$. With

$$h^\circ = \sum_{i=1}^{n} \chi_i^\circ e_i$$

it follows that $\mathcal{D}(S)$ degenerates if and only if $\sum_{i=1}^{n} \chi_i^\circ = 0$. Let S_d be defined as in (3.4) or (3.17). Denote in the following by Q_{S_d} the function

$$(4.3) \quad Q_{S_d}(z) = \overline{w_1} + \frac{\Im(w_1)}{\Im(z_1)}(z-\overline{z_1}) + (z-\overline{z_1})(z-z_1)[(S_d-z)^{-1}u_d, u_d].$$

Then the results of the preceding sections imply:

If $\sum_{i=1}^{n} \chi_i^\circ \neq 0$, the solutions of the interpolation problem (4.1) within $\mathcal{N}_0 \cup \mathcal{N}_1$ are given by

$$(4.4) \quad f(z) = Q_{S_d}(z) - (z-z_1)(z-\overline{z_1}) \cdot \frac{(\eta_0 - [(S_d-z)^{-1}u_d, a])(\overline{\eta_0} - [(S_d-z)^{-1}a, u_d])}{\tau(z) + [(S_d-z)^{-1}a, a]},$$

where the parameter τ runs through the set of all functions of $\mathcal{T}_n$ such that for the corresponding self-adjoint extension the data points z_j belong to the resolvent set. If $\sum_{i=1}^{n} \chi_i^\circ = 0$ and c_1, c_2 are as in (3.17), the solutions of (4.1) within $\mathcal{N}_0 \cup \mathcal{N}_1$ are given by

$$(4.5) \quad f(z) - Q_{S_d}(z) - (z-z_1)(z-\overline{z_1}) \cdot \frac{((c_2-z)\eta_2 - c_1\eta_0 + [(S_d-z)^{-1}u_d, a])((\overline{c_2}-z)\eta_2 - \overline{c_1\eta_0} + [(S_d-z)^{-1}a, u_d])}{\tau(z) + z(c_2-z)(\overline{c_2}-z) + [(S_d-z)^{-1}c_z, c_{\overline{z}}]},$$

where the parameter τ runs through the set of all functions of $\mathcal{T}_d$ such that for the corresponding self-adjoint extension the data points z_j belong to the resolvent set.

We will see in Section 5 that the parameters τ which have to be excluded are exactly those for which the denominator of (4.4) (or (4.5), respectively) vanishes at a data point.

In both cases the unique solution $f \in \mathcal{N}_0$ of (4.1) corresponds to the parameter $\tau(z) \equiv \infty$, that is:

Proposition 4.2. *The function Q_{S_d} in (4.3) is the unique solution of the interpolation problem (4.1) in $\mathcal{N}_0$.*

Proof. As the operator S_d acts in the positive definite inner product space $\mathcal{D}(S)$ or $\mathcal{D}_1$, respectively, we have $Q_{S_d} \in \mathcal{N}_0$ (compare [11]).
As S_d is not an extension of S we cannot apply Proposition 4.1 in order to show that Q_{S_d} is a solution, we actually have to compute the values of Q_{S_d} at $z_1, \ldots, z_n$. As the points z_i are nonreal and S_d is self-adjoint, we have $z_i \in \rho(S_d)$ for $i = 1, \ldots, n$, and $Q_{S_d}(z_1) = w_1$. If $\mathcal{D}(S)$ is nondegenerated, then for $i = 2, \ldots, n$

$$\begin{aligned} u_d + \eta_0 h^\circ &= u = e_1 = (S - z)\left(\frac{e_i - e_1}{z_i - z_1}\right) \\ &= (S_d - z)\left(\frac{e_i - e_1}{z_i - z_1}\right) + \left[\left(\frac{e_i - e_1}{z_i - z_1}\right), a\right] h^\circ \end{aligned}$$

holds. If $\mathcal{D}(S)$ is degenerated, we find for $i \in \{2, \ldots, n\}$

$$\begin{aligned} \eta_2 g + u_d + \eta_0 h^\circ &= u = e_1 = (S - z)\left(\frac{e_i - e_1}{z_i - z_1}\right) \\ &= (S_d - z)\left(\frac{e_i - e_1}{z_i - z_1} - \lambda h^\circ\right) + \left[\left(\frac{e_i - e_1}{z_i - z_1} - \lambda h^\circ\right), b\right] h^\circ \\ &\quad + \left[\left(\frac{e_i - e_1}{z_i - z_1} - \lambda h^\circ\right), a\right] g + \lambda c_1 g + \lambda c_2 h^\circ, \end{aligned}$$

where $\left(\frac{e_i - e_1}{z_i - z_1} - \lambda h^\circ\right) + \lambda h^\circ$ is the decomposition of $\frac{e_i - e_1}{z_i - z_1}$ with respect to $\mathcal{D}(S) = \mathcal{D}_1 [\dot{+}] \langle h^\circ \rangle$.
In the first case we have

$$(S_d - z)^{-1} u_d = \frac{e_i - e_1}{z_i - z_1},$$

whereas in the second case

$$(S_d - z)^{-1} u_d = \frac{e_i - e_1}{z_i - z_1} - \lambda h^\circ.$$

A straightforward computation shows that in both cases the relations $Q_{S_d}(z_i) = w_i$, $i = 1, \ldots, n$, hold. □

Remark 4.3. Proposition 4.2 gives another proof of G. Pick's result that a solution of the interpolation problem (4.1) within the class $\mathcal{N}_0$ actually exists.

Proposition 4.4. *The unique solution Q_{S_d} of (4.1) in $\mathcal{N}_0$ is holomorphic at ∞ if and only if $\mathcal{D}(S)$ is nondegenerated, and it has a simple pole at ∞ if and only if $\mathcal{D}(S)$ is degenerated.*

Proof. Let f be the unique solution of (4.1) in $\mathcal{N}_0$. Then f is a rational function and $\deg f = n-1$. From [10] we know that f is given by

$$f(z) = \frac{\sum_{i=1}^{n} \overline{\chi_i^\circ w_i} \prod_{j=1, j\neq i}^{n} (z - \overline{z_j})}{\sum_{i=1}^{n} \chi_i^\circ \prod_{j=1, j\neq i}^{n} (z - \overline{z_j})} = \frac{p(z)}{q(z)} \tag{4.6}$$

where $\sum_{i=1}^{n} \chi_i^\circ e_i = h^\circ$. As $\deg f = n-1$ (see [10]), the numerator $p(z)$ and denominator $q(z)$ in (4.6) are relatively prime.

Suppose that $\mathcal{D}(S)$ is nondegenerated, i.e., $h^\circ \notin \mathcal{D}(S)$. Equivalently, $\sum_{i=1}^{n} \chi_i^0 \neq 0$, which implies that $\deg q = n-1$. As $\deg f = \max(\deg p, \deg q) = n-1$, we find that f is analytic at ∞.

If $\mathcal{D}(S)$ is degenerated, then S_d acts in a $(n-2)$-dimensional space, i.e., it can be written as an $(n-2) \times (n-2)$-matrix. Thus

$$(S_d - z)^{-1} = \frac{1}{|S_d - z|} A^T$$

where A is the matrix of algebraic complements of $S_d - z$. The determinant $|S_d - z|$ is a polynomial of degree $n-2$, whereas each entry of A is a polynomial of degree $n-3$. Thus

$$[(S_d - z)^{-1} u_d, u_d] = \frac{p_1(z)}{q_1(z)} \tag{4.7}$$

where $\deg p_1 \le n-3$ and $q_1(z) = |S_d - z|$, i.e., $\deg q_1 = n-2$. Substituting (4.7) into the expression (4.3) for Q_{S_d}, we find that

$$Q_{S_d}(z) = \frac{p(z)}{q_1(z)}$$

where $\deg p \le n-1$ and $\deg q_1 = n-2$. As f interpolates, we must have $\deg f = n-1$ and therefore $\deg p = n-1$. Thus f has a simple pole at ∞. □

5. Explicit formulas

I. Assume that $h^\circ \notin \mathcal{D}(S)$. If $h^\circ = \sum_{i=1}^{n} \chi_i^\circ e_i$, we can assume without loss of generality that $\sum_{i=1}^{n} \chi_i^\circ = 1$.

We introduce the functions

$$N_1(z) = (\eta_0 - [(S_d - z)^{-1} u_d, a])(\overline{\eta_0} - [(S_d - z)^{-1} a, u_d])$$

and

$$M_1(z) = [(S_d - z)^{-1}a, a],$$

where a, η_0 and u_d are as in (3.4) and (3.5). As the space $\mathcal{H}$ is finite dimensional and S_d is self-adjoint, N_1 and M_1 are real rational functions.

Lemma 5.1. *Let*

$$S = \begin{pmatrix} S_d \\ [\cdot, a] \end{pmatrix} : \mathcal{D}(S) \to \mathcal{H}$$

be the matrix representation of the operator S with respect to the decomposition $\mathcal{H} = \mathcal{D}(S)\,[\dot{+}]\langle h^\circ\rangle$. Then $a = \sum_{i=1}^n \alpha_i e_i$ where the numbers α_i are the (unique) solutions of the system of linear equations

$$\text{(5.1)}\quad \sum_{i=1}^n \alpha_i \left(\frac{w_1 - \overline{w_j}}{z_1 - \overline{z_j}} - \frac{w_1 - \overline{w_1}}{z_1 - \overline{z_1}} \right) = \overline{z_j} - \overline{z_1}, \quad j = 2, \dots, n, \qquad \sum_{i=1}^n \alpha_i = 0.$$

If we put $u = e_1 = u_d + \eta_0 h^\circ$, we have

$$\eta_0 = 1 \quad \textit{and} \quad u_d = e_1 - h^\circ.$$

Proof. We have to find $a \in \mathcal{D}(S)$ such that

$$Sx = S_d x + [x, a]h^\circ, \qquad x \in \mathcal{D}(S)\,.$$

Let $x = \sum_{i=1}^n \xi_i e_i \in \mathcal{D}(S)$. Then

$$Sx = \sum_{i=1}^n z_i \xi_i e_i = \left(\sum_{i=1}^n z_i \xi_i e_i - \left(\sum_{i=1}^n z_i \xi_i \right) h^\circ \right) + \left(\sum_{i=1}^n z_i \xi_i \right) h^\circ,$$

i.e., $[x, a] = \sum_{i=1}^n z_i \xi_i$. Setting $x = e_j - e_1$ for $j \in \{2, \dots, n\}$ we obtain $[e_j - e_1, a] = z_j - z_1$ or

$$\sum_{i=1}^n \overline{\alpha_i} \left(\frac{w_j - \overline{w_1}}{z_j - \overline{z_1}} - \frac{w_j - \overline{w_1}}{z_j - \overline{z_1}} \right) = z_j - z_1, \qquad j = 2, \dots, n,$$

where additionally $\sum_{i=1}^n \alpha_i = 0$ holds as $a \in \mathcal{D}(S)$. Due to the fact that $\mathcal{D}(S)$ is nondegenerated, this system of linear equations has a unique solution $\alpha_1, \dots, \alpha_n$. As we have assumed that $\sum_{i=1}^n \chi_i^\circ = 1$, the decomposition

$$u = u_d + \eta_0 h^\circ \text{ with } u_d = e_1 - h^\circ, \qquad \eta_0 = 1$$

holds. □

Recall that if a rational function is written as a quotient of two polynomials which are relatively prime, then its degree is defined as the maximum of the degrees of the numerator and the denominator. The degree of a rational function equals the total multiplicity of its poles or its zeros (including ∞).

Proposition 5.2. *The rational function $N_1(z)$ is of degree $2n-2$, and*

$$\lim_{z\to\infty} N_1(z) = 1. \tag{5.2}$$

The zeros of $N_1(z)$ are the points $z_2,\ldots,z_n$ and $\overline{z_2},\ldots,\overline{z_n}$. The poles of $N_1(z)$ are the solutions of the equation

$$\sum_{i=1}^{n} \frac{\chi_i^\circ}{z_i - z} = 0,$$

and they are all real. Therefore

$$N_1(z) = \frac{\prod\limits_{i=2}^{n}(z-z_i)(z-\overline{z_i})}{\left[\sum\limits_{i=1}^{n}\chi_i^\circ \prod\limits_{\substack{j=1\\ j\neq i}}^{n}(z-z_j)\right]^2}. \tag{5.3}$$

The function $M_1(z)$ is of degree at most $n-1$, and

$$\lim_{z\to\infty} z^l M_1(z) = -\sum_{i=1}^{n} z_i^l \alpha_i \neq 0 \tag{5.4}$$

where $l \in \{1,\ldots,n-1\}$ is such that $S^{l-1}a \in \mathcal{D}(S)$ but $S^l a \notin \mathcal{D}(S)$. The zeros of $M_1(z)$ are the solutions of the equation

$$\sum_{i=1}^{n} \frac{\alpha_i}{z_i - z} = 0.$$

The poles of $M_1(z)$ coincide with the poles of $N_1(z)$, the multiplicity of a pole in $M_1(z)$ is half the multiplicity of the corresponding pole of $N_1(z)$. Therefore

$$M_1(z) = (-1)^{n-1}\frac{\sum\limits_{i=1}^{n}\alpha_i \prod\limits_{\substack{j=1\\ j\neq i}}^{n}(z-z_j)}{\sum\limits_{i=1}^{n}\chi_i^\circ \prod\limits_{\substack{j=1\\ j\neq i}}^{n}(z-z_j)}.$$

Proof. As S_d acts on $\mathcal{D}(S)$, we have $(S_d - z)^{-1} = \frac{L(z)}{q(z)}$ where $\deg q = \dim \mathcal{D}(S) = n-1$ and $\deg L \le n-2$. Note that $q(z) = |S_d - z|$ is a real polynomial. Thus

$$\begin{aligned} N_1(z) &= \left(1 - \frac{[L(z)u_d, a]}{q(z)}\right)\left(1 - \frac{[L(z)a, u_d]}{q(z)}\right) \\ &= \frac{q(z)^2 - q(z)([L(z)u_d, a] + [L(z)a, u_d]) + [L(z)u_d, a][L(z)a, u_d]}{q(z)^2}. \end{aligned}$$

As $\deg q^2 = 2n-2$ and the degree of the remaining terms in the numerator is at most $2n-3$, we find that $\deg N_1 \leq 2n-2$ and (5.2) holds.
From [11] we know that there exist parameters τ such that the corresponding function Q_A interpolates. Thus, as the function Q_{S_d} also interpolates, we must have $N_1(z_i) = 0$ for $i = 2, \dots, n$. As N_1 is real, it has also the zeros $\overline{z_2}, \dots, \overline{z_n}$. In particular, the degree of N_1 is in fact equal to $2n-2$.
The poles of $N_1(z)$ are obviously the eigenvalues of S_d, and therefore real. A value z is an eigenvalue of S_d if and only if there exists a vector $x = \sum_{i=1}^n \xi_i e_i \in \mathcal{D}(S)$, $x \neq 0$, such that $(S_d - z)x = 0$. We find

$$0 = (S_d - z)x = P(S-z)x = P\left(\sum_{i=1}^n (z_i - z)\xi_i e_i\right),$$

where P denotes the projection onto $\mathcal{D}(S)$ with kernel $\langle h^\circ \rangle$. Thus

$$\sum_{i=1}^n (z_i - z)\xi_i e_i = \mu h^\circ \quad \text{for some} \quad \mu \in \mathbb{C}.$$

Furthermore, as we can restrict our attention to real values of z, we must have $\mu \neq 0$. Due to this fact we find that z is an eigenvalue of S_d if and only if there exist numbers ξ_i, not all zero, such that

$$\sum_{i=1}^n (z_i - z)\xi_i e_i = h^\circ, \qquad \sum_{i=1}^n \xi_i = 0.$$

The first equation has the unique solution (observe that $z \in \mathbb{R}$ and thus $z_i - z \neq 0$)

$$\xi_i = \frac{\chi_i^\circ}{z_i - z}, \qquad i = 1, \dots, n.$$

Thus z is an eigenvalue of S_d if and only if

$$\sum_{i=1}^n \xi_i = \sum_{i=1}^n \frac{\chi_i^\circ}{z_i - z} = 0.$$

We have

$$\sum_{i=1}^n \frac{\chi_i^\circ}{z_i - z} = \frac{\sum_{i=1}^n \chi_i^\circ \prod_{\substack{j=1 \\ j \neq i}}^n (z_j - z)}{\prod_{i=1}^n (z_i - z)},$$

and therefore $\sum_{i=1}^n \frac{\chi_i^\circ}{z_i - z} = 0$ if and only if

$$\sum_{i=1}^n \chi_i^\circ \prod_{\substack{j=1 \\ j \neq i}}^n (z_j - z) = 0.$$

Now it is clear that (5.3) holds.

The assertion concerning the poles of $M_1(z)$ is obvious. Also the degree of the numerator of $M_1(z)$ is clearly at most $n-2$, whereas the degree of its denominator is $n-1$ and thus $\deg M_1 \leq n-1$.
We show that z is a zero of $M_1(z)$ if and only if $a \in \mathcal{R}(S-z)$: Put $x(z) = (S_d - z)^{-1}a$. Then

$$a = (S_d - z)x(z) = (S-z)x(z) - [x(z), a]h^\circ.$$

If $M_1(z) = 0$, we have $a = (S-z)x(z)$, i.e., $a \in \mathcal{R}(S-z)$. Conversely, suppose that $a \in \mathcal{R}(S-z)$, i.e., let $a = (S-z)x$ with some $x \in \mathcal{D}(S)$. Then

$$a = (S-z)x = (S_d - z)x + [x, a]h^\circ.$$

As $\mathcal{D}(S) \cap \mathcal{H}^\circ = \{0\}$, we find $[x,a] = 0$ and therefore $a = (S_d - z)x$, i.e., $x = (S_d - z)^{-1}a$. We conclude that $M_1(z) = 0$.
In order to determine those values z for which $a \in \mathcal{R}(S-z)$, consider the equation

$$(S-z)x = a, \qquad x \in \mathcal{D}(S).$$

With $x = \sum_{i=1}^n \xi_i e_i$ and $a = \sum_{i=1}^n \alpha_i e_i$ this equation becomes

$$(z_i - z)\xi_i = \alpha_i, \quad i = 1, \ldots, n, \qquad \sum_{i=1}^n \xi_i = 0.$$

Hence $\xi_i = \frac{\alpha_i}{z_i - z}$ if $z \neq z_1, \ldots, z_n$. Thus in this case $a \in \mathcal{R}(S-z)$ if and only if

$$\sum_{i=1}^n \frac{\alpha_i}{z_i - z} = 0.$$

If $z = z_i$ for some $i \in \{1, \ldots, n\}$ we have $a \in \mathcal{R}(S-z)$ if and only if $\alpha_i = 0$. Thus the zeros of $M_1(z)$ coincide with those of the polynomial

$$\sum_{i=1}^n \alpha_i \prod_{\substack{j=1 \\ j\neq i}}^n (z_j - z).$$

It remains to prove the relation (5.4). Let $l \in \{1, \ldots, n-1\}$ be such that $S^{l-1}a \in \mathcal{D}(S)$ but $S^l \notin \mathcal{D}(S)$. Such a number l exists as $S^k a = \sum_{i=1}^n z_i^k \alpha_i e_i$ and

$$\begin{vmatrix} 1 & z_1 & z_1^2 & \ldots & z_1^{n-1} \\ 1 & z_2 & z_2^2 & \ldots & z_2^{n-1} \\ \vdots & \vdots & \vdots & & \vdots \\ 1 & z_n & z_n^2 & \ldots & z_n^{n-1} \end{vmatrix} \neq 0.$$

Thus $S^0 a \in \mathcal{D}(S)$, but it is impossible that $S^k a \in \mathcal{D}(S)$ for $k = 0, \dots, n-1$. Consider the power series expansion of $(S_d - z)^{-1}$ outside of a sufficiently large disc:

$$(S_d - z)^{-1} = -\sum_{k=0}^{\infty} \frac{1}{z^{k+1}} S_d^k.$$

It follows that

$$M_1(z) = [(S_d - z)^{-1} a, a] = -\sum_{k=0}^{\infty} \frac{1}{z^{k+1}} [S_d^k a, a].$$

Note that $S_d^k a = S^k a$ for $k = 0, \dots, l-1$: For $k = 0$ this assertion is trivial. For $k = 1 \leq l-1$ we have

$$Sa = S_d a + [a, a] h^\circ,$$

and, as $[a, a] = \sum_{i=1}^n z_i \alpha_i = 0$, we find $Sa = S_d a$. Using induction we obtain for $k \leq l-1$,

$$S^k a = S(S^{k-1} a) = S_d(S^{k-1} a) + [S^{k-1} a, a] h^\circ = S_d(S_d^{k-1} a) = S_d^k a$$

as $[S^{k-1} a, a] = \sum_{i=1}^n z_i^k \alpha_i = 0$. Thus

$$\begin{aligned} M_1(z) &= -\sum_{k=0}^{l-1} \frac{1}{z^{k+1}} [S^k a, a] - \sum_{k=l}^{\infty} \frac{1}{z^{k+1}} [S_d^k a, a] \\ &= -\frac{1}{z^l} \left(\sum_{i=1}^n z_i^l \alpha_i \right) - \frac{1}{z^{l+1}} \sum_{k=l}^{\infty} \frac{1}{z^{k-l}} [S_d^k a, a], \end{aligned}$$

and we find

$$\lim_{z \to \infty} z^l M_1(z) = -\sum_{i=1}^n z_i^l \alpha_i.$$

The proof is complete. □

Remark 5.3. The function $N_1(z)$ has only simple zeros.

Let $\mathcal{T}_n$ be as in Definition 3.6. Proposition 4.1 and Proposition 5.2 imply:

Theorem 5.4. *Let $n \in \mathbb{N}$, $z_1, \dots, z_n \in \mathbb{C}^+$ and $w_1, \dots, w_n \in \mathbb{C}^+$ be given such that the corresponding Pick matrix $\mathbb{P}$ is positive semidefinite,* $\operatorname{rank} \mathbb{P} = n - 1$ *and $\mathbb{P} h^\circ = 0$ for a vector $h^\circ = \sum_{i=1}^n \chi_i^\circ e_i \neq 0$ with $\sum_{i=1}^n \chi_i^\circ = 1$. Let $\alpha_1, \dots, \alpha_n$ be the solutions of the system* (5.1).

The solutions f of the interpolation problem (4.1) in $\mathcal{N}_0 \cup \mathcal{N}_1$ are given by the formula

$$f_\tau(z) = \frac{\sum_{i=1}^{n} \chi_i^\circ w_i \prod_{\substack{j=1\\ j\neq i}}^{n} (z-z_j)}{\sum_{i=1}^{n} \chi_i^\circ \prod_{\substack{j=1\\ j\neq i}}^{n} (z-z_j)}$$

$$- \frac{\prod_{i=1}^{n} (z-z_i)(z-\overline{z_i})}{\sum_{i=1}^{n} \chi_i^\circ \prod_{\substack{j=1\\ j\neq i}}^{n} (z-z_j) \left[\sum_{i=1}^{n} \alpha_i \prod_{\substack{j=1\\ j\neq i}}^{n} (z-z_j) + \tau(z) \sum_{i=1}^{n} \chi_i^\circ \prod_{\substack{j=1\\ j\neq i}}^{n} (z-z_j) \right]}.$$

The parameter τ runs through those functions of $\mathcal{T}_n$ for which

$$\tau(z_k) \neq - \frac{\sum_{i=1}^{n} \alpha_i \prod_{\substack{j=1\\ j\neq i}}^{n} (z_k-z_j)}{\sum_{i=1}^{n} \chi_i^\circ \prod_{\substack{j=1\\ j\neq i}}^{n} (z_k-z_j)}, \qquad k = 1, \ldots, n,$$

holds. The function f_∞ is the unique solution in $\mathcal{N}_0$.

II. Consider now the case $h^\circ \in \mathcal{D}(S)$. Assume that $\mathbb{P}h^\circ = 0$ for $h^\circ = \sum_{i=1}^n \chi_i^\circ e_i \neq 0$ with $\sum_{i=1}^n \chi_i^\circ = 0$. Let $\gamma_1, \ldots, \gamma_n$ be the solutions of

$$\sum_{i=1}^{n} \gamma_i \left(\frac{w_i - \overline{w_j}}{z_i - \overline{z_j}} - \frac{w_i - \overline{w_2}}{z_i - \overline{z_2}} \right) = 0, \qquad j = 3, \ldots, n, \qquad \gamma_1 = 0,$$

which satisfy additionally

$$\sum_{i,j=1}^{n} \frac{w_i - \overline{w_j}}{z_i - \overline{z_j}} \gamma_i \overline{\gamma_j} = 1, \qquad \sum_{i=1}^{n} \gamma_i \in \mathbb{R},$$

and let $\alpha_1, \ldots, \alpha_n$ be the solutions of

$$\sum_{i=1}^{n} \alpha_i \left(\frac{w_i - \overline{w_j}}{z_i - \overline{z_j}} - \frac{w_i - \overline{w_2}}{z_i - \overline{z_2}} \right) = \frac{1}{\sum_{i=1}^{n} \gamma_i} (\overline{z_j} - z_2), \qquad j = 3, \ldots, n,$$

$$\alpha_1 = 0, \qquad \sum_{i=2}^{n} \alpha_i = 0.$$

Moreover, let

$$c_1 = \frac{1}{\sum_{i=1}^{n} \gamma_i} \sum_{i=1}^{n} z_i \chi_i^\circ, \qquad c_2 = z_1.$$

Let $\mathcal{T}_d$ be as in Definition 3.15 with c_1, c_2 from above. With similar computations as in the case where $\mathcal{D}(S)$ is nondegenerated we obtain:

Theorem 5.5. *Let $n \in \mathbb{N}$, $z_1, \ldots, z_n \in \mathbb{C}^+$ and $w_1, \ldots, w_n \in \mathbb{C}^+$ be given such that the corresponding Pick matrix $\mathbb{P}$ is semidefinite,* $\operatorname{rank} \mathbb{P} = n - 1$ *and that $\mathbb{P}h^\circ = 0$ for $h^\circ = \sum_{i=1}^n \chi_i^\circ e_i \neq 0$ with $\sum_{i=1}^n \chi_i^\circ = 0$.*
The solutions of the interpolation problem (4.1) *in $\mathcal{N}_0 \cup \mathcal{N}_1$ are given by the formula*

$$f_\tau(z) = \frac{\sum\limits_{i=1}^n \chi_i^\circ w_i \prod\limits_{\substack{j=1\\ j\neq i}}^n (z - z_j)}{\sum\limits_{i=1}^n \chi_i^\circ \prod\limits_{\substack{j=1\\ j\neq i}}^n (z - z_j)} - \frac{1}{\sum\limits_{i=2}^n \gamma_i \prod\limits_{\substack{j=1\\ j\neq i}}^n (z - z_j)}$$

$$\cdot \frac{\prod\limits_{i=2}^n (z - z_i)(z - \overline{z_i})}{(z - z_1)(z - \overline{z_1})(z \sum\limits_{i=2}^n \gamma_i \prod\limits_{\substack{j=1\\ j\neq i}}^n (z - z_j) + \sum\limits_{i=2}^n \alpha_i \prod\limits_{\substack{j=1\\ j\neq i}}^n (z - z_j)) + \tau(z) \sum\limits_{i=2}^n \gamma_i \prod\limits_{\substack{j=1\\ j\neq i}}^n (z - z_j)},$$

The parameter τ runs through those functions of $\mathcal{T}_d$ for which

$$\tau(z_k) \neq -\frac{(z_k - z_1)(z_k - \overline{z_1})(z_k \sum\limits_{i=2}^n \gamma_i \prod\limits_{\substack{j=1\\ j\neq i}}^n (z_k - z_j) + \sum\limits_{i=2}^n \alpha_i \prod\limits_{\substack{j=1\\ j\neq i}}^n (z_k - z_j))}{\sum\limits_{i=2}^n \gamma_i \prod\limits_{\substack{j=1\\ j\neq i}}^n (z_k - z_j)}$$

holds for $k = 1, \ldots, n$. The function f_∞ is the unique solution in $\mathcal{N}_0$.

References

[1] BRUINSMA, P.: Interpolation problems for Schur and Nevanlinna pairs; Doctoral Dissertation, Rijksuniversiteit Groningen, Groningen 1991.

[2] DIJKSMA, A., LANGER, H., DE SNOO, H.S.V.: Generalized coresolvents of standard isometric operators and generalized resolvents of standard symmetric relations in Kreĭn spaces; Operator Theory: Adv. Appl. 48 (1990), 261–274.

[3] DIJKSMA, A., LANGER, H., DE SNOO, H.S.V.: Selfadjoint Π_κ-extensions of symmetric subspaces: An abstract approach to boundary problems with spectral parameter in the boundary conditions; Integral Equations Operator Theory 7 (1984), 459–515.

[4] IOHVIDOV, I., KREĬN, M.G., LANGER, H.: Spectral theory in spaces with an indefinite metric; Mathematical Research 9, Akademie Verlag, Berlin 1982.

[5] KREĬN, M.G., LANGER, H.: Über einige Fortsetzungsprobleme, die eng mit der Theorie hermitescher Operatoren im Raume Π_κ zusammenhängen. I. Einige Funktionenklassen und ihre Darstellungen; Math. Nachr. 77 (1977), 187–236.

[6] KREĬN, M.G., LANGER, H.: Über einige Fortsetzungsprobleme, die eng mit der Theorie hermitescher Operatoren im Raume Π_κ zusammenhängen. II. Verallgemeinerte Resolventen, u-Resolventen und ganze Operatoren; J. Funct. Anal. 30 (1978), 390–447.

[7] KREĬN, M.G., LANGER, H.: On some continuation problems which are closely related to the theory of operators in spaces Π_κ. IV. Continuous analogues of orthogonal polynomials on the unit circle with respect to an indefinite weight and related continuation problems for some classes of functions; J. Operator Theory 13 (1985), 299–417.

[8] NEVANLINNA, R.: Über beschränkte analytische Funktionen; Ann. Acad. Sci. Fenn. A 22.7 (1929), 1–75.

[9] PICK, G.: Über die Beschränkungen analytischer Funktionen, welche durch vorgegebene Funktionswerte bewirkt werden; Math. Ann. 77 (1916), 7–23.

[10] WORACEK, H.: Nevanlinna-Pick interpolation: The degenerated case; Linear Algebra Appl. 252 (1997), 141–158.

[11] WORACEK, H.: An operator theoretic approach to degenerated Nevanlinna-Pick interpolation; Math. Nachr. 176 (1995), 335–350.

Institut für Analysis, Technische Mathematik und Versicherungsmathematik
Technische Universität Wien
Wiedner Hauptstr. 8–10
A-1040 Wien
Austria
hlanger@email.tuwien.ac.at, hworacek@pop.tuwien.ac.at

1991 Mathematics Subject Classification: Primary 47A57, 47B50, 30E05

Submitted: May 28, 1996

Operator Theory:
Advances and Applications, Vol. 103

Perturbation of linear semigroups

G. SCHLÜCHTERMANN

An extension of a result on the essential spectral radius of a perturbed semigroup obtained by L. Weis is presented. In addition, integral inequalities for the measure of non-compactness for strongly measurable operator-valued functions and linear semigroups are given.

1. Introduction

We will consider perturbation theorems for linear semigroups, where in particular we compare the essential spectral radii of the perturbed with the unperturbed semigroup. The main interest for this investigation arises from the fact that the asymtotic behaviour for $t \to \infty$ of a C_0-semigroup $(U(t))$ is connected with the essential spectral radius of the operators $U(t),\ t \geq 0$. The definition of the essential spectral radius is as follows (see [We, p.7] and [Vi, Vo1, We] for details). Let $T \in L(X)$. Then

$$r_e(T) := \sup\{|\lambda| : \lambda \in \sigma(T),\ \lambda \text{ is not an eigenvalue of finite algebraic multiplicity}\}.$$

The operator ideal of strictly singular operators $SS(X,Y)$ (or Kato-operators) enjoys the property that for $B \in SS(X,Y)$, $r_e(T+B) = r_e(T)$ for all $T \in L(X,Y)$ [Sche, Th.2.17, Cor.3.4].

Let us suppose that G is the generator of the C_0-semigroup $(S(t))$ and let $B \in L(X)$. Then $G+B$ generates the so called perturbed semigroup, say $(T(t))$. In applications, one is interested to know if the essential spectral radii of the perturbed and unperturbed semigroup are equal since then the asymptotic behaviour is not changed.

L. Weis gave conditions under which $r_e(T(t)) = r_e(S(t))$ [We]; we will extend his result in 3.5. I. Vidav and J. Voigt ensured this equality under certain compactness conditions [Vi, Vo1]. Their results were extended again by L. Weis to the spaces L_p for $1 \leq p \leq \infty$ and $C(K)$-spaces. Unfortunately his proof is based on a proposition given in the same paper whose proof contains a gap as J.Voigt showed (see [Vo2, 1.2]).

Here we will show the invariance of the essential spectral radius for a larger class of spaces which includes that of L. Weis. The assumption L. Weis used for this result was "$BS(s_1)B \cdots BS(s_n)B$ is strictly singular" for all finite sets $\{s_1, \dots, s_n\} \subset \mathbb{R}_+$. This hypothesis can be modified to the condition that the products of the operators are compact (weakly compact, conditionally weakly compact, respectively), and these conditions can be verified more easily in applications, cf. [We].

Throughout the paper X, Y are Banach spaces, $B(X)$ and $S(X)$ denote the unit ball and the unit sphere of X, respectively. $L(X, Y)$ is the space of linear and bounded operators, $W(X, Y)$ ($CW(X, Y)$) is the space of weakly compact (conditionally weakly compact, respectively) operators. (Ω, Σ, μ) stands for a finite measure space, and $U : \Omega \longrightarrow L(X, Y)$ is called strongly measurable if $U(\cdot)(x)$ is measurable for all $x \in X$.

First we give some results on the essential spectral radius, which are used later, and we will develop the connection to the measure of non-compactness.

Remark 1.1.

a) For $T \in L(X.Y)$, the quantity

$$c(T) := \inf\{\|T|_Z\| : Z \text{ is a closed subspace with } \operatorname{codim} Z < \infty\}$$

is called the *measure of non-compactness.* We note that $c(T) = 0$ if and only if T is compact [LeS, p.9] The connection to the essential spectral radius is given by $r_e(T) = \lim_{n\to\infty} c(T^n)^{\frac{1}{n}}$ (see e.g. [We, p.7]).

b) For a C_0-semigroup $(S(t))_{t\geq 0}$ let $\omega := \inf\{t^{-1} \ln \|S(t)\| : t > 0\}$ and $\omega_e := \inf\{\gamma \leq \omega : \exists\, M \in \mathbb{R} : \ c(S(t)) \leq Me^{\gamma t}\}$. Then

$$r_e(S(t)) = e^{\omega_e t} \quad \text{for all} \quad t > 0.$$

ω (ω_e, respectively) is called the type (essential type, respectively) of the semigroup (see [We, Prop.2.1]).

c) Let $|||\cdot|||$ be an equivalent norm on X with

$$\forall\, x \in X : \quad m\|x\| \leq |||x||| \leq M\|x\|, \qquad m, M > 0 \text{ suitable.}$$

If c_{eq} denotes the measure of non-compactness for this equivalent norm, then we have

$$\forall\, T \in L(X) : \quad c(T) \leq \frac{M}{m} c_{eq}(T) \quad \text{and} \quad c_{eq}(T) \leq \frac{M}{m} c(T).$$

Thus, for a C_0-semigroup $(S(t))$, the essential spectral radius is invariant for different equivalent norms on X, using a).

d) $\mathcal{S} := \{Z \subset X : \operatorname{codim} Z < \infty\}$ will denote the set of all subspaces with finite co-dimension. We have $Z \in \mathcal{S}$ if and only if there are $x_1^*, \ldots, x_n^* \in X^*$ so that $Z = \bigcap_{i=1}^n x_i^{*-1}(\{0\})$. Hence

$$Z_1, Z_2 \in \mathcal{S} \quad \Longrightarrow \quad Z_1 \cap Z_2 \in \mathcal{S}.$$

e) Let $U : \Omega \longrightarrow L(X, Y)$ be strongly measurable and bounded. We suppose that there is a decreasing sequence $(Z_n) \subset \mathcal{S}$ such that $\omega \mapsto \|U(\omega)|_{Z_n}\|$ is measurable and $c(U(\omega)) = \lim_{n\to\infty} \|U(\omega)|_{Z_n}\|$ for almost all $\omega \in \Omega$.

Then we have:

$$c\left(\int_\Omega U(\omega)\,d\mu(\omega)\right) \le \int_\Omega c(U(\omega))\,d\mu(\omega).$$

Proof (of e)). By the definition of the measure of non-compactness one has

$$\begin{aligned} c\left(\int_\Omega U(\omega)\,d\mu(\omega)\right) &\le \limsup_{n\to\infty} \sup_{x\in B(Z_n)} \left\|\int_\Omega U(\omega)(x)\,d\mu(\omega)\right\| \\ &\le \limsup_{n\to\infty} \int_\Omega \|U(\omega)|_{Z_n}\|\,d\mu(\omega) = \int_\Omega c(U(\omega))\,d\mu(\omega) \end{aligned}$$

by the Lebesgue domination theorem. Thus e) is proved. □

2. Essential spectral radius for perturbed semigroups

Before we start with the investigation of the spectral radius of a perturbed semigroup, we will state some auxiliary results which may be of independent interest. In the following proposition we will present a "multiplicative" version of Remark 1.1 e). This proposition will be used in the proof of the main theorem in which we state the equality of the essential spectral radius for the unperturbed and perturbed semigroup. The next proposition also improves [KS, Th. 3.7 a)], provided a certain assumption is imposed on the operator-valued function.

Proposition 2.1. *Let $U_1,\dots,U_k : \Omega \longrightarrow L(X)$ be bounded and strongly measurable. Let us suppose that for all $i = 1,\dots,k$ there exists a decreasing sequence $(Z_n^i) \subset \mathcal{S}$ so that for all $i = 1,\dots,k$,*

$$\omega \longmapsto \|U_i(\omega)|_{Z_n^i}\| \text{ is measurable and } c(U_i(\omega)) = \lim_{n\to\infty} \|U_i(\omega)|_{Z_n^i}\| \quad \text{a.e. on } \Omega.$$

Then we have:

a) *If X does not contain ℓ_1, then*

$$c\left(\int_\Omega U_1(\omega)\cdots U_k(\omega)\,d\mu(\omega)\right) \le \int_\Omega c(U_1(\omega))\cdots c(U_k(\omega))\,d\mu(\omega).$$

b) *If X is separable, then*

$$c\left(\int_\Omega U_1(\omega)\cdots U_k(\omega)\,d\mu(\omega)\right) \le 2\int_\Omega c(U_1(\omega))\cdots c(U_k(\omega))\,d\mu(\omega).$$

c) *If X is arbitrary, then*

$$c\left(\int_\Omega U_1(\omega)\cdots U_k(\omega)\,d\mu(\omega)\right) \le 4\int_\Omega c(U_1(\omega))\cdots c(U_k(\omega))\,d\mu(\omega).$$

Proof. a) For $n \in \mathbb{N}$ we define

$$Z_n := \bigcap_{i=1}^{k} Z_n^i.$$

Then with Remark 1.1 c) we conclude that $Z_n \in \mathcal{S}$ for all $n \in \mathbb{N}$. For $\omega \in \Omega$, $n \in \mathbb{N}$ and $i = 1, \ldots, k$ we define $\phi_{n,i}(\omega) := \inf_{j=1,\ldots,k} \|U_i(\omega)|_{Z_n^j}\|$. The functions $\phi_{n,i}$ are measurable for $n \in \mathbb{N}$ and $i = 1, \ldots, k$. Keeping in mind that $c(U_i(\omega)) \leq \|U_i(\omega)|_{Z_n}\| \leq \phi_{n,i}(\omega) \leq \|U_i(\omega)|_{Z_n^i}\|$ for all $n \in \mathbb{N}$, $i = 1, \ldots, k$ and $\omega \in \Omega$, according to the assumption we get

$$\forall\, i = 1, \ldots, k: \quad c(U_i(\omega)) = \lim_{n\to\infty} \phi_{n,i}(\omega) \quad \text{a.e. on } \Omega.$$

Since the U_i are bounded for all $i = 1, \ldots, k$, we may assume without loss of generality that

$$\sup_{i=1,\ldots,k} \sup_{\omega\in\Omega} \|U_i(\omega)\| \leq 1.$$

Let $0 < \epsilon < \frac{1}{\mu(\Omega)}$. The Egorov theorem provides us with a set $\Omega_\epsilon \in \Sigma$, $\mu(\Omega \setminus \Omega_\epsilon) < \frac{\epsilon}{2}$, such that

$$\sup_{i=1,\ldots,k} \sup_{\omega\in\Omega_\epsilon} \left|\phi_{n,i}(\omega) - c(U_i(\omega))\right| \to 0, \qquad n \to \infty.$$

For simplicity we assume that $k = 2$. The argument can easily be transferred to any $k \in \mathbb{N}$. We can find an $n_0 \in \mathbb{N}$ such that for all $n \geq n_0$

$$\sup_{i=1,2} \sup_{\omega\in\Omega_\epsilon} \left|\phi_{n,i}(\omega) - c(U_i(\omega))\right| < \frac{\epsilon}{8\mu(\Omega)}.$$

Because $c(U_i(\omega)) \leq \phi_{n,i}(\omega) \leq 1$ for all $\omega \in \Omega$ and $i = 1, 2$, we conclude

$$\begin{aligned}
&\sup_{\omega\in\Omega_\epsilon} |\phi_{n,1}(\omega)\cdot\phi_{n,2}(\omega) - c(U_1(\omega))\cdot c(U_2(\omega))| \\
&< \sup_{\omega\in\Omega_\epsilon} \left|\phi_{n,1}(\omega)\cdot\phi_{n,2}(\omega) - \left(\phi_{n,1}(\omega) - \frac{\epsilon}{8\mu(\Omega)}\right)\cdot\left(\phi_{n,2}(\omega) - \frac{\epsilon}{8\mu(\Omega)}\right)\right| \\
&\leq \left|\frac{\epsilon}{8\mu(\Omega)} + \frac{\epsilon}{8\mu(\Omega)} - \frac{\epsilon^2}{64\mu(\Omega)^2}\right| \\
&< \frac{\epsilon}{4\mu(\Omega)}.
\end{aligned} \tag{2.1}$$

Using (2.1), for $n \geq n_0$ we deduce

$$\int_{\Omega_\epsilon} \phi_{n,1}(\omega)\cdot\phi_{n,2}(\omega)\,d\mu(\omega) \leq \int_{\Omega_\epsilon} c(U_1(\omega))\cdot c(U_2(\omega))\,d\mu(\omega) + \frac{\epsilon}{4}.$$

Since $\|U_i(\omega)\| \leq 1$ for all $\omega \in \Omega$ and $i = 1, 2$, for $n \geq n_0$ we have

$$\int_{\Omega} \phi_{n,1}(\omega)\cdot\phi_{n,2}(\omega)\,d\mu(\omega) \leq \int_{\Omega} c(U_1(\omega))\cdot c(U_2(\omega))\,d\mu(\omega) + \frac{\epsilon}{2}. \tag{2.2}$$

Let $P : X \longrightarrow Z_{n_0}$ denote the projection onto Z_{n_0}. Then $I - P$ is a finite rank operator and, thus, $V(\omega) := (I-P)\cdot U_2(\omega)$ is compact for all $\omega \in \Omega$. Since $\ell_1 \not\subset X$, by [S1, Th. 3.2] we know that $\overline{V} : X \longrightarrow L_1(\mu, X)$ is compact, too. Therefore by Remark 1.1 b) there is a subspace $\widetilde{Z} \subset Z_{n_0}$, $\widetilde{Z} \in \mathcal{S}$, such that $\|\overline{V}|_{\widetilde{Z}}\| \le \frac{\epsilon}{4}$. For $x \in B(\widetilde{Z})$ we estimate

$$\begin{aligned}
(2.3)\qquad & \int_\Omega \|U_1(\omega)\cdot U_2(\omega)(x)\|\,d\mu(\omega) \\
&= \int_\Omega \|U_1(\omega)\cdot P\cdot U_2(\omega)(x) + U_1(\omega)\cdot(I-P)\cdot U_2(\omega)(x)\|\,d\mu(\omega) \\
&\overset{\|U_1(\omega)\|\le 1}{\le} \int_\Omega \|U_1(\omega)\cdot P\cdot U_2(\omega)(x)\|\,d\mu(\omega) + \frac{\epsilon}{4} \\
&\overset{\|U_1(\cdot)|_{Z_{n_0}}\|\le \phi_{n_0,1}(\cdot)}{\le} \int_\Omega \phi_{n_0,1}(\omega)\cdot\|P\cdot U_2(\omega)(x)\|\,d\mu(\omega) + \frac{\epsilon}{4} \\
&= \int_\Omega \phi_{n_0,1}(\omega)\cdot\|\big(U_2(\omega) - (I-P)\cdot U_2(\omega)\big)(x)\|\,d\mu(\omega) + \frac{\epsilon}{4} \\
&\overset{\widetilde{Z}\subset Z_{n_0}}{\le} \int_\Omega \phi_{n_0,1}(\omega)\cdot\phi_{n_0,2}(\omega)\,d\mu(\omega) \\
&\quad + \int_\Omega \phi_{n_0,1}(\omega)\cdot\|(I-P)\cdot U_2(\omega))(x)\|\,d\mu(\omega) + \frac{\epsilon}{4} \\
&\le \int_\Omega \phi_{n_0,1}(\omega)\cdot\phi_{n_0,2}(\omega)\,d\mu(\omega) + \frac{\epsilon}{2}.
\end{aligned}$$

Using (2.2) and (2.3), we get

$$\begin{aligned}
c\Big(\int_\Omega U_1(\omega)\cdot U_2(\omega)\,d\mu(\omega)\Big) &\le \sup_{x\in B(\widetilde{Z})} \Big\|\int_\Omega U_1(\omega)\cdot U_2(\omega)(x)\,d\mu(\omega)\Big\| \\
&\le \sup_{x\in B(\widetilde{Z})} \int_\Omega \|U_1(\omega)\cdot U_2(\omega)(x)\|\,d\mu(\omega) \\
&\overset{(2.3)}{\le} \int_\Omega \phi_{n_0,1}(\omega)\cdot\phi_{n_0,2}(\omega)\,d\mu(\omega) + \frac{\epsilon}{2} \\
&\overset{(2.2)}{\le} c(U_2(\omega))\,d\mu(\omega) + \epsilon.
\end{aligned}$$

Since $\epsilon > 0$ is arbitrary, the proof of a) is done. b) and c) are direct consequences of the submultiplicity of the measure of non-compactness c (cf. [LeS]) and of [KS, Th.3.7 b), c)]. The measurability of $c(U_i(\cdot))$, $i = 1, \dots n$, is given by the assumption. □

Remark 2.2. Up to now it is not known to us whether the constants in the integral equations of Proposition 2.1 can be lowered or even removed. But as we will see later, for the application to perturbed semigroups this is of no significance.

The next lemma will give conditions under which it is possible to find a suitable sequence $(Z_n) \subset \mathcal{S}$ needed for Remark 1.1 e) and Proposition 2.1, respectively.

Lemma 2.3. *Let $T > 0$ and let $U : [0,T] \longrightarrow L(X,Y)$ be strongly measurable and bounded such that either*

i) $[0,T] \ni t \longmapsto \|U(t)|_Z\|$ *is decreasing for all* $Z \in \mathcal{S}$ *or*

ii) $[0,T] \ni t \longmapsto \|U(t)|_Z\|$ *is increasing for all* $Z \in \mathcal{S}$.

Then there is a decreasing sequence $(Z_n) \subset \mathcal{S}$ so that

$$c(U(t)) = \lim_{n\to\infty} \|U(t)|_{Z_n}\| \quad \textit{for almost all} \;\; t \in [0,T].$$

Proof. For $Z \in \mathcal{S}$ we consider the function $\|U(\cdot)|_Z\| : [0,T] \ni t \longmapsto \|U(t)|_Z\|$ as an element in the order complete Banach lattice $L_1([0,T])$ (see [LT, p.4]). Then there exists $\phi := \inf_{Z\in\mathcal{S}} \|U(\cdot)|_Z\|$ in the Banach lattice $L_1([0,T])$ (see [LT, p.4, 1.a.3]). Hence we find a sequence $(Z_n) \subset \mathcal{S}$ such that $\phi = \inf_{n\in\mathbb{N}} \|U(\cdot)|_{Z_n}\|$ in $L_1([0,T])$. According to Remark 1.1 d), the sequence may be chosen to be decreasing.

Let us define $\psi(t) := \inf_{n\in\mathbb{N}} \|U(t)|_{Z_n}\|$, $t \in [0,T]$. The definition of ψ reveals that for all $Z \in \mathcal{S}$, $\psi(t) \le \|U(t)|_Z\|$ for almost all $t \in [0,T]$. For $Z \in \mathcal{S}$ we define $N_Z := \{t \in [0,T] : \psi(t) > \|U(t)|_Z\|\}$. We claim that $M := \bigcup_{Z\in\mathcal{S}} N_Z$ has measure zero.

We will give the proof of this claim in the case i); the proof of ii) is similar. For $k \in \mathbb{N}$ we define $M_k := \{t \in [0,T] : \exists\, Z \in \mathcal{S} : \|U(t)|_Z\| + \frac{1}{k} < \psi(t)\}$. Then evidently $M = \bigcup_{k\in\mathbb{N}} M_k$. We prove that $\lambda(M_k) = 0$ for all $k \in \mathbb{N}$.

Let $k \in \mathbb{N}$. We suppose that $\lambda^*(M_k) > 0$ where λ^* denotes the outer measure. We define $a := \inf M_k$ and $b := \sup M_k$. Then we can find a $t_1 \in (a,b) \cap M_k$ such that

$$\lambda^*((t_1, b) \cap M_k) > 0. \tag{2.4}$$

Such a t_1 exists according to the definition of a, b and since $\lambda^*(M_k) > 0$. Because $t_1 \in M_k$, the definition of M_k provides a $\widetilde{Z}_1 \in \mathcal{S}$ so that

$$\left\|U(t_1)|_{\widetilde{Z}_1}\right\| + \frac{1}{k} < \psi(t_1). \tag{2.5}$$

Since $\psi(t) \le \|U(t)|_{\widetilde{Z}_1}\|$ for almost all $t \in [0,T]$, we can find a $t_2 > t_1$, $t_2 \in (t_1, b) \cap M_k$, $\lambda^*((t_2, b) \cap M_k) > 0$, so that $\psi(t_2) \le \|U(t_2)|_{\widetilde{Z}_1}\|$. The general assumption that $t \longmapsto \|U(t)|_{\widetilde{Z}_1}\|$ is decreasing finally gives us

$$\psi(t_2) + \frac{1}{k} \le \left\|U(t_2)|_{\widetilde{Z}_1}\right\| + \frac{1}{k} \le \left\|U(t_1)|_{\widetilde{Z}_1}\right\| + \frac{1}{k} < \psi(t_1).$$

We now proceed in identical fashion, applying (2.4) and (2.5) to t_2. We obtain a sequence of strictly increasing numbers $(t_m) \subset M_k$ such that $\psi(t_{m+1})+\frac{1}{k} < \psi(t_m)$, which is a contradiction since $\psi(t) \geq 0$ for all $t \in [0,T]$. Hence $\lambda(M) = 0$. This means that there is a set $N \subset [0,T]$ of Lebesgue measure zero such that for all $t \in [0,T] \setminus N$ and all subspaces $Z \in \mathcal{S}$ of finite co-dimension

$$\psi(t) \leq \|U(t)|_Z\|.$$

Hence, the proof of the lemma is completed. □

Remark 2.4.

a) Since we are interested in the application of the results stated in Proposition 2.1 and Lemma 2.5, we will select a special situation where especially the assumption of Lemma 2.5 is satisfied. For this purpose we suppose that for a dense set $(x_n)_{n\in\mathbb{N}} \subset S(X)$ the map $t \longmapsto \|U(t)(x_n)\|$ is decreasing for all $n \in \mathbb{N}$. Then the assumptions of the preceding lemma evidently are fulfilled. The same is true if "decreasing" is replaced by "increasing".

b) Let $(S(t))$ be a C_0-semigroup. Then

$$\exists\, \omega \in \mathbb{R}\ \exists\, M > 0\ \forall\, t \in [0,\infty): \quad \|S(t)\| \leq Me^{\omega t}.$$

With the help of a classical result (see e.g. [Pa, p.19]), we find an equivalent norm $|||\cdot|||$ on X such that $|||S(t)||| \leq e^{\omega t}$. According to Remark 1.1 c) the essential spectral radius is not affected by changing to an equivalent norm. Hence, we will consider a C_0-semigroup $(S(t))$ in the next lemma such that $\|S(t)\| \leq e^{\omega t}$.

The preceding results are used to deduce an important inequality for operators of a C_0-semigroup.

Lemma 2.5. *Let $(S(t))_{t\geq 0}$ be a C_0-semigroup on X with $\|S(t)\| \leq e^{\omega t}(t \in [0,\infty))$ and suitable $\omega \in \mathbb{R}$. Then:*

$$\exists\, M > 0\ \ \forall\, t > 0\ \ \forall\, m \in \mathbb{N}\ \ \forall\, B \in L(X):$$

$$c\left(\int_{\Omega_m} S(s_1)BS(s_2)\cdots BS(s_m)BS(t-s_1-\cdots-s_m)\,ds_1\cdots ds_m\right)$$

$$\leq c(B)^m M \int_{\Omega_m} c(S(s_1))\cdots c(S(s_m))c(S(t-s_1-\cdots-s_m))\,ds_1\cdots ds_m$$

where $\Omega_m := \{(s_1,\ldots,s_m) \in [0,t]^m : \sum_{j=1}^m s_j \leq t\}$.
If $\ell_1 \not\subset X$, then $M = 1$, if X is separable, then $M = 2$, and we have $M = 4$ otherwise.

Proof. We define

$$\widehat{S}(s) := e^{-\omega t}S(s).$$

Then obviously $(\widehat{S}(s))_{s\geq 0}$ is a C_0-semigroup with $\sup_{s\geq 0}\|\widehat{S}(s)\|\leq 1$. For $i=1,\dots,m$ let us define operator-valued maps

$$\begin{aligned} U_i \;:\; \Omega_m &\longrightarrow L(X)\\ \omega &\longmapsto U_i(\omega):=\widehat{S}_{s_i}\circ B, \qquad \omega=(s_1,\dots,s_m)\in\Omega_m. \end{aligned}$$

We need the existence of a decreasing sequence $(Z_n^i)\subset\mathcal{S}$ with the property that for $i=1,\dots,m$, $\omega\longmapsto\|U_i(\omega)|_{Z_n^i}\|$ is measurable and $c(U_i(\omega))=\lim_{n\to\infty}\|U_i(\omega)|_{Z_n^i}\|$ a.e. on Ω_m to be able to conclude the inequality of Proposition 2.1. To this end, we apply Lemma 2.3 to $[0,t]\ni s\longmapsto\widehat{S}(s)\circ B$. The assumptions are satisfied because for all subspaces Z and all $s,h>0$,

$$\|\widehat{S}(s+h)|_Z\|\leq\|(\widehat{S}(h)\circ\widehat{S}(s))|_Z\|\leq\|\widehat{S}(h)\|\cdot\|\widehat{S}(s)|_Z\|\leq\|\widehat{S}(s)|_Z\|,$$

hence in particular $s\longmapsto\|\widehat{S}(s)\circ B|_Z\|$ is also decreasing for all $Z\in\mathcal{S}$. As a consequence there is a decreasing sequence $(Z_n)\subset\mathcal{S}$ such that for $i=1,\dots,m$:

$$c(U_i(\omega))=c(\widehat{S}(s_i)\circ B)=\lim_{n\to\infty}\|\widehat{S}(s_i)\circ B|_{Z_n}\|=\lim_{n\to\infty}\|U_i(\omega)|_{Z_n}\|\quad\text{f.a.a. }\ \omega\in\Omega_m.$$

Furthermore, $s\longmapsto\|\widehat{S}(t-s)|_Z\|$ is increasing. Hence, according to Lemma 2.3, there is a decreasing sequence $(Z_n^{m+1})\subset\mathcal{S}$ so that

$$c(\widehat{S}(t-s))=\lim_{n\to\infty}\|\widehat{S}(t-s)|_{Z_n^{m+1}}\|\quad\text{f.a.a. }\ s\in[0,t].$$

Thus we can deduce f.a.a. $\omega=(s_1,\dots,s_m)\in\Omega$:

$$c(\widehat{S}(t-(s_1+\cdots+s_m)))=\lim_{n\to\infty}\|\widehat{S}_{t-(s_1+\cdots+s_m)}|_{Z_n^{m+1}}\|.$$

We can apply Proposition 2.1 since the assumptions are verified, and we find an $M>0$ such that

$$\begin{aligned} &c\Big(\int_{\Omega_m}\widehat{S}(s_1)B\widehat{S}(s_2)\cdots B\widehat{S}(s_m)B\cdot\widehat{S}(t-s_1-\cdots-s_m)\,ds_1\cdots ds_m\Big)\\ &\leq M\int_{\Omega_m}c(\widehat{S}(s_1)\circ B)\cdots c(\widehat{S}(s_m)\circ B)\cdot c(\widehat{S}(t-s_1-\cdots-s_m))\,ds_1\cdots ds_m\\ &\leq c(B)^mM\int_{\Omega_m}c(\widehat{S}(s_1))\cdots c(\widehat{S}(s_m))\cdot c(\widehat{S}(t-s_1-\cdots-s_m))\,ds_1\cdots ds_m. \end{aligned}$$

The last inequality is due to the fact that the measure of non-compactness is submultiplicative [LeS, p.9,Lemma 3.2.]. But this means for the original semigroup $(S(t))_{t\geq 0}$ that

$$c\Big(\int_{\Omega_m}S(s_1)BS(s_2)\cdots BS(s_m)BS(t-s_1-\cdots-s_m)\,ds_1\cdots ds_m\Big)$$

$$= c\left(\int_{\Omega_m} e^{\omega s_1}\widehat{S}(s_1)Be^{\omega s_2}\widehat{S}(s_2)\cdots Be^{\omega s_m}\widehat{S}(s_m)B\cdot e^{\omega(t-s_1-\cdots-s_m)}\right.$$
$$\left.\cdot\widehat{S}(t-s_1-\cdots-s_m)\,ds_1\cdots ds_m\right)$$
$$= e^{\omega t}c\left(\int_{\Omega_m}\widehat{S}(s_1)B\widehat{S}(s_2)\cdots B\widehat{S}(s_m)B\widehat{S}(t-s_1-\cdots-s_m)\,ds_1\cdots ds_m\right)$$
$$\leq e^{\omega t}c(B)^m M\int_{\Omega_m} c(\widehat{S}(s_1))\cdots c(\widehat{S}(s_m))\cdot c(\widehat{S}(t-s_1-\cdots-s_m))\,ds_1\cdots ds_m$$
$$= c(B)^m M\int_{\Omega_m} e^{\omega s_1}c(\widehat{S}(s_1))\cdots e^{\omega s_m}c(\widehat{S}(s_m))$$
$$\cdot e^{\omega(t-s_1-\cdots-s_m)}c(\widehat{S}(t-s_1-\cdots-s_m))\,ds_1\cdots ds_m$$
$$= c(B)^m M\int_{\Omega_m} c(S(s_1))\cdots c(S(s_m))c(S(t-s_1-\cdots-s_m))\,ds_1\cdots ds_m.$$

The value of M is given according to Proposition 2.1. □

Remark 2.6. If $(S(t))_{t\geq 0}$ is a C_0-semigroup, there is an $\widetilde{M}>0$ and an $\omega\in\mathbb{R}$ so that for all $t\geq 0$, $\|S(t)\|\leq\widetilde{M}e^{\omega t}$ (see [Pa, p.4, Th.2.2]). According to [Pa, p.19] there is an equivalent norm $|||\cdot|||$ on X such that

$$\forall\, x\in X:\quad \|x\|\leq|||x|||\leq\widetilde{M}\|x\|\quad\text{and}\quad |||S(t)|||\leq e^{\omega t}.$$

Hence using Lemma 2.5 and the inequality of Remark 1.1 c) we get for an arbitrary C_0-semigroup

$$\forall\, t>0\;\;\forall\, m\in\mathbb{N}\;\;\forall\, B\in L(X):$$
$$c\left(\int_{\Omega_m} S(s_1)BS(s_2)\cdots BS(s_m)BS(t-s_1-\cdots-s_m)\,ds_1\cdots ds_m\right)$$
$$\leq M\widetilde{M}^{2(m+1)}c(B)^m\overline{\int_{\Omega_m} c(S(s_1))\cdots c(S(s_m))\,c(S(t-s_1-\cdots-s_m))\,ds_1\cdots ds_m}$$

where $\Omega_m := \{(s_1,\ldots,s_m)\in[0,t]^m : \sum_{j=1}^m s_j\leq t\}$.

We are done with the required preliminary results and can derive the main theorem which is concerned with the essential spectral radius of a perturbed semigroup. The proof follows from the Dyson-Phillips-expansion (see [Kat, p.497, IX, 2.11]).

Theorem 2.7. *We suppose that X has the property that either $W(X,X)\subset SS(X,X)$ or $CW(X,X)\subset SS(X,X)$. Let $B\in L(X)$. If for some integer $m\in\mathbb{N}$ and all m-tupels $(t_1,t_2,\ldots,t_m)$ of positive numbers the operator $BS(t_1)BS(t_2)\cdots BS(t_m)B$ is either weakly compact or conditionally weakly compact, then*

$$\forall\, t\geq 0:\quad r_e(S(t))=r_e(T(t)),$$

where $(T(t))$ is the perturbed semigroup.

Proof. Let us recall the representation of the perturbed semigroup $(T(t))_{t\geq 0}$ according to the Dyson-Phillips expansion (see e.g. [Kat, p.497]). By iterating

$$T(t) = S(t) + \int_0^t S(s)BT(t-s)\,ds$$

we get

$$T(t) = \sum_{j=0}^{m} W_j(t) \; + \; R_m(t)$$

where

$$\begin{aligned} W_0(t) &= S(t), \\ W_j(t) &= \int_{s_1+\cdots+s_j\leq t} S(s_1)BS(s_2)\cdots BS(s_j)BS(t-s_1-\cdots-s_j)\,ds_1\cdots ds_j, \\ &\qquad\qquad j=1,\ldots,m. \\ R_m(t) &= \int_{s_1+\cdots+s_{m+1}\leq t} S(s_1)BS(s_2)\cdots BS(s_{m+1})B \\ &\qquad\qquad \cdot\, T(t-s-s_1-\cdots-s_{m+1})\,ds_1\cdots ds_{m+1} \end{aligned}$$

for $t \geq 0$. By [S2, Th.2.2], $R_m(t)$ is weakly compact (conditionally weakly compact, respectively), hence strictly singular. For the proof of $r_e(T(t)) = r_e(S(t))$, we basically follow the lines of the proof of [We, p.11, 3.1.Th.].
Let ω_e and $\widetilde{\omega}_e$ be the essential type of $(S(t))$ and $(T(t))$. First we show that $\widetilde{\omega}_e \leq \omega_e$. Let ω', ω'' so that $\omega_e < \omega' < \omega''$ and let M be according to Remark 1.1 b) such that

$$c(S(t)) \leq \widehat{M}e^{\omega' t}.$$

Since $R_m(t)$ is strictly singular, $R_m(t)$ does not change the essential spectral radius (see [Sche, p.1067, Th.2.17, p.1069, Cor.3.4]). Thus we obtain

$$r_e(T(t)) = r_e\left(\sum_{j=0}^{m} W_j(t)\right) \leq c\left(\sum_{j=0}^{m} W_j(t)\right).$$

By Remark 1.1 b) and Remark 2.6 we have

$$\begin{aligned} r_e(T(t)) &\leq \widehat{M}e^{\omega' t} + \sum_{j=1}^{m} \widehat{M}_j \int_{s_1+\cdots+s_j\leq t} c(S(s_1))\cdots c(S(s_j)) \\ &\qquad\qquad \cdot\, c(S(t-s_1-\cdots-s_j))\,ds_1\cdots ds_j \\ &\leq \widehat{M}e^{\omega' t} + \sum_{j=1}^{m} \widehat{M}^j M_j t^j e^{\omega' t} \\ &\leq C\cdot e^{\omega'' t}, \end{aligned}$$

with $M_j := M\widetilde{M}^{2(j+1)}c(B)^j$, M according to Lemma 2.5 and $\widetilde{M}$ as in 2.6, where C is sufficiently large. Because we have chosen $\omega'' > \omega_e$ arbitrarily, Remark 1.1 b) gives

$$\widetilde{\omega}_e \leq \omega_e.$$

Showing the reverse inequality $\omega_e \leq \widetilde{\omega}_e$, we use the expansion

$$S(t) = \sum_{j=0}^{m} \widetilde{W}_j(t) + \widetilde{R}_m(t)$$

which is obtained by interchanging the roles of $(S(t))$ and $(T(t))$ and replacing B by $(-B)$. Hence

$$\widetilde{R}_m(t) = \int_{s_1+\cdots+s_{m+1}} (-1)^{m+1} T(s_1)BT(s_2)\cdots BT(s_{m+1})B \cdot S(t-s-s_1-\cdots-s_{m+1})\, ds_1 \cdots ds_{m+1}.$$

We can replace the $T(s_k)$ in the representation of $\widetilde{R}_m(t)$ by $\sum_{j=0}^{m} W_j(s_k)+R_m(s_k)$. Thus $\widetilde{R}_m(t)$ can be written as a set of multiple integrals of operator products, each containing a factor of the form

$$BS(s_1)B\cdots BS(s_m)B.$$

These factors are weakly compact (conditionally weakly compact, respectively) according to the assumption. Since those operators form an operator ideal, $\widetilde{R}_m(t)$ is weakly compact (conditionally weakly compact, respectively). Now proceed as in the proof for the reverse inequality, checking that Proposition 2.1 and Lemma 2.5, respectively, can be applied to $(T(t))$ since $(T(t))$ is again a C_0-semigroup. This procedure leads to $\omega_e = \widetilde{\omega}_e$ and hence, Remark 1.1 b) tells us that $r_e(T(t)) = r_e(S(t))$. □

Corollary 2.8. *Assume that X has the property that either $SS(X,X) = W(X,X)$, $CW(X,X) = SS(X,X)$ or $SS(X,X) = K(X,X)$. Let $B \in L(X)$. If for $m \in \mathbb{N}$ and all tupels $(t_1, t_2, \ldots, t_m)$ of positive numbers the operator $BS(t_1)BS(t_2)\cdots BS(t_m)B$ is strictly singular , then*

$$\forall\, t \geq 0: \quad r_e(S(t)) = r_e(T(t)).$$

In the following remark we will point out, that Theorem 2.7 and Corollary 2.8 are substantial extensions of the result of L. Weis.

Remark 2.9.

a) If X has the Dunford-Pettis Property, i.e., each $T \in W(X,Y)$ maps weakly convergent sequences into norm convergent sequences, then X fulfills the condition

in Theorem 2.7. This is the case if either $X = L_1(\nu)$ for a positive measure space (Γ, Θ, ν) or $X = C(K)$ for a compact Hausdorff space K (see [DU, p.176, 5.Notes]). These spaces even satisfy the assumption of Corollary 2.8. For these cases, Theorem 2.7 and Corollary 2.8 extend the result of L. Weis [We, p.11, 3.1.Th.].
If every infinite dimensional subspace of X contains a subspace isomorphic to ℓ_1, then $CW(X,X) = SS(X,X)$.
b) An analysis of the proof of Theorem 2.7 also gives the following result:
Let us suppose X has the property that for each non-strictly singular operator $T \in L(X)$, there is a complemented subspace $Z \subset X$ so that $T|_Z$ is an isomorphism and Z is isomorphic to ℓ_p, $1 < p < \infty$. The L_p-spaces, $1 < p < \infty$, have this property (see e.g. [We, p.10, Prop.2.4]). Then, if for some $m \in \mathbb{N}$ and all choices of $t_1, t_2, \dots, t_m > 0$ the operator $BS(t_1)B(t_2) \cdots BS(t_m)B$ is strictly singular, we obtain

$$\forall\, t \geq 0: \quad r_e(S(t)) = r_e(T(t)).$$

The proof uses the fact that on ℓ_p the strictly singular operators coincide with the compact operators (see [GMF])]. The most important part in the proof of Theorem 2.7 is the fact that $R_m(t)$ and $\widetilde{R}_m(t)$, respectively, are strictly singular. Since on an infinite dimensional Banach space every compact operator is strictly singular, we can deduce with the help of [KS, Cor. 3.8] that this is fulfilled on ℓ_p, here. Thus Theorem 2.7 is also true for L_p-spaces, $1 < p < \infty$.
c) We remark that recently J. Voigt [Vo3] has obtained a result similar to Theorem 2.7, but with a different approach.
d) The gap in the original proof of [We, p.11, 3.1.Th.] is filled by Proposition 2.1 and [KS, Th. 3.7], respectively. It has turned out that the underlying integral inequality for the measure of non-compactness for operator-valued functions is more sophisticated.
e) The inequality which we stated in Remark 1.1 d) is not true if we change codim $Z < \infty$ to dim $Z = \infty$. For this purpose, we define for an operator $T : X \longrightarrow Y$

$$u(T) := \inf\{\|T|_Z\| : Z \text{ is a closed subspace with } \dim Z = \infty\}.$$

Let $(\Omega, \Sigma, \mu) = ([0,1], \mathcal{L}, \lambda)$ and $(r_n)_{n \in \mathbb{N}}$ be the sequence of Rademacher functions on $[0,1]$ with values in $\{0,1\}$. We define for $t \in [0,1]$:

$$\begin{aligned} U(t) \quad : \quad & \ell_1 \longrightarrow \ell_1, \\ & e_n \longmapsto 2e_n r_n(t). \end{aligned}$$

Then for $x = \sum_{n \in \mathbb{N}} a_n e_n \in S(\ell_1)$,

$$\left\| \int_{[0,1]} U(t)(x)\, d\lambda(t) \right\| = 2 \left\| \int_{[0,1]} \sum_{n \in \mathbb{N}} a_n e_n r_n(t)\, d\lambda(t) \right\|$$

$$
\begin{aligned}
&= 2\left\|\sum_{n\in\mathbb{N}}\int_{[0,1]} a_n e_n r_n(t)\,d\lambda(t)\right\| \\
&= 2\sum_{n\in\mathbb{N}}\left\|\int_{[0,1]} a_n e_n r_n(t)\,d\lambda(t)\right\| \\
&= 2\sum_{n\in\mathbb{N}}\frac{1}{2}|a_n| = 1
\end{aligned}
$$

Thus

$$
\inf_{Z\subset X,\dim Z=\infty}\left\|\left(\int_{[0,1]} U(t)d\lambda(t)\right)\Big|_Z\right\| = 1. \tag{2.6}
$$

Now let $t\in[0,1]$ be irrational, $t=\sum_{n\in\mathbb{N}}\epsilon_n 2^{-n}$. Then $\left(r_n(t)\right)_{n\in\mathbb{N}} = (\epsilon_n)_{n\in\mathbb{N}}$. We define $Z(t) := \overline{\operatorname{span}}\{e_n : \epsilon_n = 0\}$. Then dim $Z(t)=\infty$ since t is irrational. By construction of $Z(t)$, $U(t)|_{Z(t)} = 0$. Thus

$$
\int_{[0,1]} \inf_{Z\in X,\dim Z=\infty} \|U(t)|_Z\|\,d\lambda(t) = 0.
$$

Comparing this with (1) we see that the same inequality as in Remark 1.1 d) does not hold for u.

f) Application for this perturbation theorem can be found in the theory of neutron transport equations (see e.g. [Vi, Vo1, We]).

We close by deducing a useful result concerning time evolution of the perturbation. The succeeding result is implicitly contained in Theorem 2.7 and extends a result given by J. Voigt [Vo1].

Theorem 2.10. *We suppose that for all choices of $t_1,t_2>0$ the operator $S(t_1)BS(t_2)$ is weakly compact (conditionally weakly compact, respectively). Then for all $t\geq 0$ the operators*

$$
\begin{aligned}
T(t)-S(t) &: X \longrightarrow X,\\
\overline{T}-\overline{S} &: X \longrightarrow L_1([0,t],X),\\
&\quad x \longmapsto (T(\cdot)-S(\cdot))x,
\end{aligned}
$$

are weakly compact (conditionally weakly compact, respectively).

Proof. From

$$
T(t) = S(t) + \int_0^t S(t)BT(t-s)\,ds
$$

we get

$$
\forall\, t\geq 0:\quad T(t) = S(t) + \int_0^t S(s)BT(t-s)\,ds
$$

$$= S(t) + \int_0^t S(s)BS(t-s)\,ds + \int_0^t \left(\int_0^{t-s_1} S(s_1)BS(s_2)BT(t-s_1-s_2)\,ds_2 \right) ds_1.$$

We note that the above representation implies that for all $t \geq 0$ the operator $T(t) - S(t)$ is weakly compact (conditionally weakly compact, respectively) by [S2]. Again by [S2] (with a slight modification of the proofs), we deduce for all $t \geq 0$ that $\overline{T} - \overline{S}$ is weakly compact (conditionally weakly compact, respectively). □

References

[DU] Diestel, J.,Uhl, J.: Vector measures; AMS Math. Surveys 15 (1977).

[GMF] Gohberg, I.C., Markus, A., Feldman, I.: Normally solvable operators and ideals associated with them; Izv. Akad. Nauk. Moldav. USSR 10 (1960), 51–69 (in Russian), English translation: Amer. Math. Soc. Transl. 61 (1967), 63–84.

[Kat] Kato, T.: Perturbation of linear operators, Springer Verlag, Berlin New York 1974.

[KS] Kunze, M., Schlüchtermann, G.: Measures of non-compactness and structures of Banach spaces; Math. Nachr. (to appear).

[LeS] Lebow, A., Schechter. M.: Semigroups of operators and measure of non-compactness; J. Funct. Anal. 7 (1971), 1–26.

[LT] Lindenstrauss, J., Tzafriri, L.: Classical Banach spaces II; Springer Verlag, Berlin New York 1977.

[Pa] Pazy, A.: Semigroups of linear operator and applications to partial differential equations; Springer Verlag, Berlin 1983.

[Pe1] Pelczynski, A.: Strictly singular and cosingular operators; Bull. Polish Acad. Sci. Math. 13 (1965), 31–41.

[Pe2] Pelczynski, A.: On Banach spaces containing L_1; Studia Math. 30 (1968), 231–246.

[Pi] Pietsch, A.: Operator ideals; North Holland, Amsterdam 1980.

[Sche] Schechter, M.: Quantities related to strictly singular operators, Indiana Univ. Math. J. 21:11 (1972), 1061–1071.

[S1] Schlüchtermann, G.: Hereditary properties for duals of Bochner L_p–function spaces; Rocky Mount. J. Math. (to appear).

[S2] Schlüchtermann, G.: On weakly compact operators; Math. Ann. 292 (1992), 263–266.

[Vi] Vidav, I.: Spectra of perturbed semigroups with applications to linear transport theory; J. Math. Anal. Appl. 30 (1970), 264–279.

[Vo1] VOIGT, J.: A perturbation theorem for the essential spectral radius of strongly continuous semigroups; Monatsh. Math. 90 (1980), 153–161.

[Vo2] VOIGT, J.: The convex compactness property for the strong operator topology; Note Mat. XII (1992), 259–269.

[Vo3] VOIGT, J.: Stability of the essential type of strongly continuous semigroups; preprint, 1992.

[We] WEIS, L.: A generalization of the Vidav-Jörgens perturbation theorem for semigroups and its application to transport theory; J. Math. Anal. Appl. 129 (1988), 6–23.

Mathematisches Institut
Ludwig-Maximilians-Universität
Theresienstrasse 39
80333 München
Germany
schluech@rz.mathematik.uni-muenchen.de

1991 Mathematics Subject Classification: Primary 47A55; Secondary 46G10

Submitted: December 20, 1995

Operator Theory:
Advances and Applications, Vol. 103

On the approximation of operators and the convergence of the spectra of the approximants

M.P.H. Wolff

Let the bounded linear operator T be approximated discretely by a sequence of operators (S_n). Then the distance $\text{dist}(\sigma_{ap}(T), \sigma_\varepsilon(S_n))$ between the approximative point spectrum $\sigma_{ap}(T)$ of T and the ε–pseudospectra $\sigma_\varepsilon(S_n))$ of (S_n) converges to 0. A somewhat related theorem on the pseudospectrum of Toeplitz operators was proved by L. Reichel and L.N. Trefethen [ReTr].

1. Introduction

The spectrum $\sigma(T)$ of a bounded linear operator T on the Banach space E is an upper semicontinuous function of T (see [K], 208–210). More precisely, for bounded sets A, B of the complex plane $\mathbb{C}$ we define $\text{dist}(A, B) := \sup_{a\in A}\{\inf\{|b - a| : b \in B\}\}$. If $\lim_n \|T - S_n\| = 0$ holds for a sequence $(S_n)_{n\in\mathbb{N}}$ in the Banach algebra $\mathcal{L}(E)$ of all bounded linear operators on E, then $\lim_n \text{dist}(\sigma(S_n), \sigma(T)) = 0$, but in general

$$\lim_{n\to\infty} \text{dist}(\sigma(T), \sigma(S_n)) = 0. \tag{1.1}$$

does not hold. It is the main aim of our paper to show that a modification of (1.1) does even hold in the case of discrete convergence.

Throughout the whole paper by an operator we always mean a bounded linear operator on a Banach space. By $\rho(T)$ we denote the resolvent set $\rho(T) = \{\lambda \in \mathbb{C} : (\lambda - T) \text{ is bijective}\}$. For $\varepsilon > 0$ we define $\rho_\varepsilon(T) := \{\lambda \in \rho(T) : \|(\lambda - T)^{-1}\| < 1/\varepsilon\}$. Then the ε–pseudospectrum, introduced by Trefethen (cf. [Tr]), is the complement $\sigma_\varepsilon(T) := \mathbb{C} \setminus \rho_\varepsilon(T)$. Let us also recall the notion of the approximative point spectrum of the operator T which is the set

$$\sigma_{ap}(T) := \{\lambda \in \sigma(T) : \inf_{\|x\|=1} \{\|(\lambda - T)x\|\} = 0\}.$$

Finally we have to explain the notion of discrete convergence (see [Vai], §§1–2, [Fat], 306–329, or [R], 124–132, cf. also [PS], 114–118, for the related notion of "P-convergent sequences of operators"). Let E and F_n, $n \in \mathbb{N}$, be Banach spaces and let P_n be operators from E into F_n. The sequence $(F_n, P_n)_{n\in\mathbb{N}}$ *approximates* E if $\lim_n \|P_n u\| = \|u\|$ holds for all $u \in E$. From the uniform boundedness theorem it follows then that (P_n) is uniformly bounded. A sequence $(u_n)_n$ with $u_n \in F_n$ for all n *converges discretely* to $u \in E$ (in symbols $u_n \xrightarrow{d} u$) if $\lim_n \|u_n - P_n u\|_n = 0$. A sequence $(S_n)_n$ of operators S_n on F_n *converges discretely* to T on E (in symbols $S_n \xrightarrow{d} T$) if $\lim_n \|P_nTu - S_nP_nu\|_n = 0$ for all $u \in E$. Note that then $(S_n)_n$ is bounded by [Vai], (8), p.25. Concerning examples to this concept we refer to the references cited above. Obviously, strong convergence is a special case of this general notion.

2. The main result

Before we give our main result let us show that it is in a certain sense optimal by discussing two examples.

Example 2.1.
(i) Let $E = \ell^2(\mathbb{N})$ be the usual Hilbert space and consider the shift T given by $(Tf)(k) = f(k+1)$. Then $\sigma(T) = \sigma_{ap}(T) = \{\lambda \in \mathbb{C} : |\lambda| \leq 1\}$. Moreover, every λ with modulus strictly less than 1 is an eigenvalue. Now set $F_n = E$, $P_n = I$ (the identity operator) and

$$(S_nf)(k) = \begin{cases} f(k+1), & k \leq n-1, \\ 0, & k \geq n. \end{cases}$$

Then $(S_n)_n$ converges strongly to T and $\sigma(S_n) = \{0\}$ for every n, but for all $\varepsilon > 0$ we have $\lim_n \operatorname{dist}(\sigma(T), \sigma_\varepsilon(S_n)) = 0$ (cf. (1.1)).
(ii) Let T be as before and take $S = T^*$. Then $\sigma_{ap}(S) = \{\lambda \in \mathbb{C} : |\lambda| = 1\}$. Set $F_n = \mathbb{C}^n$ and take $S_n(x_1, \ldots, x_n) = (x_n, x_1, \ldots, x_{n-1})$. Then S_n is a unitary operator, hence $\sigma_\varepsilon(S_n) \subset \{\lambda : 1 - \varepsilon \leq |\lambda| \leq 1 + \varepsilon\}$. Obviously, $S_n \xrightarrow{d} S$ but $\overline{\lim}_n \operatorname{dist}(\sigma(S), \sigma_\varepsilon(S_n)) = 1 - \varepsilon$. However, $\lim_n \operatorname{dist}(\sigma_{ap}(S), \sigma_\varepsilon(S_n)) = 0$.

After having illuminated the general situation a little we now formulate our main result:

Theorem 2.2. *Let $(F_n, P_n)_{n\in\mathbb{N}}$ approximate the Banach space E. Let T be a bounded linear operator on E, and let $(S_n)_{n\in\mathbb{N}}$ be a sequence of operators S_n on F_n converging discretely to T. Then*

$$\lim_{n\to\infty} \operatorname{dist}(\sigma_{ap}(T), \sigma_\varepsilon(S_n)) = 0 \tag{2.1}$$

holds for every $\varepsilon > 0$.

Let $f(z) = \sum_{n=0}^\infty a_n z^n$ be a holomorphic function with absolut convergent series of coefficients. Moreover, let A be the Toeplitz operator with symbol f, and let A_n be its n-th section. Then Theorem 3.3 in [ReTr] says $\lim \operatorname{dist}(\sigma_\varepsilon(A), \sigma_\varepsilon(A_n)) = 0$. This theorem is obviously related to our one. The investigation of the exact relation between this result and our theorem will be postponed to another publication.

Corollary 2.3. *If all F_n are Hilbert spaces and if moreover all S_n are unitary operators, then* $\sigma_{ap}(T) \subset \{\lambda : |\lambda| = 1\}$.

Corollary 2.4. *If E is a Hilbert space and if in addition T is a normal operator, then* $\lim_{n\to\infty}(\sigma(T), \sigma_\varepsilon(S_n)) = 0$.

3. Auxiliary results and proofs

We use the notion and techniques of ultraproducts (cf. e.g. [Schae] or see [RaeWo]). Let $\mathcal{U} \subset \mathcal{P}(\mathbb{N})$ be a free ultrafilter and denote by $m_\mathcal{U}$ the finitely additive $\{0,1\}$-valued measure on $\mathbb{N}$ given by $m_\mathcal{U}(A) = 1$ iff $A \in \mathcal{U}$.

Let (F_n) be a sequence of Banach spaces over $\mathbb{C}$. Denote by F_∞ the Banach space of all bounded sequences $\widetilde{x} = (x_n)_n$ with $x_n \in F_n$ for all n, equipped with the norm $\|\widetilde{x}\| = \sup_n(\|x_n\|_n)$. Then $F_0 := \{\widetilde{x} : \lim_\mathcal{U} \|x_n\|_n = 0\}$ is a closed subspace of F_∞ and the quotient space is called the ultraproduct $\widehat{F}$ of (F_n) with respect to $\mathcal{U}$. The norm on $\widehat{F}$ is given by $\|\widehat{y}\| = \|\widetilde{y} + F_0\| = \lim_\mathcal{U} \|y_n\|_n$. If all F_n are equal to some single Banach space E, then the ultraproduct $\widehat{E}$ is called the ultrapower of E with respect to $\mathcal{U}$. In this case E is isometrically embedded into $\widehat{E}$ by $x \mapsto (x, x, x, \ldots) + F_0$.

If (E_n), (F_n) are two sequences of Banach spaces, and if (S_n) is a bounded sequence of operators, $S_n : E_n \to F_n$, then by $\widetilde{S}\widetilde{x} = \widetilde{(S_n x_n)}$ there is defined a bounded linear operator from $\widetilde{E} \to \widetilde{F}$ which maps E_0 into F_0. Thus it uniquely induces a bounded linear operator $\widehat{S} : \widehat{E} \to \widehat{F}$, the ultrapower of (S_n) with respect to $\mathcal{U}$. If all E_n and F_n are equal to some single Banach space E, and if moreover all S_n are equal to some operator T, then $\widehat{T}$ is called the ultrapower of T. After identification of E with its isometric canonical subspace in $\widehat{E}$, $\widehat{T}$ becomes an extension of T.

We need the following result 2.2 from [RaeWo].

Lemma 3.1. *Let $(S_n)_{n\in\mathbb{N}} =: \mathcal{S}$ be a bounded sequence of operators S_n on the Banach space F_n, and for every $\varepsilon > 0$ set*

$$^0\sigma_\varepsilon(\mathcal{S}) = \{\lim_\mathcal{U} \lambda_n : \lambda_n \in \sigma_\varepsilon(S_n)\ m_\mathcal{U}\text{-a.e.}\}.$$

Then the following assertions hold:

(1) $\sigma_{\varepsilon'}(\widehat{S}) \subset {}^0\sigma_\varepsilon(\mathcal{S}) \subset \sigma_\varepsilon(\widehat{S})$ *for all* $0 < \varepsilon' < \varepsilon$.

(2) $\sigma(\widehat{S}) = \bigcap_{\varepsilon>0} {}^0\sigma_\varepsilon(\mathcal{S})$.

In order to make the proof of Theorem 2.2 more transparent we isolate the following lemma:

Lemma 3.2. *Let E, (F_n, P_n), T, and (S_n) be as in Theorem 2.2. Moreover, let $\mathcal{U}$ be a free ultrafilter on $\mathbb{N}$ and denote by $\widehat{E}$, etc. the ultrapowers, ultraproducts,* *respectively, of the given objects. Then $\widehat{P}$ maps E isometrically onto a closed subspace of $\widehat{F}$. Moreover, for each $x \in E$ one has $\widehat{S}\widehat{P}x = \widehat{P}Tx$.*

Proof. By hypothesis we have $\|x\| = \lim_n \|P_n x\|_n$, which implies the first assertion. The second one follows similarly. □

It should be known that $\sigma_{ap}(T)$ is closed. We include a proof for the sake of convenience.

Lemma 3.3. *$\sigma_{ap}(T)$ is always closed.*

Proof. If $\inf\{\|(\lambda_0 - T)x\| : \|x\| = 1\} =: \delta > 0$, then for $|\lambda - \lambda_0| < \delta/2$ and x of norm 1, it follows $\delta/2 \leq \|(\lambda_0 - T)x\| - |\lambda_0 - \lambda|\|x\| \leq \|(\lambda - T)x\|$. So the complement of $\sigma_{ap}(T)$ is open. □

Proof (of Theorem 2.2). Suppose that the assertion does not hold. Then there exists an $\varepsilon > 0$ and an ultrafilter $\mathcal{U}$ on $\mathbb{N}$ such that $\lim_{\mathcal{U}} \operatorname{dist}(\sigma_{ap}(T), \sigma_\varepsilon(S_n)) = 2\delta > 0$. By definition of the function dist, there exists $(\lambda_n) \subset \sigma_{ap}(T)$ and for each n an element $\mu_n \in \sigma_\varepsilon(S_n)$ such that $|\lambda_n - \mu_n| = \inf\{|\lambda_n - \mu| : \mu \in \sigma_\varepsilon(S_n)\} > \delta$ $m_{\mathcal{U}}$–a.e.. But then $\lambda := \lim_{\mathcal{U}} \lambda_n \in \sigma_{ap}(T)$, $\mu := \lim_{\mathcal{U}} \mu_n \in {}^0\sigma_\varepsilon(\mathcal{S})$, and

$$\delta \leq |\lambda - \mu| = \operatorname{dist}(\{\lambda\}, {}^0\sigma_\varepsilon(\mathcal{S})) \tag{3.1}$$

holds. Since $\lambda \in \sigma_{ap}(T)$ there exists a sequence (x_n) of normalized vectors x_n in E such that $\lim_{\mathcal{U}} \|(\lambda - T)x_n\| = 0$. Lemma 3.2 implies $\|\lim_{\mathcal{U}}(\lambda - \widehat{S})\widehat{P}x_n\| = \lim_{\mathcal{U}} \|\widehat{P}(\lambda - T)x_n\| = 0$, and since $\|\widehat{P}x_n\| = 1$ by Lemma 3.2, we obtain $\lambda \in \sigma(\widehat{S}) \subset {}^0\sigma_\varepsilon(\mathcal{S})$, a contradiction to equation (3.1). □

Proof (of Corollary 2.3). Using the spectral representation of the unitary operator V, say, we obtain $\sigma_\varepsilon(V) \subset D_\varepsilon := \{\lambda \in \mathbb{C} : 1 - \varepsilon \leq |\lambda| \leq 1 + \varepsilon\}$. Hence $\sigma_{ap}(T) \subset D_\varepsilon$ for all $\varepsilon > 0$ by Theorem 2.2. □

Proof (of Corollary 2.4). If T is normal, then $\sigma_{ap}(T) = \sigma(T)$. □

References

[Fat] FATTORINI, H.O.: The Cauchy Problem; Addison-Wesley, Reading, MA 1983.

[K] KATO, T.: Perturbation Theory of Linear Operators; 2nd ed., Springer Verlag, Berlin Heidelberg New York 1980.

[PS] PRÖSSDORF, S., SILBERMANN, B.: Numerical Analysis for Integral and Related Operator Equations; Akademie Verlag, Berlin 1991.

[RaeWo] RAEBIGER, F., WOLFF, M.P.H.: On the approximation of positive operators and the behaviour of the spectra of the approximants; Integral Equations Operator Theory (to appear).

[R] REINHARDT, H.J.: Analysis of Approximation Methods for Differential and Integral Equations; Springer Verlag, Berlin Heidelberg New York 1985.

[ReTr] REICHEL, L., TREFETHEN, L.N.: Eigenvalues and pseudo-eigenvalues of Toeplitz matrices; Linear Algebra Appl. 162/164 (1992), 153–185.

[Schae] SCHAEFER, H.H.: Banach Lattices and Positive Operators; Springer Verlag, Berlin Heidelberg New York 1974.

[Tr] TREFETHEN, L.N.: Pseudospectra of matrices; in: Griffiths, D.F. (ed.): Numerical Analysis, Proceedings of the 14th Dundee Conference 1991; Pitman, London 1992, 234 –264.

[Vai] VAINIKKO, G.: Funktionalanalysis der Diskretisierungsmethoden; B.G. Teubner Verlag, Leipzig 1976.

Mathematisches Institut
Universität Tübingen
Auf der Morgenstelle 10
72076 Tübingen
manfred.wolff@uni-tuebingen.de

1991 Mathematics Subject Classification: Primary 47A10; Secondary 47A58, 41A35

Submitted: May 30, 1996

Titles previously published in the series

OPERATOR THEORY: ADVANCES AND APPLICATIONS

BIRKHÄUSER VERLAG

Edited by

I. Gohberg,

School of Mathematical Sciences, Tel-Aviv University, Ramat Aviv, Israel

This series is devoted to the publication of current research in operator theory, with particular emphasis on applications to classical analysis and the theory of integral equations, as well as to numerical analysis, mathematical physics and mathematical methods in electrical engineering.

75. **C.B. Huijsmans, M.A. Kaashoek, B. de Pagter**: Operator Theory in Function Spaces and Banach Lattices. The A.C. Zaanen Anniversary Volume, 1994 (ISBN 3-7643-5146-2)
76. **A.M. Krasnosellskii**: Asymptotics of Nonlinearities and Operator Equations, 1995, (ISBN 3-7643-5175-6)
77. **J. Lindenstrauss, V.D. Milman** (Eds): Geometric Aspects of Functional Analysis Israel Seminar GAFA 1992–94, 1995, (ISBN 3-7643-5207-8)
78. **M. Demuth, B.-W. Schulze** (Eds): Partial Differential Operators and Mathematical Physics, 1995, (ISBN 3-7643-5208-6)
79. **I. Gohberg, M.A. Kaashoek, F. van Schagen**: Partially Specified Matrices and Operators: Classification, Completion, Applications, 1995, (ISBN 3-7643-5259-0)
80. **I. Gohberg, H. Langer** (Eds): Operator Theory and Boundary Eigenvalue Problems. International Workshop in Vienna, July 27–30, 1993, 1995, (ISBN 3-7643-5275-2)
81. **H. Upmeier**: Toeplitz Operators and Index Theory in Several Complex Variables, 1996, (ISBN 3-7643-5282-5)
82. **T. Constantinescu**: Schur Parameters, Factorization and Dilation Problems, 1996, (ISBN 3-7643-5285-X)
83. **A.B. Antonevich**: Linear Functional Equations. Operator Approach, 1995, (ISBN 3-7643-2931-9)
84. **L.A. Sakhnovich**: Integral Equations with Difference Kernels on Finite Intervals, 1996, (ISBN 3-7643-5267-1)

85/ **Y.M. Berezansky, G.F. Us, Z.G. Sheftel**: Functional Analysis, Vol. I + Vol. II, 1996,
86. Vol. I (ISBN 3-7643-5344-9), Vol. II (3-7643-5345-7)

87. **I. Gohberg, P. Lancaster, P.N. Shivakumar** (Eds): Recent Developments in Operator Theory and Its Applications. International Conference in Winnipeg, October 2–6, 1994, 1996, (ISBN 3-7643-5414-5)
88. **J. van Neerven** (Ed.): The Asymptotic Behaviour of Semigroups of Linear Operators, 1996, (ISBN 3-7643-5455-0)
89. **Y. Egorov, V. Kondratiev**: On Spectral Theory of Elliptic Operators, 1996, (ISBN 3-7643-5390-2)
90. **A. Böttcher, I. Gohberg** (Eds): Singular Integral Operators and Related Topics. Joint German-Israeli Workshop, Tel Aviv, March 1–10, 1995, 1996, (ISBN 3-7643-5466-6)
91. **A.L. Skubachevskii**: Elliptic Functional Differential Equations and Applications, 1997, (ISBN 3-7643-5404-6)
92. **A.Ya. Shklyar**: Complete Second Order Linear Differential Equations in Hilbert Spaces, 1997, (ISBN 3-7643-5377-5)
93. **Y. Egorov, B.-W. Schulze**: Pseudo-Differential Operators, Singularities, Applications, 1997, (ISBN 3-7643-5484-4)
94. **M.I. Kadets, V.M. Kadets**: Series in Banach Spaces. Conditional and Unconditional Convergence, 1997, (ISBN 3-7643-5401-1)
95. **H. Dym, V. Katsnelson, B. Fritzsche, B. Kirstein** (Eds): Topics in Interpolation Theory, 1997, (ISBN 3-7643-5723-1)
96. **D. Alpay, A. Dijksma, J. Rovnyak, H. de Snoo**: Schur Functions, Operator Colligations, and Reproducing Kernel Pontryagin Spaces, 1997, (ISBN 3-7643-5763-0)
97. **M.L. Gorbachuk / V.I. Gorbachuk**: M.G. Krein's Lectures on Entire Operators, 1997, (ISBN 3-7643-5704-5)
98. **I. Gohberg / Yu. Lyubich** (Eds): New Results in Operator Theory and Its Applications The Israel M. Glazman Memorial Volume, 1997, (ISBN 3-7643-5775-4)
99 **T. Ayerbe Toledano / T. Dominguez Benavides / G. López Acedo**: Measures of Noncompactness in Metric Fixed Point Theory, 1997, (ISBN 3-7643-5794-0)
100 **C. Foias / A.E. Frazho / I. Gohberg / M.A. Kaashoek**: Metric Constrained Interpolation, Commutant Lifting and System, 1998, (ISBN 3-7643-5889-0)
101 **S.D. Eidelman / N.V. Zhitarashu**: Parabolic Boundary Value Problems, 1998, (ISBN 3-7643-2972-6)
102 **I. Gohberg / R. Mennicken / C.Tretter** (Eds): Differential and Integral Operators. International Workshop on Operator Theory and Applications, IWOTA 95, in Regensburg, July 31–August 4, 1995, 1998, (ISBN 3-7643-5890-4)

MIX
Papier aus verantwortungsvollen Quellen
Paper from responsible sources
FSC® C105338

If you have any concerns about our products,
you can contact us on
ProductSafety@springernature.com

In case Publisher is established outside the EU,
the EU authorized representative is:
Springer Nature Customer Service Center GmbH
Europaplatz 3, 69115 Heidelberg, Germany

Printed by Libri Plureos GmbH
in Hamburg, Germany